City on the Edge

City on the Edge

Technology, Politics, and the Fight for the Soul of San Francisco

Jonathan Weber

ATRIA BOOKS
New York Amsterdam/Antwerp London
Toronto Sydney/Melbourne New Delhi

ATRIA
BOOKS

An Imprint of Simon & Schuster, LLC
1230 Avenue of the Americas
New York, NY 10020

First Atria Books hardcover edition June 2026

ATRIA BOOKS and colophon are registered trademarks of Simon & Schuster, LLC

Map by Betsy Streeter

Manufactured in the United States of America

1 3 5 7 9 10 8 6 4 2

Library of Congress Cataloging-in-Publication Data has been applied for.

ISBN 978-1-6680-7491-6
ISBN 978-1-6680-7493-0 (ebook)

Contents

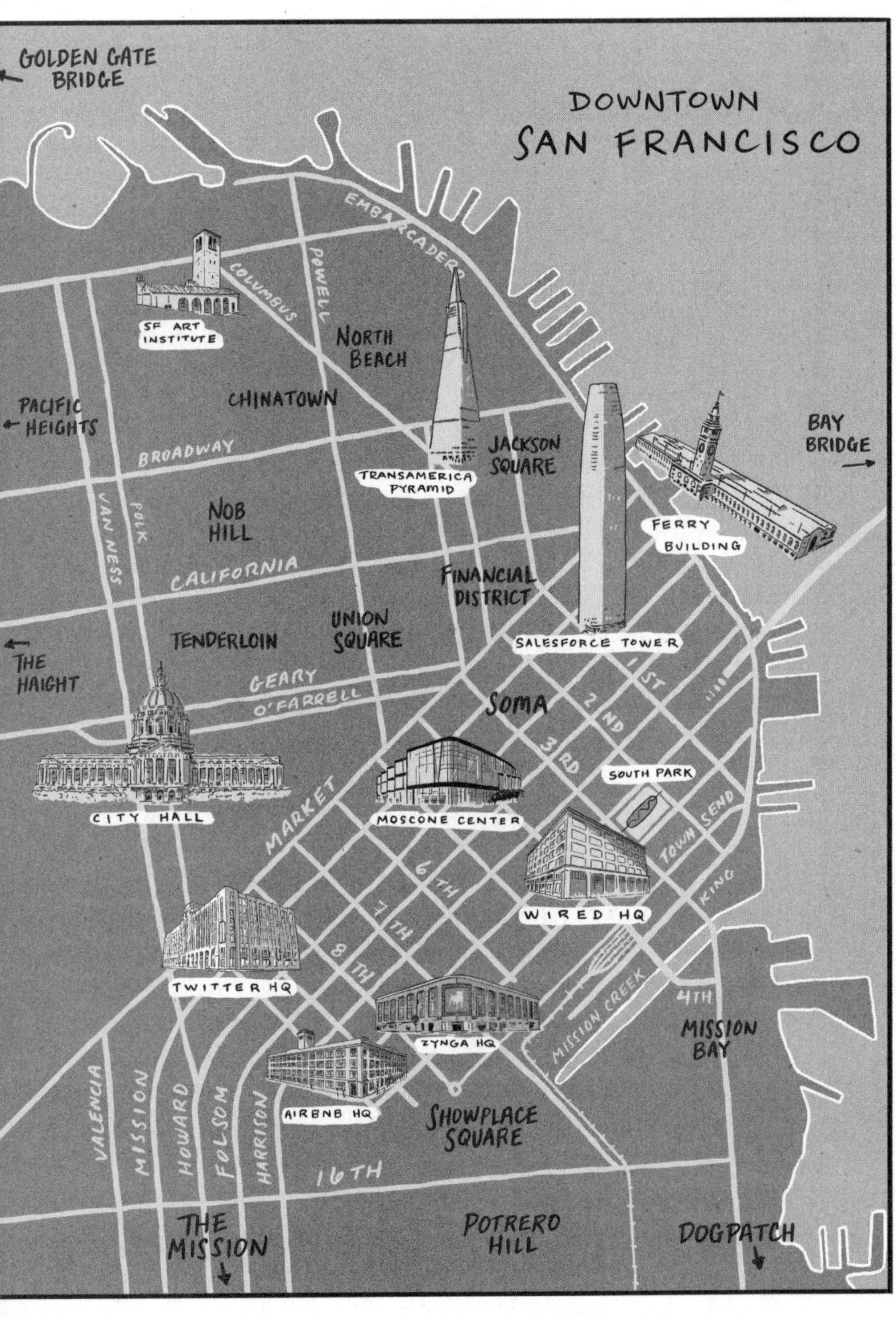

DOWNTOWN SAN FRANCISCO
GOLDEN GATE BRIDGE
EMBARCADERO
POWELL
COLUMBUS
SF ART INSTITUTE
NORTH BEACH
CHINATOWN
PACIFIC HEIGHTS
BROADWAY
JACKSON SQUARE
TRANSAMERICA PYRAMID
BAY BRIDGE
FERRY BUILDING
VAN NESS
POLK
NOB HILL
CALIFORNIA
FINANCIAL DISTRICT
UNION SQUARE
TENDERLOIN
THE HAIGHT
SALESFORCE TOWER
GEARY
O'FARRELL
SOMA
1ST
2ND
3RD
SOUTH PARK
CITY HALL
MARKET
MOSCONE CENTER
TOWNSEND
KING
6TH
7TH
8TH
WIRED HQ
TWITTER HQ
MISSION CREEK
4TH
MISSION BAY
ZYNGA HQ
AIRBNB HQ
SHOWPLACE SQUARE
VALENCIA
MISSION
HOWARD
FOLSOM
HARRISON
16TH
THE MISSION
POTRERO HILL
DOGPATCH

Foreword

In the summer of 1990 I was named the first "Silicon Valley" correspondent for the *Los Angeles Times*, based out of the bureau in San Francisco. The paper's business editor, Marty Baron, who'd later rise to fame as editor of *The Washington Post*, was ahead of the curve in creating the position, which only a handful of big national publications had at the time. Technology news was an afterthought in the grand firmament of the global media, a dry and complicated topic that was important for businesspeople, maybe, but rarely destined for the front page. Personal computers had been around for a decade, and Steve Jobs was famous, but outside of corporate offices, PCs were mainly used as word processors. Cell phones, let alone smartphones, were still years from being commonplace.

The city of San Francisco didn't have a technology industry at all. That was the domain of Palo Alto and Santa Clara and the other small cities that made up Silicon Valley, about an hour to the south.

The *Times* bureau was at Fox Plaza, a hideous twenty-nine-story rectangle on a ghostly center-city corridor called Mid-Market, not far from City Hall. Built in the place of the historic Fox Theatre, a glorious Rococo movie palace that was torn down in the redevelopment frenzy of the early 1960s, Fox Plaza might have been the most reviled building in the whole city, with a creepy warren of mostly empty retail spaces making up its so-called plaza, and dim hallways of faded offices and apartments up above. People jumped to their deaths from the upper floors with disconcerting frequency. Anton LaVey, a local musician and performer who'd later conjure up the Church of Satan, famously put a curse on the new building during the last show at the Fox Theatre, and whatever God he was praying to seemed to have delivered.

Directly across Market Street from Fox Plaza was the Western Furniture Exchange and Merchandise Mart, a hulking, twelve-story Art Deco building that took up an entire block but showed little life outside twice-yearly trade shows, its broad sidewalks grimy and deserted. The once-elegant theaters and office buildings nearby had seen much better days too, their ground-floor spaces shuttered if they weren't occupied by liquor stores or smoke shops, the doorways populated by people toting sleeping bags and glass pipes. Mid-Market wasn't completely free of regular commerce: One big newer building, also notable for its ugliness, housed a windowless Bank of America data center, and there was an insurance company down the street. The Orpheum and Golden Gate Theatres, and the Civic Auditorium, would bring out crowds for performances. With City Hall closed for repairs after the 1989 Loma Prieta earthquake, the city government was renting office space in the annex that was adjacent to the Furniture Mart, which kept a few office workers around.

Mostly, though, Mid-Market was becoming a showcase for the city's ballooning homelessness problem, exacerbated by the loss of blue-collar jobs and single-room occupancy hotels in the adjacent industrial district known as South of Market, or SoMa. It would be twenty years before the Furniture Mart would become the centerpiece of a high-profile effort to revive the neighborhood, and a symbol of what the city hoped to become.

This part of town wasn't representative of San Francisco as a whole, certainly. But in the pre-Internet era, the city's ever-volatile economy was in a tough spot. Born of the 1849 gold rush and made truly rich by a silver bonanza a decade later, "Imperial San Francisco" (the title of not one but two books) was a colossus: master of a regional natural resources empire, tip of the spear of American colonial ambitions in the Pacific, home to the great railroads and shipping companies of the West and wellspring of the state's top politicians and business oligarchs. Its power and influence reached new heights during World War II, with the city serving as headquarters of the Pacific Fleet, primary mustering point for the war against Japan, and a key contributor to the "arsenal of democracy."

The postwar period was much less kind. San Francisco had been an industrial hub, with steel mills and food-packaging and shipyards, and it all ramped up dramatically during World War II—only to be deconstructed, first suddenly, with the end of the war, and then gradually, with globalization and the nationwide flight of manufacturing to non-union states and cheaper countries. The mines of the Sierras were long-since depleted, the redwoods of the North Coast thoroughly harvested, the railroads no longer the profit machines of yore. The city was increasingly dependent on tourism, and the banks and corporate headquarters of the financial district, but many of them were going away too.

There were a lot of political scars from the violent 1970s, when Mayor George Moscone and Supervisor Harvey Milk were assassinated, the preacher Jim Jones led nine hundred people (many of them city residents) to their deaths in a South American jungle, the "Zebra killers" randomly shot white people on the streets (including future Mayor Art Agnos) and the heiress Patty Hearst was kidnapped by a group of disturbed criminals who called themselves the Symbionese Liberation Army. Then there was the AIDS epidemic, which in 1990 hadn't yet begun to wane and would ultimately kill some ten thousand San Franciscans. The Black community had never really recovered from the destruction of the Fillmore district, the lively center of Black life that was mostly bulldozed in the '60s and '70s in the name of redevelopment. Severe poverty stalked some areas, including pockets of Chinatown as well as Hunters Point and the Bayview.

Dianne Feinstein, the mayor from 1978 to 1988, had won the city's affection with her poise and empathy in the wake of the twin assassinations—though she was much more conservative than her predecessor—and she did much to reinforce the downtown financial district and push the transition to a white-collar economy. At the same time, a powerful slow-growth movement, born of environmental and neighborhood preservation activism dating to the 1960s, had been fighting the "Manhattanization" of the city. Feinstein, though no friend of the left on most issues, had sympathy for those sentiments, backing a citywide rezoning in the late 1970s that all but shut down housing development in much of San Francisco.

Even as the slow-growthers rallied against new office buildings, the corporate pillars of downtown were decamping to cheaper, safer suburban pastures, or succumbing to mergers, with oil giant Chevron in 1989 following construction giant Bechtel and the Southern Pacific Railroad out the door. San Francisco wasn't the money-center of the West it had been, and its flagship Bank of America, and later Wells Fargo, would eventually be gone too. Art Agnos, who succeeded Feinstein as mayor and sat well to her left, didn't have a lot of answers on what would power the city's economy, especially after the 1989 earthquake and then the 1990 recession, which pushed unemployment to 10 percent. Real estate prices were rising, but that was partly because there had been so little construction, and the homelessness problem was quickly getting worse. The city was still a magnet for the young and restless, and entertainment and tourism were healthy. But that wouldn't be enough.

Yet even with the economic challenges—and in some ways because of them—San Francisco culture was bounteous in all kinds of ways. With its breathtaking natural beauty and inviting spirit, and a history that encouraged breaking the mold, it remained a powerful beacon for creatives, renegades and entrepreneurs of every type. The influx of hippies and gays in the 1960s and 1970s, though resisted at first by the city's Irish and Italian old guard, had helped imbue it with an appreciation for different ways of living, and for the politics of social freedom and social justice. There were a lot of cheap spaces to rent, offering plenty of room for the punk rockers and alt-rockers and ravers who'd kept the music scene lively, and for the experimental artists of the cultural underground. A rich Latino culture bloomed in the Mission, marked by colorful murals splashed across buildings and some of the best burritos anywhere. The Black neighborhoods of the Western Addition, Hunters Point and the Bayview retained a powerful sense of local pride, even amid poverty, racism and violence that weighed heavily. Chinatown, and the numerous Asian neighborhoods in the western and southern parts of the city, imbued the streets with a rich variety of sights and scents and social scenes, as did the Russians and Ukrainians of the Richmond and numerous smaller tribes.

But unbeknownst to almost everyone at the time, this unique cultural brew was part of a confluence of political, technological and financial forces that over the next several decades would vault the city, and the world, into a future that bends our imaginations still. San Francisco was about to become the test-bed for the Internet, and the spiritual and financial capital of the digital technology revolution.

It's not that the Internet was invented in San Francisco, or even in nearby Silicon Valley. Computer scientists working for institutions including MIT, UCLA and UC Berkeley in the 1970s and 1980s were mostly responsible for that, creating technical protocols and an overall architecture for an open global network that could connect any computer, anywhere. Big telecom companies built the infrastructure and helped operate it. A British computer scientist working in Switzerland pioneered the World Wide Web, a means of navigating the emerging network that made it accessible to mere mortals. The Internet, like most massively successful human undertakings, has many parents.

It was in San Francisco, though, and especially in the former industrial district known as SoMa, that emerging digital technology met the unique creative zeitgeist of the city—and it was the technical innovations, cultural flowering and political idealism that arose from that alchemy that would come to define the Internet era.

Electronic publishing, social media, video streaming, virtual reality, cloud software, programmatic advertising, online marketplaces, the "sharing economy" and even the architecture of the website itself all had roots in San Francisco. Alongside that, the city nurtured a kaleidoscope of political beliefs about how technology might change the world for the better. As the journalist John Markoff has shown, the personal computer revolution was inspired by the anti-authority, individual empowerment, proto-environmentalists of the1960s. The Internet as we know it was animated in its earliest days by a vibrant successor: the novel creative culture of San Francisco in the 1990s.

The Internet economy would transform the city into one of the wealthiest and most dynamic places on the planet—a destination for the best and brightest, a political powerhouse that spawned the likes of Nancy Pelosi, Kamala Harris, Dianne Feinstein, and Gavin Newsom, and a diverse mecca of unrivaled food, incredible parks, first-rate arts institutions, and every sort of lifestyle choice. San Francisco was a font of invention and opportunity, yet still charming, approachable, inescapably beautiful, and happy to let everyone do their thing. It would even birth an idealized doppelganger in the extraordinary annual event that is Burning Man.

Yet somehow, the Internet era was not the happy story that it should have been for San Francisco.

A host of urban ills lurked in the shadows even as the Internet industry remade the local economy, and when the Covid pandemic emptied the downtown skyscrapers and exposed the severity of the problems in the streets, the city's decades-long ascent was abruptly and shockingly reversed. Long after the pandemic eased, San Francisco remained a shadow of its former self, with downtown streets eerily quiet, long blocks occupied mostly by the homeless, drug-addled and mentally ill, and the office and retail sectors only barely recovering from a years-long freefall.

Idealism about what the Internet might bring had taken a big hit too, for a host of reasons that were at once political, economic and social. By 2025, even as artificial intelligence was emerging as a major economic force, the hopes that a "new economy" might introduce more collaborative, egalitarian, and inspirational modes of work—and support "San Francisco values" on immigration, civil rights and the environment—were all but dead. Donald Trump's second ascent to the White House, assisted in critical ways by a cohort of far-right tech industry figures who'd long been on the fringe of the San Francisco Internet scene, proved an emphatic nail in the coffin. A different kind of era was at hand.

San Francisco has always occupied a special place in the national psyche—the mythic, fog-shrouded city on the edge of the continent, mesmerizing in its beauty, founded by fortune seekers and freethinkers, the birthplace of the counterculture and pretender to the title of "greatest city in the world." That all but assured it would become a leading symbol in the national culture wars. It's a place that's at once completely unique and also an exemplar of both the glory and the troubles of all American metropolises.

For the right, San Francisco is exhibit number one for the argument that progressive policies on crime and drugs have ruined Democratic-run cities. Among moderate Democrats, San Francisco is a great example of the failures of anti-growth liberalism, as outlined in the "abundance" theory popularized by journalists Ezra Klein and Derek Thompson. The progressive left has its diagnosis too: If only the tech companies and the billionaires weren't so greedy, there'd be plenty of resources to solve social problems like homelessness.

There is certainly some truth in each of these. Yet as I revisited the remarkable history of the past three decades and interviewed more than two hundred people at the center of events, I found that none of the pat explanations really captured what had happened in San Francisco. The deeper story, in my view, is only partly about the various ways that city government fumbled the ball, or that the tech industry's heedlessness created needless strife. It's also about how the intensifying forces of techno-capitalism operate by their own logic. Considering the city and its tech industry as protagonists in a wider drama leads to more nuanced and richer interpretations.

With the artificial intelligence boom developing extraordinarily quickly and San Francisco well-ensconced as its headquarters, the city is recovering from the worst of its post-pandemic woes. A new political regime is in place, and its focus on accountability and law and order has a lot of commonsense appeal. It's easy to envision a cleaner, brighter, better-run city, one where market forces rule freely and there's less tolerance for bad behavior. Tech innovations, from self-driving cars to robot food delivery to medical interventions for extending human life, would be welcomed without a fuss. It would be a nicer, safer place, better for tourists, conventioneers and tech companies, the envy of many other cities, and undoubtedly preferable in the eyes of many residents.

But would it be San Francisco?

The story I aim to tell in these pages begins in the early 1990s, the start of what I refer to as the "Internet era" in the city. A simple statistic frames the period well: In 1991, just 2 percent of city jobs were in technology, according to city economist Ted Egan. By 2019, that number was 35 percent. The Internet era ended with the pandemic, when many of the companies that powered the Internet economy either closed their San Francisco offices or shrunk them down dramatically. The 2024 election provides a fitting coda. The AI era has now begun.

The rise of the Internet industry, which took place in several distinct phases over the course of thirty years, is one key thread in this tale of San Francisco's transformation. The evolution of city politics over the same period, where four different mayors from the same faction wrestled with an increasingly left-wing Board of Supervisors, is another. How the culture of the city first nourished the tech industry and then changed alongside it is the third major strand of the tapestry.

I've endeavored to weave together a big story from the many small dramas that played out on the ground, with the goal of offering an intimate view of San Francisco that will prove more illuminating, and more interesting, than the countless hot takes and formulaic news stories that have done so much to shape perceptions of the city. It aims to shed fresh light on how things really happened in one of the world's most fascinating places during a pivotal period in its history, and help explain both the city's shocking Covid-era decline and the political shift in a tech industry that had been comfortably aligned with the Democratic Party. The story of San Francisco offers a unique and valuable perspective on some of the most consequential political, social, cultural and economic developments of our times.

Focusing on the city proper, as distinct from "Silicon Valley," might sometimes seem artificial when talking about the Internet industry. The difference is real though, and I think the city serves as a useful prism in considering the momentous rise of technology in modern society. Conversely, looking at the city through the lens of the Internet revolution illuminates crucial dimensions of the modern urban experience.

This is also at times a personal narrative, born of my good fortune in being able to witness the city's remarkable transformation from a number of journalistic vantage points. In some cases I was a direct participant in events. More often I was working as a reporter and editor, trying to understand it all and help my readers do the same. My sincere hope is that this version of San Francisco's history offers not just a sense of what happened—and why—over the course of three decades, but an idea of how it felt to live and work in this magical place as it was reshaped by the forces of history.

It all happened very fast.

PART ONE

The Rise of the Internet Economy (1990–2001)

CHAPTER 1

South Park and the Birth of the Web

Brian Behlendorf was only a week into his freshman year at UC Berkeley, in 1991, when he logged onto a campus computer and stumbled upon the future.

Growing up in the Los Angeles suburb of La Cañada, adjacent to the engineering meccas of Caltech and NASA's Jet Propulsion Laboratory, Behlendorf had gotten interested in tech early. An amiable, broad-faced kid with stringy blond hair and a passion for music and math, he was part of the first generation of Americans to come of age with personal computers and had been going online by the time he was in high school. He'd explored the world of electronic bulletin boards—computer systems established mostly by hobbyists in the 1980s that allowed people with PCs to connect via a phone line and chat with others, play games and share software. He'd dabbled with an early IBM service called Prodigy, one of the first national computer networks.

Behlendorf was also a fan of techno music, and he'd gone to Berkeley to study computer science—and hopefully partake in the party scene.

He considered himself pretty computer-savvy, but the machines he found at UC Berkeley were much more sophisticated than the IBM-compatible personal computers he'd used in high school. They were powered by a software system called Unix, first developed at the famed AT&T Bell Laboratories, and they were connected directly to similar machines all around the world. There was no need to fuss with dialing in, or worry about the cost of long-distance phone calls, a big issue in those days. All sorts of fascinating information and interesting people were just a few keystrokes away, twenty-four hours a day.

"I'm like, what is this?" he recalled years later, a twinkle of wonder still in his eyes. "I'd had the experience of messaging people and that kind of thing, but this was much more open, much more raw—more clunky, but also distributed. There was clearly no company at the center of this. I fell down a rabbit hole of 'where did this come from,' and started to learn some things."

Behlendorf didn't quite realize it at the time, but he'd discovered the Internet.

Launched in the late 1970s as a US Department of Defense research project, the Internet by the early 1990s connected scores of university, government and corporate research labs around the world. Berkeley was an important hub and the birthplace of a new version of Unix, written by a graduate student named Bill Joy, that would become a technical pillar of the network. The Internet at the dawn of the 1990s was still noncommercial, sometimes aggressively so, with many denizens in the science and research communities determined to shield it from the big telecom and tech companies that might turn it into something different. It was still all but unknown to the general public.

For Behlendorf, the big attraction at first was free messaging and the absence of a governing authority. The bulletin boards he'd visited, and online services like Prodigy, were centralized systems, subject to the whims of their owners and easy for governments to monitor and control. The Internet was something else.

"That was the same year there was a coup in Russia," Behlendorf recalled, referring to the failed 1991 effort by the Russian military to depose then-president Boris Yeltsin. All regular international telecom links had been cut off, leaving a few Russian universities that were connected to the Internet as the only portal for communication with the rest of the world. Behlendorf found it amazing that something developed by and for computer researchers—and heavily populated by technically curious "hackers" who liked testing and tinkering—could become a humanitarian lifeline. It suddenly seemed obvious: "Maybe there's a sociological angle to this," he mused. The Internet was clearly going to be about much more than computers.

Behlendorf immediately found a practical purpose for it, albeit one rather less high-minded than saving Russians from the abyss: He wanted to find the coolest underground raves.

This new genre of ecstatic dance party had originated in the UK in the 1980s, bringing together electronic music, mind-altering drugs and all-night revelry, often in hidden places. Behlendorf was among the many early computer enthusiasts who were entranced with the experimental and subversive aspects of rave culture, not to mention the mechanical-engineering challenge of throwing huge parties in secret locations.

"I was struggling to find the better parties that I knew existed out there, where I could get in," Behlendorf recalled. But he couldn't go to a lot of the events, because he didn't want to get a fake ID. He was only nineteen.

Behlendorf started to think about how the Internet might help him better plug into the burgeoning rave circuit. Electronic mailing lists were already the standard means of sharing information among computer mavens, so that was an obvious place to start. In 1992, he created a list called SF Raves. This was well before most people had email, limiting its potential influence, but the list was an immediate hit, drawing hundreds of enthusiastic subscribers in its first few weeks. One thing led to another.

"I thought, 'I should put the archives of this list somewhere public so people can catch up with the conversations,'" he recalled. "'And gee, there's all these rave fliers, why don't I scan them in? There's a scanner in school that I could use. There's these DJ sets I have on audio cassette. Why don't I see if I can rip them to this new file format called MP3?'" Sharing music over the Internet seemed like a promising frontier.

His timing could not have been better. A couple of years earlier, a London-born computer scientist named Tim Berners-Lee, working at the European nuclear research laboratory in Geneva known as CERN, had developed the initial protocols for what he called the World Wide Web. Now his big ideas were starting to become a reality. Berners-Lee described the Web as "a wide-area hypermedia information retrieval initiative aiming to give universal access to a large universe of documents." It would bring to life "hypertext," first described by the computer visionary Ted Nelson, where documents were connected by "links" that could be clicked on with a mouse.

Curious young techies like Behlendorf—often in collaboration with university researchers supported by government grants—were now starting to use the new World Wide Web specifications to make things. The SF Raves mailing list begat HyperrealSF, an email server and website that Behlendorf had built to host the global rave community. It was among the first nonacademic websites on the Internet, and it served its purpose very well: Behlendorf was getting to know a lot of people, many of whom shared his overlapping interest in music, psychedelic drugs and the new frontiers of online information technology. It would eventually lead him to South Park, in San Francisco, where an advance guard of hackers, designers, engineers, artists and creative rebels of one sort or another was growing fast, and laying the groundwork for the Internet era.

A grassy oval lined with row houses and low-slung brick buildings, South Park sits in the southeastern corner of the precinct of SoMa, short for "south of Market Street." Originally built in the 1850s as an exclusive, gated redoubt for the prosperous merchants of the Gold Rush, South Park's initial heyday was brief, as was so often the case for ambitious real estate projects in this boom-and-bust town. A clever entrepreneur and engineer named Andrew Smith Hallidie had invented the cable car in the early 1870s, and with an influx of immigrants starting to crowd the blocks surrounding South Park after Second Street was leveled, the wealthy shifted their affections to newly accessible Nob Hill.

Like most of the city, South Park was rebuilt from scratch after the 1906 earthquake and fire, but by the second half of the century it had been subsumed into the light manufacturing and warehouse economy of SoMa, with several residential hotels owned by Japanese Americans and another serving a growing Filipino community, alongside clusters of rooming houses and small workshops. Longshoremen waiting for the call from a nearby Sailor's Union hall began warming themselves over a bonfire in the park in the 1930s, and it would burn almost continuously for forty years, with drug dealers and assorted vagrants and homeless people gradually replacing the laborers. By the end of the 1980s, the garment workshops, printing plants, metal works, warehouses and breweries that once dotted SoMa had mostly faded away, part of a post–World War II wave of deindustrialization that had staggered older cities around the country.

Yet something new was starting to simmer amid the derelict warehouses and abandoned factories of South Park and SoMa. At the southern edge of the district was China Basin Landing, where a sprawling six-story structure built by the Southern Pacific Railroad in the 1920s was now being upgraded for a new generation of companies. Printing and photo-processing had long had a home in SoMa, and there were a couple of sizable companies that created corporate videos and made copies of tapes, a labor-intensive process at the time. A handful of trade publications that covered the computer industry—employing writers, designers and photographers—were based in SoMa. There were art, architecture and design studios scattered about, often animated by graduates of the prestigious SF Art Institute, or the California College of the Arts, or the film school at SF State. The "No Nothing Cinema," a makeshift theater with scavenged seats inside an old milk-bottling plant on Berry Street—where a baseball stadium would later be built—featured indie bands and independent movies.

"A lot of us worked in the 'video ghetto' of SoMa and China Basin," recalled Will Kreth, who had a job doing film-to-tape transfer and video duplication for big companies like Pacific Bell and Levi's. In the early 1990s, Kreth caught on with a software firm called PF Magic, one of the first companies dedicated to creating "multimedia" computer programs based on the silvery discs known as CD-ROMs. He'd later be among the first employees of a new magazine that would play a central role in the rise of San Francisco's Internet economy.

The tech business in the city proper—as opposed to Silicon Valley down south—was tiny and artisanal. Jasmine Technologies, reputed to be the first computer hardware company based in San Francisco, made a good living for a few years in the late 1980s selling external hard drives for the Apple Macintosh, the cutting-edge machine of the day. The drives gave the Mac far more storage capacity, which was crucial for handling imagery and sound—key ingredients of the incipient enthusiasm for multimedia computing. Jasmine had a modest building next to Anchor Brewing in the industrial neighborhood of Potrero Hill, just south of SoMa.

Will Kreth had worked there too. So did Michael Mikel, later known as Danger Ranger, one of the founders of the annual reverie in the Nevada desert known as Burning Man. Mikel's true passion lay in organizing stunts for the Cacophony Society, a free-form group of urban adventurers and performance artists that was just finding its groove. But he'd learned electrical engineering in the military and community college, and was working as a manufacturing consultant when he was hired by Jasmine.

"I walk in and there were three guys sitting on benches assembling these things with hand screwdrivers. I went out and bought electric screwdrivers and increased production by 300 percent in one day," he deadpanned many years later. "We were selling by mail, and we would get the order and then we'd buy the parts and assemble it. That was basically how it worked." The company's edge, he said, was customer service. "Half the people working there were basically hippies, you know, idealistic, and they were really focused on customer satisfaction." Rob Schmitt, who had some computer skills when he arrived in the city from Minnesota in the late 1980s, was quickly hired at Jasmine, and then recruited by Mikel into Cacophony. He'd go on to invent some of the most celebrated Cacophony stunts, including the extant happening known as Santa Con.

Jasmine's drives were very popular, and it was a talented bunch: Its president at the dawn of the decade was Barry Schuler, who'd go on to be a top executive at America Online Inc. and a successful venture investor. But the company would flame out in a blaze of customer complaints and lawsuits as small manufacturers struggled to keep up in the burgeoning personal computer business.

Tom Jennings, a hacker, skateboarder and queer activist who'd be instrumental in building the city's first Internet service provider, lived in an old warehouse on Shipley Street in SoMa. Like many buildings in the area, it was destined for the wrecking ball as the old economy declined, and in the meantime could be had for dirt cheap. It was a group house, mostly gay skate punks and bike messengers, and they called themselves the Shred of Dignity Skaters' Union. "We had our shit together," Jennings recalled fondly. "We allowed cheap beer, some pot, no drugs. There were a lot of needle drugs in San Francisco at the time, and we were not part of it. The house was like half runaway teenagers, but super-responsible." Jennings published a gay punk zine called *Homocore*, which attracted a following, and he never shied from expressing his views to the authorities. "We did political protests. We put on shows. We did police protests, where we out-megaphoned the fucking police—we had a pickup truck with a full-scale sound system on it, 5,000 watts. We had a great time."

Jennings already had some considerable accomplishments under his belt as a self-taught programmer. He'd played an important role in the development of the Phoenix BIOS, a critical if unsung bit of personal computer technology. He was the creator of FidoNet, which allowed dial-up bulletin boards to talk with one another and functioned as a sort of proto-Internet in the 1980s and early 1990s. Yet Jennings and the smattering of online tinkerers in SoMa were outliers in the neighborhood: The technology industry of the time was still firmly implanted in suburban Silicon Valley.

But a convergence of new technology and San Francisco's creative energy was starting to change that. The Apple Macintosh, introduced with the legendary "Big Brother" television ad in 1984, brought visuals to the personal computer, making it much easier to use and far more expressive—not just what a document might say, but how the text would look on a page, together with pictures and other images. John Battelle, later a prolific media entrepreneur, recounted his astonishment in using the Mac for the first time, in a Berkeley dorm room in 1985. "I turned it on and started mucking around with it, and I was blown away. I just couldn't believe it, compared to the PC . . . it was like night and day," Battelle recalled. "I was like, oh fuck, this is going to change everything. This is the greatest artifact in the history of mankind."

The early Macs were far too underpowered, and expensive, to deliver on that dream. But the machine built a fervent following among the small but growing cohort of technologists, artists, futurists and out-of-the-box thinkers who believed computers could change the world for the better. Apple had opened a small multimedia lab in the city's ritzy Presidio Heights district to develop educational software for the Macintosh—Will Kreth, of Jasmine and PF Magic, was one early employee—and it became an important talent laboratory for multimedia innovation. The annual Macworld expo evolved into a festival for the fledgling San Francisco tech set, overflowing from Brooks Hall, the old conference venue near City Hall, to the new Moscone Center in central SoMa. A large contingent of the crowd looked less like the pallid, suit-cladded attendees of a typical trade show than those you might find at, well, a rave.

A lot of those people were talking to one another on The WELL, an electronic bulletin board founded in 1985 by *Whole Earth Catalog* impresario Stewart Brand, together with a medical doctor and creative thinker named Larry Brilliant. Based across the Golden Gate Bridge in Sausalito, it had become one of the first virtual communities worthy of the name, an intimate but diverse gathering place that nurtured conversations among hackers, journalists, environmentalists, Grateful Dead followers, organic food enthusiasts and everything in between. By 1990, The WELL was home to an effervescent group-chat of the technologically and socially curious who'd do much to shape the Internet era.

Brand, who grew up in Illinois, had arrived in San Francisco in the early 1960s by way of Stanford, drawn by the bustling bohemian culture of North Beach. He caught on with the new photography program at the SF Art Institute, established by the famed Yosemite chronicler Ansel Adams. He'd also been part of then-legal experiments with LSD that were happening around Stanford and was enamored with its mind-expanding potential; he soon found himself helping to organize acid trips with the writer Ken Kesey and his merry pranksters, which were later immortalized in Tom Wolfe's 1968 epic, *The Electric Kool-Aid Acid Test*. Brand seemed to be everywhere as the 1960s counterculture exploded, even assisting the legendary inventor Douglas Engelbart in what became known as "the Mother of All Demos," held at Brooks Hall in 1968. Engelbart, a solid fifteen years ahead of his time, showed how an easy-to-use computer could work, with a mouse and overlapping windows and pull-down menus and most of the other features that are central to the computers we know today.

Brand would win fame and fortune with the *Whole Earth Catalog*, conceived as a guide to new tools that could enable people to live better and closer to nature. It was Brand, as much as anyone, who forged the early connections between 1960s counterculture and emerging digital technologies. Now his pioneering online service was doing the same for the emerging Internet era.

It was on The WELL that John Perry Barlow, a bluff, handsome Wyoming cattle rancher and polymath who wrote lyrics for the Grateful Dead and was early to the idea of "cyberspace," recounted an unnerving visit from a federal agent. The officer was investigating an alleged theft of some Apple computer code, but seemed to know nothing at all about technology. Around the same time, the US Secret Service had been conducting nationwide raids on a group called Legion of Doom, which the government thought was a criminal hackers' den but consisted mostly of curious and mischievous teenagers. Barlow saw danger in what looked like an ignorant government crackdown on the nascent Internet.

Barlow's missive was spotted on The WELL by Mitch Kapor, a left-leaning software entrepreneur from Brooklyn who'd done so well as the founder of Lotus Development Corp., the creator of a seminal PC software program, that he had a private jet. He reached out to Barlow, and a few days later dropped in on him at the Wyoming ranch. "As a late spring snowstorm swirled outside my office, we spent several hours hatching what became the Electronic Frontier Foundation," Barlow wrote later. Shortly after that, Barlow told his story to *The Washington Post*, and support poured in.

John Gilmore, a libertarian-minded engineer who was among the most influential programmers of the early Internet and had made his own fortune as the fifth employee of Sun Microsystems, offered a six-figure check. Steve Wozniak, the Apple cofounder, got in touch. Stewart Brand was on board too. So was Esther Dyson, who produced PC Forum, the top gathering for the tech elite, along with an influential newsletter, and was the only woman among the early cohort. After a series of dinners in San Francisco in the summer of 1990, the EFF, destined to be a powerful player in shaping the laws governing the Internet, was officially born.

The group would help nurture a novel strand of politics, rooted in the counterculture, that married a belief in technology's potential to connect and empower people with libertarian ideas about self-sufficiency and suspicion of centralized authority. One branch of the left-wing politics that had arisen out of the 1960s was hostile to computers, which were viewed as tools of oppressive corporations and the capitalist state. But Barlow, Brand and their fellow travelers were turning that idea on its head. Stanford author and sociologist Fred Turner labeled them the New Communitarians, using technology to build self-actualizing communities and creative cultures without rejecting the mechanisms of capitalism, or expecting the government to do much besides leave them alone.

The cheap industrial spaces of 1990s SoMa were a magnet for upstart businesses and cultural renegades of every sort. At the former Hamm's brewery on Bryant Street, punk rockers had commandeered the abandoned beer vats to use for rehearsal space. Survival Research Laboratories, a collective led by a performance artist named Mark Pauline, staged elaborate shows of hacked-together industrial machines spewing fire and battling one another, with a famous 1989 performance beneath the 280 freeway titled "Illusions of Shameless Abundance" reaching a fiery climax that enshrouded drivers in smoke. Pauline had also studied at the SF Art Institute, and his marriage of mechanical tinkering and high-voltage performance was an inspiration among the growing band of creatives working at the intersection of art and technology.

In 1991, at the corner of Ninth and Folsom, party impresario Joegh

Bullock created the Anon Salon, where an ensemble of artists, technologists and performers experimented with electronic music, fire art, outré fashion, sexual expression, every kind of theater—and emerging digital technologies. The idea of creating a "virtual reality" in the digital world was an animating principle, and the Australian American engineer Mark Pesce would lead the creation of a computer language for virtual spaces, known as VRML. Bullock and his partner, Marcia Crosby, were "the epicenter of the hipster underground art scene at the time," Harley Dubois, one of the Burning Man cofounders, said when Bullock died in 2022.

Anon Salon was where all the cool kids were going.

The local rave circuit got a big jolt when a British DJ and singer named Mark Heley came to San Francisco at the dawn of the decade and birthed a roving party called ToonTown. A 1991 New Year's Eve bacchanal at a four-story SoMa warehouse, featuring virtual reality booths, video chats and intense laser light shows accompanying the techno beats, would put raves at the center of the emerging cultural map. There was even something of a house magazine in the form of *Mondo 2000*, led by a refugee from upstate New York named Ken Goffman, which celebrated the marriage of technology, music and psychedelic drugs; Timothy Leary, the LSD guru, made frequent appearances in its pages. The raves, and *Mondo 2000*, were both a magnet and an inspiration for the tech subcultures that were starting to flourish in the byways of SoMa.

Around the same time, in the basement of John Gilmore's house in the Haight-Ashbury neighborhood, once the epicenter of the Summer of Love, San Francisco's first general-purpose consumer Internet provider was taking shape. Tom Jennings, the SoMa hacker and skate punk, would come aboard to help.

Local bulletin boards like The WELL, along with the big national network America Online, were growing fast in the early 1990s. But it was still hard to get on the Internet proper—the free, open, global network where you might email anyone around the world, or talk with others via something known as Internet relay chat, or participate in discussion groups on what was called Usenet. It was a noncommercial network controlled by an assortment of university, government and telecom industry administrators. You couldn't get to it from The WELL, or anyplace else, unless you had a friend in the right place, or a lot of money for a special phone line. John Gilmore, almost by accident, would come up with a solution for that.

Gilmore was a singular figure in the tech business, a self-taught programmer from Pennsylvania who'd moved to San Francisco in the late 1970s and quickly proved himself an exceptional talent, and a freethinker. Working for a computer-terminal company in Silicon Valley, he'd found his way to the Internet's predecessor, known as the Arpanet, and discovered a mailing list devoted to the topic of microprocessors, one of his areas of interest. He eventually decided to track down the Stanford engineer who was doing the most advanced work on a powerful new processor, which turned out to be a German graduate student named Andy Bechtolsheim.

In 1984, after finding Bechtolsheim's phone number through an old Arpanet function called "finger," he called. "Andy isn't here," said a distracted Vinod Khosla, then a Stanford professor, who picked up the phone. Khosla said he was busy setting up a company and didn't have time to talk. Gilmore mentioned what he'd been working on—the complicated job of getting software written for one chip to work on a different one—and Khosla thought again. Gilmore was soon hired by what would be called Sun Microsystems, working alongside founders Bechtolsheim, Khosla and Scott McNealy, all from Stanford, plus Bill Joy, the Unix wizard from Berkeley. He would stay three and a half years, and earn permanent financial independence. For their part, the four founders would become billionaires, and influential figures in the industry for many years to come.

A lanky, blue-eyed man who still sports the scraggly beard and shaggy long hair of the Grateful Dead follower that he is, Gilmore was one of those kids who couldn't wait to get away from home and followed his own beat even after he did. While working at Sun, he'd been an active member of the Suicide Club, a secret society that traced its roots to the 1960s free-school movement at San Francisco State, and whose exploits included taking over abandoned military bunkers for elaborate secret parties, or climbing the Golden Gate Bridge in formal wear and eating a meal at the top. Gilmore didn't have a lot of patience for top-down authority, be it the government or the telephone company—the nemesis of many early hackers for its controlling ways and costly services. He loved the tech world in part because you didn't need a license, or a degree, to do your thing.

The emerging Internet could be a place for unfettered free expression and experimentation, and Gilmore was very happy to help people connect.

After leaving Sun, Gilmore had started a company called Cygnus Solutions to provide technical support for "free" software—programs created by a growing cadre of technologists who believed in open, collaborative development. Cygnus needed good network connections, and Gilmore eventually installed an expensive, dedicated communications line at his house in the Haight. It had much more capacity than he needed for himself. "Then I have friends who are saying, 'Hey John, can I put a modem in your basement and dial in and be on the Internet?' " recalls Gilmore. "So we did it that way." It was the beginning of The Little Garden. You just had to buy two modems, connect one to your PC, bring the other one to Gilmore's house and call the phone company to run another line. Then you could make a local phone call to your own special number, and you were on the Internet.

The basement-modem setup quickly became a lot to manage, what with the need to collect the checks and reboot the machines when they died and haggle with the phone company, which didn't quite know what to make of this group of oddballs who kept ordering more lines. Tom Jennings had already developed FidoNet, an Internet precursor, and he needed to make some money—and The Little Garden needed him. He and Gilmore did a deal where Jennings would get half the revenue for any new customers he signed up at the base rate of $80 a month. He was off and running, and The Little Garden rented an office in an old bank building on Sixteenth Street in the Mission, just as the city's new tech economy was beginning to take shape.

Gilmore was proud of The Little Garden's approach. "Because it was started as a cooperative network, by libertarians, we didn't impose any controls over what you could do with your Internet connection," he said later. "Almost all of the existing ISPs [Internet service providers] came with rules that said you can only use this for your own personal use: you can't use it for business, you can't resell it to anybody, blah, blah, blah. We basically said: 'We don't care what you do with this. Try not to hurt anybody.' "

Internet fever was now sweeping university laboratories and hobbyist garages in the Bay Area and beyond, and there were competing ideas on where it might go next; Berners-Lee's World Wide Web wasn't the only game in town. A tool called Gopher, released in 1991, also provided a simple means of browsing documents on the Internet, solving some of the same problems as the Web protocols. An MIT spin-out called Thinking Machines had led the development of "Wide Area Information Servers" for searching connected computers, and a Grateful Dead–loving programmer named Brewster Kahle had moved to San Francisco and cofounded a company to build it. WAIS was based in Menlo Park, "as far north as I could put a company in 1992," he recalled later. Ted Nelson, the hypertext visionary, was still toiling away on what he called Project Xanadu, which promised everything the Web could do and more.

Everyone involved agreed on one thing though: This was the beginning of something very big. What it might mean for the city of San Francisco was a question for another day.

For Louis Rossetto, an itinerant writer and editor with big ideas and no money, the digital technology revolution was nothing less than the greatest story of modern times. He just needed the chance to tell it to the world.

Then living in Amsterdam, Rossetto flew to San Francisco in early 1991 for the Macworld trade show, on a mission to recruit collaborators for a new magazine that would lead, and chronicle, the massive upheaval he saw coming. He was messianic in his conviction that new technologies would force a top-to-bottom reordering of culture, politics, business and every type of human relationship. This was something to be celebrated, not feared: It was nothing less than an epochal opportunity to throw off the chains of the mundane modernity imposed by soulless corporations and corrupt governments, and empower the people with magical new tools. With his flowing brown hair, brooding gaze and gift for grand pronouncements, Rossetto presented like a prophet, or at least a poet, of an emerging new world, clear in his vision and eager to enlighten the masses.

Rossetto had created a house magazine for an Amsterdam company that was developing tech tools for language translation, which was then sold to another Dutch publisher and renamed *Electric Word*. It was a vibrant title, expounding on the implications of digital technologies, and was among the first to be produced with the desktop publishing techniques made possible by the Apple Macintosh. But it had its limitations. "It was difficult to shoehorn all my discoveries into *Electric Word*," Rossetto said many years later. He and his partner, Jane Metcalfe, an ebullient blonde whom he'd met in Paris a couple of years earlier, formulated a plan for a new magazine, and decided they needed to make a go of it in America. They made a good team. Rossetto, who grew up on Long Island and earned his libertarian bona fides as a student at Columbia, tended to be stubborn and impossibly single-minded, carrying people along on the power of his ideas and his spectacular skills as a storyteller. Metcalfe, a Kentucky native who'd graduated from the University of Colorado, was cheerful, practical, ambitious and not afraid of work, or much else really, a crucial counterbalance to her brilliant but often difficult partner.

The Macworld trip was a success, and Rossetto and Metcalfe made the move to San Francisco soon after, cadging some workspace on the first floor of a crumbling warehouse building at the corner of Second Street and South Park, courtesy of a kindly Montana native named Randy Stickrod. He'd founded an ahead-of-the-curve trade magazine called *Computer Graphics World* that had been a success, and was now lending his expertise, and some spare real estate, to the burgeoning creative community of South Park. The neighborhood could still be a dangerous place to walk around, but changes were afoot, and Rossetto and Metcalfe were enamored with San Francisco. "There were so many artists and so many musicians and people from all walks of life, and every little street corner had a gallery," Metcalfe recalled later. "There was a lot of underground art and parties with intellectual themes."

The aspiring publishers still needed to raise money. But they hadn't had any luck with the pin-striped executives in the high-rise office suites of Time Inc. and Condé Nast and the other New York conglomerates that dominated the magazine business. The suits hadn't gotten it at all. In their view, computer publications were about machines and software, not politics and culture; in any case they were a boring niche that the big publishers were happy to leave to the smooth-talking Bill Ziff, the magazine heir who'd bet it all on *PC Week*, or to Pat McGovern, the odd bird up in Boston who'd built *Computerworld* into an empire. But the computer trade titans didn't see Rossetto and Metcalfe's vision either. New national magazines were the product of years of careful market research and multimillion-dollar direct mail campaigns. This strange young couple with a fluorescent brochure and a lot of unhinged rhetoric about the digital revolution wasn't going anywhere in this world.

"They were like, 'Yeah, have fun with your little hobby magazine there,'" Rossetto recalled. His vision was far grander than that.

In California, Rossetto and Metcalfe were among their people. An unlikely promoter named Richard Saul Wurman, an architect by training, had also seen that the future of technology was about culture and commerce, not bits and bytes, and he'd somehow persuaded prominent young techies and assorted creative thinkers to come to an event he called Technology, Entertainment and Design. TED, with its signature TED Talks, would later become an international media franchise, but in its first incarnation it was a once-a-year, four-day event in the out-of-the-way coastal town of Monterey. Rossetto credits the second annual TED gathering, in 1991, with persuading him that there really were digital revolutionaries out there who needed a magazine. At TED the next year, he sought out Nicholas Negroponte, founder of the MIT Media Lab, who offered him a seed investment on the spot (and secured for himself the coveted back-page column slot).

It was Metcalfe who gave their project a name: *Wired.*

Some of their ideas, and their future writers, were already embedded in Ken Goffman's *Mondo 2000*, but Goffman, who went by the pen name R. U. Sirius, was publishing on a shoestring, and his title was too edgy for the mainstream. *Wired*, on the other hand, was very much about bringing the ideas to the masses, with all the design and marketing horsepower it could muster. Goffman recalls the dawning realization in the early 1990s that "the Internet is a place where people can have businesses and feel safe, it's not just for weirdos. Like [author and tech-culture critic] Doug Rushkoff said, 'the Internet is going to be serious, not R. U. Sirius.'" *Mondo* would gradually fade away as the 1990s wore on.

Rossetto had convinced designer John Plunkett and his wife and partner, Barbara Kuhr, to join the venture as creative directors, and they produced a radically distinctive look, with Day-Glo colors and creative type treatments that at their best were bold, fun and elegant all at once, if sometimes hard to read. Kevin Kelly, whose sunny, Christian-inflected positivity about the tech future would prove a good complement to Rossetto's political certitudes, was recruited from the *Whole Earth Review* as executive editor. John Battelle, then a twenty-five-year-old Berkeley journalism grad, was brought on as managing editor, getting the job despite a clean-cut look and earnest vibe that almost disqualified him—*Wired* hired for cool from its earliest days.

Battelle had grown up in Pasadena, and he'd learned a little programming on his parents' Apple II—enough that when he got to UC Berkeley and needed a part-time job, he found one helping an entrepreneur create software for the newfangled IBM personal computer. After graduation and a brief stint at the trade publication *MacWEEK*, he'd returned to Berkeley for the graduate program in journalism, and though he was subsequently offered a coveted entry-level job at a suburban edition of the *Los Angeles Times*, he wanted to cover tech. The *Times* told him they already had someone on the Silicon Valley beat. (That was me.)

Louis Rossetto called over Labor Day weekend in 1992. "He basically pitched me my idea for my master's thesis at Berkeley that I never got to do—a magazine on digital technology with broad cultural appeal, a *Rolling Stone* of tech," Battelle recalled later. Still, he wasn't totally convinced when he met Rossetto a few days later at the decrepit South Park warehouse. "He had this prototype of the magazine that he had done with Plunkett and I thought, 'This is a little too *Mondo 2000* for my taste, a little too weird,'" Battelle recalled. "But the story selection was good, and then we just started talking about stories."

They'd get the first issue of *Wired* out in early 1993, distributing it the old-fashioned way.

"We took the boxes and went to Macworld and stood at the top of the escalators and handed them out," recalls Battelle. His wife, Michelle, who worked in the TV business, had a connection at CNN, and she persuaded the network to come out, which turned out to be a boon. "The piece was like three or four minutes long, and they ran it over and over again because CNN was really young and they didn't have a lot of content," says Battelle. "Kind of a cool piece, with a bunch of weird-looking people in San Francisco, right?" Crucially, the magazine had succeeded in securing national newsstand distribution. There were also ads on the sides of buses in San Francisco and New York: "Get Wired."

It was an instant sensation. Subscriptions poured in unsolicited through the mail as the magazine touched the large universe of people who were fascinated by new technologies and believed in their potential to change everything, but had never realized there was a whole world of like-minded people out there. It also arrived just at the moment when multimedia software based on CD-ROMs—primarily games, educational programs, electronic books and magazines—was taking off. The CD had been around for a while for recorded music, but only recently had the technology improved enough for it to easily handle pictures and video. A company later known as Macromedia had set up shop in SoMa in 1989 and created crucial tools for making interactive CD-ROM software. The new medium was spurring enough activity that people were starting to call the area around South Park "Multimedia Gulch."

The building at Second and South Park became an incubator of sorts for emerging media. *Wired* moved to the third floor, making way for writers and cartoonists Mark Frauenfelder and Carla Sinclair and their culture zine, *Boing Boing*, which anticipated some of *Wired*'s enthusiasm for topics like cyber science fiction and brain-machine technology. Zines, traditionally aimed at niche enthusiasms, were having a moment: *Boing Boing* sold 17,500 copies of its most popular edition. A literary magazine called *Might*, cofounded by a young writer named Dave Eggers who'd later pen best-selling novels on topics including the city's Internet culture, would move into the building too. The old warehouse was awaiting an earthquake retrofit, and the rent was exceptionally cheap.

At the other end of South Park, a sturdy onetime printing factory at the corner of Third and Bryant was starting to see small multimedia companies replace the garment workshops where Chinese seamstresses still toiled. One of those companies, Vivid Studios, founded by a group of designers and working at first on CD-ROM software, took it upon itself to promote the neighborhood's new brand. Nathan Shedroff, one of the Vivid founders, printed up maps showing all the multimedia businesses in the neighborhood. When *Wired* needed bigger and more permanent space, he suggested the company move into the Third Street building, which would become a headquarters first for the multimedia industry and then for the Internet industry that quickly supplanted it. The building's denizens nurtured a party culture that fit the moment: When *Wired* won a National Magazine Award in its first year, the Vivid crew swarmed its office with Jello shots—"We were very generous with alcohol," Shedroff said later. The antics built a lot of bonds among the companies and contributed to the collaborative spirit bubbling across Multimedia Gulch.

Soon a fresh group of offbeat young talents, many inspired by *Wired*, began descending on SoMa from around the world, looking for their way into the new digital world. The *Wired* offices would become an obligatory stop for politicians, business executives and celebrities who wanted to burnish their tech credentials. Even Vice President Al Gore showed up, touting the "information superhighway." Everyone was trying to figure out what the quick advances in digital technology might mean, for the city and the country and the future of humanity.

"We were trying to define digital culture," recounted Daniel Carter, then a young magazine designer who'd moved west after a stint with the legendary but short-lived New York magazine *Spy*. "Before, if you were interested in technology, you were just some shlumpy person with a weird interest in something no one understood. Silicon Valley was a bunch of semiconductor companies. The Internet blew up that stereotype. Now there seemed to be a new type of person. It was . . . us." In the new tech era, design and creative skills were the equal of technical excellence and a necessary complement. San Francisco was the natural home for such a marriage.

San Francisco's media economy, not yet disrupted by the Internet, was still fairly robust. For half a century, the city's self-conscious sensibilities had been defined by Herb Caen, the nonpareil columnist for the *San Francisco Chronicle* who wrote six days a week for more than fifty years and established the mood and the mores of the city with authority, humor and verve. He'd given his crucial blessing to the gays and the hippies back in the '60s, and mandated that his town was never to be called "Frisco" or "San Fran"—prohibitions that stand to this day. He never tired of celebrating the city's delights, even in tough times, and his sympathy for slow-growth politics, civil rights and environmental protection were a major boost for the local left, even if he wasn't of the left himself. He developed an especially close bond with Willie Brown, the extraordinary politician who'd come to dominate local affairs, and the two of them would have lunch together every week for thirty years.

Over at the *Examiner*, the original flagship of William Randolph Hearst's tabloid empire, the swashbuckling editor Phil Bronstein was in charge, bringing the populist flair of the publication's founder and news-making abilities of his own: He'd later marry the actress Sharon Stone, of *Basic Instinct* fame, and make headlines getting bitten by a giant lizard at the Los Angeles Zoo. Bronstein appreciated out-of-the-box writers and characters like Warren Hinckle, who'd helped animate the city's literary scene since the 1960s, when he ran the influential left-wing magazine *Ramparts*. The afternoon *Examiner* shared business operations with the staid morning *Chronicle*, founded back in 1865 and still owned by the de Young family, but it wasn't a happy marriage. Will Hearst III, the grandson of the founder and publisher of the *Examiner* in the 1980s, would lament later that the so-called joint operating agreement between the two papers had handcuffed him in his efforts to meet the digital moment.

The *Bay Guardian*, founded in 1966 by Bruce Brugmann and Jean Dibble, had followed New York's *Village Voice* in creating the blend of left-wing politics and entertainment listings that defined the alternative weekly. Brugmann was still in the chair, and he and his longtime right-hand, editor Tim Redmond, were setting the agenda for the political left. The *Bay Guardian* could easily make or break candidates with its endorsements, and it ceaselessly beat the drum on pet issues, especially the perfidy of PG&E, the local power company. Along with an upstart rival, *SF Weekly*, the paper was thick with ads for bars and bands and movies, a go-to choice for young people looking for something to do.

"My primary source of information was the *Bay Guardian*," recalls Sharky Laguana, who'd arrived in the city in 1990 as a homeless teenager and would eventually become a business owner and a member of the city's Homelessness Commission. Laguana was a foster kid from Cincinnati, who after aging out had followed a girl to a spiritual retreat center near Santa Cruz called Mt. Madonna, and after she broke his heart he made his way to the city with nothing but a cheap guitar and dreams of making it in the music business. "I was poor, I was a renter, I was intelligent but I was not a target for any of the things that businesses were looking for," he said. "My politics, I just believed anything the *Bay Guardian* wrote."

Community newspapers, including two serving Chinese speakers, had strong followings too. The *Bay Area Recorder*, chronicling the gay community, entered journalism lore in 1992 with a cover depicting then–San Francisco Sheriff Richard Hongisto obscenely handling his baton, dubbing it his "big stick." It was a reference to Hongisto's ham-handed roundup of protesters and journalists during the local version of the riots that followed the Rodney King verdict in Los Angeles. Hongisto responded to the stinging satire by ordering his deputies to confiscate the newspapers from their distribution boxes on the streets—an impressive display of bad judgment that quickly cost him his job.

The *Los Angeles Times*, for its part, with a newsroom 1,200 strong and profits running at $200 million a year, was bigger than its two San Francisco counterparts combined and took great pride in its journalistic excellence. In 1990, the digital future was already visible on the horizon, and "Los Angeles" was not a brand that played very well in San Francisco, but the corporate bosses at the paper—then still controlled by the founding Chandler family—had nonetheless decided that this was the moment to advance a long-standing aspiration to be the "newspaper of the West."

Just a few months after I'd arrived that year, the bureau moved from desolate Fox Plaza, near City Hall, to a shiny new building clad in dark red marble at 388 Market Street, in the heart of the financial district. It was a huge upgrade: We had a whole floor with curved, floor-to-ceiling windows, outfitted for a newsroom of forty and envisioned as the home base for a new Northern California edition of the paper. But we had barely unpacked when the brass thought better of it, spooked by the recession of 1990 and the obvious risks, and abruptly pulled the plug on the project. The bureau of a half dozen reporters and columnists would rattle around the fancy space for the next five years.

For me, an awkward question remained: If I was the Silicon Valley correspondent, what was I doing in San Francisco?

The sunny tech heartlands of Palo Alto, Santa Clara, Menlo Park and Cupertino, with their sprawling, low-slung office complexes and strip malls hugging traffic-clogged suburban boulevards, were a whole different world. Thirty, forty and fifty miles south of the city, global brand names like Hewlett-Packard and Intel employed tens of thousands of people, while up-and-comers like Sun Microsystems nipped at their heels. The older companies had emerged from a very different kind of tech culture, one where the military loomed large, and there was still a lot of soldering and metal-bending going on inside faceless factories built of prefab walls. Some of the biggest operations—like the IBM disk drive plant in San Jose—were owned by the behemoths from New York and Boston that still ruled a very corporate computer industry. There were a lot of good jobs in Silicon Valley and pleasant suburban homes that people could buy with the wages they earned. With cutting-edge research coming out of Stanford and a startup economy supported by the venture capital firms lining Sand Hill Road, on the border of Palo Alto and Menlo Park, there was plenty of innovation too.

Yet even in Silicon Valley, the early 1990s were not a very optimistic moment. Japanese companies had emerged as formidable competitors in the semiconductor business that had given the Valley its name, and were threatening to upend the rest of the computer business too. That—not what was happening with the hackers and punks of SoMa—was the story of the moment for a technology business press corps that consisted of just a few dozen people across all of the mainstream newspapers and magazines, both national and local.

Andy Grove, the Intel CEO, had sent a fiddle to Washington a few years earlier, a message about what he thought the government was doing while the US chip industry burned. Hewlett-Packard, the original garage startup and paterfamilias of the Valley, was struggling in a highly competitive and volatile computer business. Steve Jobs, ousted from Apple, was desperately trying to get the doomed NeXT computer out the door, reduced at times to virtually begging reporters for ink. (He tracked me down one morning at my girlfriend's apartment—"Hey Jonathan, it's Steve. Steve Jobs"—chatting like we were old friends.) Microsoft was ascendant, though despised by many in Silicon Valley as a copycat for its Windows software, and a bully for squeezing out competition. An Apple spin-out called General Magic was the hot startup of the moment, but would prove too far ahead of its time.

Compared with today, the tech industry was small, and strikingly personal. Bill Gates, usually disheveled and always argumentative, was happy to mix it up with critics in the hallways of Esther Dyson's PC Forum, or to lecture me, in a memorable interview, about how stupid I was to think that a federal antitrust investigation of his company was even news. (Microsoft's protracted battle with the US government, which began in 1992, would handicap its business for a decade.) Computer and software companies occupied only a modest corner of the global business landscape, certainly not on par with the big banks and oil companies and the other stalwarts of the old economy, and wielding none of the power and influence they'd later enjoy.

Tech was also very cyclical, and the recession that took hold in 1990 solidified a down cycle in the industry that wouldn't reverse for a couple of years. It didn't look obvious at all that tech was going to be the savior for San Francisco's faltering economy.

■ ■ ■

Brian Behlendorf was at the Great American Music Hall, a storied concert venue in the heart of the city's Tenderloin district, when he ran into Jonathan Steuer, who'd just been hired to run *Wired*'s new online venture, soon to be named HotWired. It had been a year since Behlendorf had launched SF Raves and his hyperreal.org website.

Steuer, a Wisconsin native, had moved to the city for graduate studies in digital media at Stanford, and was living in a group house on Ramona Street in the Mission. He'd joined The Little Garden, and was developing his own weekly dinner salon, called Cyborganic, to bring the burgeoning crowd of digital enthusiasts together. Now he'd been tapped to create an electronic expression of *Wired*, first on America Online—the go-to choice for nontechies—and then on the amorphous new medium of the Web. Steuer knew Behlendorf from SF Raves, and they were both big fans of the band that was playing that night, a rave ensemble with the too-perfect moniker Space Time Continuum. Steuer invited Behlendorf to come by the *Wired* offices and have a look.

Behlendorf took him up on it, and soon he started helping out, getting paid a pittance but reveling in the creative swirl, and the possibilities. *Wired*, as a company, was blowing up, hiring scores of people for the magazine and the new digital venture and a host of other projects as Louis Rossetto pursued his expansive vision—a TV show, a book publishing imprint, Wired cafés, maybe even hotels. As a brand, *Wired* had captured a moment, and a story of monumental import, and it was ready to double down. Music always pumped through the *Wired* offices, and now the old printing factory was throbbing with optimistic energy—smart young rebels with brightly colored hair and too many tattoos, merrily creating the future. It was all a world away from the New York media establishment. The aspirations were boundless.

"This Internet thing could be how you actually bring society together," Behlendorf recalled thinking at the time. "I got so wild-eyed about that."

Rossetto had been jolted into action when a September 1993 *Wired* cover story about Singapore had led to the issue being banned in that country, and some Singaporean engineers who were fans of the magazine had taken it upon themselves to replicate the issue on the fledgling World Wide Web. The success of *Wired* had led Condé Nast chairman Si Newhouse to invest $2.5 million in the company at the beginning of 1994 to build up its print circulation, but since subscribers were coming in on their own, Rossetto decided to use the money for the Web project instead.

"We were making *Wired* and realizing what the opportunities were in this Internet space beyond AOL, and we realized that [the Web] was basically the future of media, period. The Web was the place where individuals could create media and build businesses outside of the control of the normal channels. You didn't need a gatekeeper like Condé Nast or Ziff Davis to tell you whether you could make a magazine, you could just make something on the Web and go directly to people." It was natural for *Wired* to be out front.

"We didn't just think of ourselves as reporters chronicling something," Rossetto added. "We felt ourselves as players. We were eating our own dog food. We were going to make our own media and pioneer this new space."

Rossetto had set up HotWired as a separate entity, and everyone agreed that it needed to be its own thing, a new kind of media. But the consensus ended there. Steuer had introduced Rossetto to Howard Rheingold, a big personality on The WELL and a pioneering thinker on virtual communities, and he'd been brought on as executive editor. Rheingold wanted HotWired to focus on building community, and two-way relationships among readers and staffers. Rossetto wasn't interested in that at all. He was a magazine guy, and an evangelist; people were supposed to listen to him, not the other way around. The community was the readers and advertisers, Rossetto explained later, not amateurs creating content. "He was like, this is going to be a slick, polished thing with an attitude," Steuer recounted. "And we don't care what anyone thinks about it. The last thing we want to do is have conversation, community, interactivity—fuck all that."

The questions didn't end there: What kind of stories would be offered on HotWired beyond what was in the magazine? Would people be required to register? Would they have to pay? How would it make money?

And then there was the small matter of how to make anything at all. Website software barely existed.

Behlendorf was hired as perhaps the world's first "webmaster," and he would find the solution. From the beginning of SF Raves, he'd been engaged with a group of researchers at the National Center for Supercomputing Applications at the University of Illinois-Champaign, led by a graduate student named Rob McCool. The NCSA was developing software for hosting websites, aiming to improve on Tim Berners-Lee's original code. Like most early Internet work, it was a collaborative project, underwritten by the federal government, with people voluntarily contributing and the results made freely available—an approach later to be labeled "open source." The group would share information via the project's mailing list and send one another software fixes, or "patches," and then try them out. If they were good they'd be integrated into the code. The NCSA software, known as a Web server, would power HotWired.

At the very same time, another group at the University of Illinois, led by a combative, hyper-intelligent twenty-eight-year-old named Marc Andreessen, was building a Web browser, called Mosaic. Andreessen had grown up in rural Wisconsin and had big ambitions, one of which was to make money, so he was all ears when he got a call from Jim Clark, a computer graphics expert who'd made a fortune with his first company, Silicon Graphics, before being pushed out by his investors. Like Andreessen, who resented the pretentiousness and privilege of the Stanford and MIT grads, Clark had a huge chip on his shoulder, having never gotten the respect from the Silicon Valley elite that he thought he deserved. Clark persuaded Andreessen to leave Champaign and come west, and together they formed Mosaic Communications, soon to be renamed Netscape, to commercialize Web software.

Shortly thereafter, Rob McCool and the other Illinois grad students Behlendorf had been working with remotely had some news: "They tell the list, sorry, we're not gonna be able to release any more updates. We're all leaving school and going to join this startup in Silicon Valley," Behlendorf recalled.

McCool's NCSA server software was good, and the university's license allowed anyone to develop their own versions, as long as they shared the code. Since the university no longer looked like a good partner, Behlendorf proposed to his collaborators on the list that they do it themselves, and about twenty people joined in. Behlendorf had two motivations: He wanted a good Web server for his own purposes—HotWired already needed features beyond what the NCSA software offered—and he believed strongly in the idea of open source. Personal computer software was becoming a Microsoft monopoly, limiting innovation and keeping costs high. Maybe the Internet could do better.

"We thought it'd be really nice if the Web stayed open and didn't go the way of the desktop," Behlendorf says. "A browser is a lot harder to build than a server. But we felt if we could keep the server side open, we'd have a shot." He'd just seen a documentary about a Native American tribe, and its name had a nice resonance with the patches they shared to improve their code. The Apache Group was formed, with twenty-one initial members, and would prove an extraordinary success: The free, open-source Apache Web server would run a majority of all websites on the global Internet for more than fifteen years.

Back at HotWired, Behlendorf and his software solved one key problem: providing the technical foundation that would organize and control the content and interactions that are at the core of a website. But that was just the beginning. Rossetto at first insisted on requiring registration for HotWired's users, so Behlendorf and the tech team had to figure out how such a registration would work. How could the Web server know whether a particular visitor was registered or not? They came up with the idea of a digital token that would be discreetly left on a visitor's machine, so the website would recognize them the next time they visited—something we now know as a cookie.

If the technical hurdles were challenging, the business questions were no less daunting. John Battelle wasn't in charge of HotWired, but he was responsible for the overall editorial budget, and HotWired was supposed to have its own editorial staff of several dozen digital-savvy journalists, writing for topic "verticals" like travel and culture. He raised the obvious questions at a weekly editorial meeting.

"How are we going to pay these people?" Battelle asked. "We don't have enough money in the company. What's the business model? We can't charge subscriptions, everybody is already paying fourteen bucks a month for dial-up service and no one is going to want to pay more for, like, information." Advertising was the obvious alternative, but there was no template for Internet advertising. What would the ads look like? How would they work?

Tim O'Reilly, a computer book publisher based in Sebastopol, a town in rural Sonoma County about fifty miles north of the city, had already seen that this new construct of the website could effectively be an advertisement in itself. The company had also built Global Network Navigator, an early tool for finding things on the Web, and it had an ad across the bottom of the screen. Another example was Prodigy, the IBM service, which had a blinking advertisement that would take over the whole page when it was clicked. Battelle suggested a horizontal graphical advertisement, and Rossetto had the idea of putting it at the top of the page, where it would disappear as people scrolled down. When clicked on, rather than taking over the page, it would take you to the website of the advertiser.

And so the banner ad, a core Internet advertising format to this day, was born.

There was still a problem though: If clicking an ad took you to the advertiser's website, that advertiser would need to, um, have a website. And nobody had a website.

Steuer's childhood friend and now-housemate Jonathan Nelson, a music producer and event promoter as well as a techie, saw the opportunity. He'd been following the emergence of the Web, and was especially keen on figuring out how the Internet might help the music business. Now he recognized that companies would want websites. So together with his brother Matthew, Brian Behlendorf and a programmer named Cliff Skolnick, he started a company called Organic Online and rented an office just upstairs from *Wired*. Behlendorf began moonlighting for Organic, shuttling between the two floors as he built the technical foundations for both HotWired *and* its advertisers. Nelson lugged around a computer with Behlendorf's software on it to show marketers what they were talking about. When HotWired launched in the fall of 1994, five of the seven advertisers had websites built by Organic.

HotWired got off to a splashy start, but quickly learned the lesson that countless publishers have confronted since: It's hard to make money from Internet banner ads, and all-but-impossible to earn enough to pay a substantial editorial staff. At first, big advertisers were happy to pony up $10,000 a month for a banner—*Wired* was the "hot book." But online audiences were small, and slow Internet speeds and the janky browsers of the day made it hard to do anything fancy. Wired Digital, as the new company formed to house HotWired was called, was spending money like mad and not bringing in very much, and soon there were layoffs and a changing of the guard. Andrew Anker, a young investment banker who'd been hired as the CEO, pushed out Steuer. Rheingold soon left too.

Steuer joined a San Francisco startup called CNET, where Halsey Minor, an aristocratic young entrepreneur from Virginia, together with his friend Shelby Bonnie, were creating cable TV shows about computers and technology out of an old warehouse building near Fisherman's Wharf. Now they were eyeing the Web. They'd had the foresight to acquire some very valuable Internet addresses—or "domain names," as they were known—including news.com. Before long CNET was mainly an Internet site devoted to tech industry news, with a little TV on the side. It was a good niche, and CNET would go on to be a major player in tech media.

Jim Clark and Marc Andreessen's Netscape launched its Web browser in the fall of 1994—you could buy it in a box for $49—and by the next year websites were springing up all over as the Internet caught the attention even of those who didn't care about tech. In August 1995, Netscape sold stock to the public for the first time in an initial public offering and saw its shares quadruple in price. It's a day still referred to in Silicon Valley as the "Netscape moment," when venture capitalists and others in the money business realized that the Internet was not only the next big thing but maybe the biggest thing ever.

Yahoo, a simple, category-based index of Web content founded in 1994 by Stanford graduate students David Filo and Jerry Yang, had at first vowed never to take advertising (they told me as much directly when they came to visit the *LA Times* in early 1995). Many of the pre-Web Internet pioneers, mostly academics and researchers, were still fighting the commercialization of the Internet. But the writing was now very clearly on the wall. Before the year was out Filo and Yang changed their tune and raised money from Sequoia Capital, king of the venture capital firms, and would go public in 1996.

It wasn't all about the money though, not yet anyway: The idealistic vision of the Internet as a liberating tool of personal empowerment and community connection, open to use and build upon, and free of censorship and profiteering, still burned bright, especially in San Francisco.

Those beliefs would soon be put to the test.

CHAPTER 2

The "Motor City" of the Internet

Willie Brown, dapper as always in an elegant dark-gray suit, grinned broadly as he surveyed the star-studded crowd that had gathered for his inauguration, gesturing with delight as he acknowledged friends in the audience and rivals too.

It was a chilly, foggy day on January 8, 1996, but 7,500 people had turned out for his swearing-in at the outdoor plaza of the Yerba Buena Center for the Arts, an unprecedented showing for such an event. The sixty-one-year-old Brown bantered with President Bill Clinton, who had joined the festivities by phone, teasing the president for making him wait and then boasting of the glory of a city with "no snow and no Republicans," accompanied by his trademark cackle.

He warmly called out his predecessors, who together made up a tableau of the tumultuous political history of recent years: Frank Jordan, the genial but awkward former police chief who'd always seemed out of place in City Hall, and lost to Brown badly in his run for reelection; Art Agnos, adored by the left in this deep blue town but so ineffectual that he was a one-term mayor too; and of course Dianne Feinstein, who'd led the city in the wake of the horrific murders of Mayor George Moscone and Supervisor Harvey Milk at City Hall in 1978. Barbara Boxer, who together with Feinstein constituted California's historic all-female delegation in the US Senate, was there, and Representative Maxine Waters, among the country's most senior Black elected officials, was up from Los Angeles. Representative Nancy Pelosi, not yet Speaker of the House but clearly a rising star, was stuck in a Washington snowstorm, "in tears," Brown reported, to be missing the festivities.

Gina Moscone, the late mayor's widow, along with her children, was a guest of honor, an enduring symbol of a trauma that still shadowed the city's politics. Her husband had been a liberal's liberal, devoted to civil rights, a union man from a working-class Italian family, handsome, charming and a bit of a rogue, a partyer in a town that liked to party. Like his close friend and mentor Phil Burton, a legendary liberal lion and power broker who'd held the congressional seat later occupied by Pelosi, Moscone had been aligned with the progressive wing of a national party that had nominated George McGovern in 1972, and locally he tilted against the "downtown" corporate interests in the name of defending the neighborhoods. Moscone had won the 1975 mayor's race by a whisker over an opponent who represented the white, ethnic old guard of San Francisco, which had for years been fighting the hippies and gays and lefties and artists who'd made the city the headquarters of the global counterculture. Moscone, along with Milk—one of the country's first openly gay public officials and a hero in his Castro district—was assassinated by another member of that old guard, former Supervisor Dan White, who was then acquitted of murder and would serve just five years for manslaughter. The trial is remembered nationally for the "Twinkie defense," with White's lawyers arguing that eating so much of the ersatz pastry was a sign of his mental illness. Locally, though, it was seen as something far darker: politicized prosecutors and corrupt policemen joining in an unholy alliance to protect one of their own.

Willie Brown and George Moscone were kindred spirits, people lovers and men-about-town who reveled in the push-and-pull of politics and wanted to stick it to the old-boy establishment. Moscone's political path had been as straightforward as it gets: He grew up in the Marina, met his future wife in grade school, attended the prestigious St. Ignatius Catholic high school and went to law school at Hastings, the traditional training ground for local politicos. He became friends with Phil Burton and his brother John, later a congressman as well, who together were building a potent, progressive Democratic political machine. They recruited Moscone to the cause.

Willie Brown had gone to Hastings too, but he'd had to work as a janitor to pay the bills.

Brown's career was nothing less than a tour de force of personal and political entrepreneurship, perhaps the greatest up-by-your-bootstraps story in American government since Alexander Hamilton. Born very poor in a violent, segregated small town in Texas, Brown was industrious from childhood, shining shoes and doing odd jobs and keeping his head down, before leaving for San Francisco at seventeen to live with an uncle—a hustler of the benevolent kind, by Brown's description. Working his way through college at San Francisco State, and then Hastings, Brown's exceptional intelligence and formidable capacity for work, combined with charm, humor and a killer sense of style, were hard not to notice even amid the deep prejudices of the day. He quickly gained notoriety as a young criminal defense lawyer, representing the streetwalkers and drug dealers that no one else wanted to take on, and then plunging into the burgeoning civil rights movement in the city. He helped lead a sit-in against racist landlords at a housing development called Forest Knolls, and organized the defense of hundreds of SF State students arrested in an antiwar protest. He'd later represent Mario Savio, the leader of the Berkeley free speech movement.

Brown first ran for state assembly in 1962 after being tapped by Phil Burton, who was determined to oust a long-serving conservative Democrat. He lost, but ran again two years later and prevailed, and would spend the next three decades representing San Francisco in Sacramento, wielding enormous power as the wily, indefatigable Speaker of the assembly for most of that time. His hold on the body was such that he was able to flip two votes and retain the speakership even after Republicans won the majority in the 1994 election, an almost inconceivable feat.

Brown honed a highly personal and transactional approach to government, tending to the needs of legislators—and also lobbyists—and demanding loyalty in return, while retaining a lucrative legal practice on the side. It wasn't everyone's idea of good government, to put it mildly, but Brown made no apologies, securing jobs and ceremonial appointments for political allies and seeking deals with business interests and Republicans across the aisle, all in the name of getting things done. It was classic machine politics in many ways, with jobs and other favors for allies, and creative punishments for those who strayed; assembly members who had a committee chairmanship, or even a nice office in the capital, knew very well what they had to do to keep them.

The target of more than one federal corruption investigation, Brown liked his Italian suits, exotic sports cars and pretty women, but he laughed off allegations of self-dealing. The FBI in the late 1980s had mounted a massive sting operation involving a fake shrimp processing business that ensnared a dozen people, but not the then Speaker of the assembly, widely assumed to be the target. Willie Brown wasn't considered the smartest person in California politics for nothing, and he knew what lines not to cross. In his 2006 autobiography, Brown dismisses the episode as an elaborate entrapment set up by unethical—and bumbling—careerists at the FBI, who wanted the glory that would come with bagging a top Democratic politician, and a Black man to boot.

Though the Feds never got anywhere in their pursuit of Brown, their efforts emboldened Republicans who'd been driven to despair by the speaker's iron grip in Sacramento, and in 1990 they successfully promoted a term-limits measure conceived primarily to end his dominance. Now the "Ayatollah of the Assembly," as he once called himself, would need a new gig. The mayor's office seemed like an obvious step in hindsight, though Brown would say later that it didn't look that way at the time.

Truth be told, Brown didn't love the more mundane aspects of retail politics and hadn't faced a competitive election in thirty years. "I have never shaken as many hands as I had to shake this time around in all of my life," he joked to the crowd on Inauguration Day. "I'm not sure I ever want to see another bus stop at seven thirty in the morning." He'd also have a lot of work to do in understanding the complicated mechanics of San Francisco's city government, which operated according to a 1932 City Charter that put numerous limitations on the powers of the mayor, and had been amended by voters so many times that it was impenetrable to all but the most committed insiders. The city was a complicated place; with its exceptional diversity, unique geography and boom-and-bust economy, it was often deemed ungovernable.

Brown seemed to be trying to convince himself as he waxed repeatedly on the need for unity now that the election was over. "The differences are no longer there. They must be blurred. We must unite," he implored. And on this day at least, that hope didn't seem too far-fetched, with the celebration capped by the mayor taking a turn as the conductor for the middle school band that had come out to play. That evening, he'd trade his suit for a regal outfit featuring a cape and tails—over the top even for Brown, a legendary clothes horse who sometimes changed four times a day—and be serenaded by the rock icon Santana and other local heroes, along with a hundred thousand citizens, at a "Soul of the City" party on Pier 45. Willie had star power, and he was leaning into it hard. And for the moment at least, the city was swept up in the Willie show, ready to believe that it was all one big happy family now that their proud papa was at the head of the table.

Brown knew very well, though, that the divides were deep, and maybe unbridgeable.

Dianne Feinstein had done much to push the transition to a white-collar economy, and plenty of big companies still called the city home. But anti-growth sentiment ran strong. Led by Haight-Ashbury activists Calvin Welch and Sue Bierman, the slow-growth movement grew out of the "freeway revolt" of the 1960s, when neighborhood activists succeeded in stopping interstate connections that would have ripped through the middle of Golden Gate Park and marred the northern waterfront. Voters approved limits on new office buildings in 1986, and another ballot measure banned any construction that would cast a shadow on a city park. These were expressions of what the political scientist Richard Deleon would call a progressive "anti-regime" that sought to stall further development by any means necessary, on the grounds that it benefited only monied interests. It was a dynamic playing out in cities around the country as environmental regulations and the "new urbanism," popularized by Jane Jacobs in her landmark 1961 book *The Death and Life of Great American Cities*, brought stiff opposition to redevelopment projects. In San Francisco especially, it had gotten very difficult to build anything at all.

San Francisco's progressive coalition was a reflection of the city's extraordinary diversity, with scores of political clubs and community groups representing a proverbial rainbow of ethnic, racial and cultural perspectives; they often converged around neighborhood protection and preservation. Classic NIMBYs held sway in the city's western residential precincts, from the fog-shrouded mansions of Sea Cliff, near the Presidio, to the neat row houses of the Richmond and Sunset districts, and the leafy precincts of West Portal and Park Merced; those neighborhoods were not interested in change, and they were especially not interested in apartment buildings that might block their views and bring a different population. Then there were the community organizers in the poorer, denser districts, especially the Mission, SoMa and Chinatown, who were against development on the grounds that it pushed out low-income residents. They effectively allied themselves with the wealthy Westside homeowners fighting change in their neighborhoods.

Brown was a civil rights warrior and an ally of labor unions, and he'd even been a leader of the freeway protests back in the 1960s. But after many years in Sacramento, he was no lefty when it came to economic issues. He'd evolved into a pro-development "business Democrat," friendly with the local oligarchs, and he liked to talk about infrastructure: stadiums, train lines, office towers and hospitals, and the remaking of old industrial sites into twenty-first-century housing. They were the sort of ambitious, legacy-making projects that get many an urban leader going, and Brown had a long list. Most urgent was a new baseball stadium to keep the Giants in town; Candlestick Park, on a picturesque waterfront spit at the city's southern border, was too cold and windy to draw crowds, and the team was constantly threatening to scurry off to San Jose. There were the massive brownfields of the former naval facilities at Hunters Point and Treasure Island, seemingly good sites for lots of new housing, but poisoned with toxic waste. And there was Mission Bay, where the old Southern Pacific rail yard offered the biggest parcel of developable land in the city. Politically, these sorts of developments were generally popular, uniting bankers and developers and the construction trade unions, and offering the promise of fresh tax revenues along with new civic venues and better transit. But the anti-growth contingent hated them. There were no easy compromises here.

San Francisco was a small city, but its government was outsized, in part for structural reasons: It was both a city and a county, for one, meaning that core responsibilities that reside with counties in California, especially public health, were part of the city's brief. It directly owned San Francisco International Airport, which generated healthy profits, as well as the Hetch Hetchy dam and aqueduct network, which brought water from the Sierra Nevada mountains that the city both distributed to its own residents and sold to its neighbors. San Francisco also spent more than most cities on salaries and benefits for civil servants, with its public employee unions deeply embedded in the political system. Comparisons were tricky: Municipal budgeting in California is very complicated by nature, with tax revenues and payments flowing back and forth among Sacramento and local governments and school districts according to arcane formulas dictated by the legislature, or sometimes voters. But any way you measured it, San Francisco was spending far more per resident than most cities.

The political coalitions among unions, community groups, political clubs and business interests were intricate too—and crucial to master, in Brown's view, before even thinking about policy. On Inauguration Day, he appointed the city's first Asian American police chief and first Black fire chief—bold and not-strictly-legal moves that Brown remained very proud of years later. He proceeded to diversify city government while installing political cronies in key positions and showering favors on allies in labor and elsewhere, all the while maneuvering to undercut the power of the Board of Supervisors, an eleven-member body of elected officials that functioned as the city's legislative branch and could block the mayor's initiatives, or take up their own. Brown meant it on bringing diversity to city government and getting his big projects done. He meant it on the patronage too.

Brown quickly set to work on what would be his signature initiatives: restoring City Hall, a glorious masterwork of French Renaissance architecture that had been badly damaged in the Loma Prieta earthquake; renovating the Ferry Building and reclaiming the Embarcadero waterfront, long disfigured by an elevated freeway that was rendered unusable by the quake and torn down; winning voter approval for the baseball stadium, which had eluded his predecessors; and getting Mission Bay off the ground.

Brown dominated a room, radiating energy with a mirthful grin, grabbing an arm here, whispering in an ear there, always the best-dressed, always prepared. And now he was dominating San Francisco. Willie was everywhere—at City Hall, fussing over the renovations or posing for a photo-op or persuading somebody of something, but also at John's Grill or Sam's or Perry's, drinking and dealing and hamming it up with constituents, or at the memorial for earthquake survivors, or welcoming a conference to town. He held weekly office hours at Le Central, a French restaurant near Union Square, where anyone could approach him with their pet peeves, or ask him for a job. He was working, always, but making it look like a whole lot of fun. And the city loved him back, at least at the beginning, with approval ratings north of 60 percent. The straight talk was refreshing for everyone.

The digital economy, which would soon upend the city, wasn't on Brown's radar. For all its promise, it was still small beer: Ted Egan, before being named city economist in 2004, had written an academic paper on Multimedia Gulch in the mid-1990s, estimating that the industry employed twelve to fifteen thousand people at the time—less than 5 percent of the city's workforce.

The fledgling industry didn't have a local political agenda. It mostly wanted to be left alone, to make new things and throw big parties and play with drugs and technology without worrying about the Man. The city didn't want anything from the industry either, not yet. "I didn't pay any attention to the tech world," Brown recounted later. "I was trying to figure out local government."

He'd have to pay attention soon enough.

Netscape's 1995 IPO had awakened the venture capitalists of Silicon Valley and the bankers of Wall Street to the promise of the Internet, and they were beginning to stoke a financial tsunami that would sweep across the city in the late 1990s as the consumer Internet was born.

Modern venture capital had come into its own in Silicon Valley in the 1960s, with pioneers such as Arthur Rock, who'd back Intel and Apple, seeing an opportunity in funding California tech companies that the East Coast bankers of the day considered too risky and too far away. They developed a straightforward playbook: raise money from institutional investors such as pension funds and university endowments, and put it into select technology startups a little bit at a time. Though Rock and some others would make huge fortunes, venture capital for decades occupied a modest nook in the palace of high finance, invisible even to much of the business world. In Silicon Valley, it wasn't always respected by the technologists who were taking the money; Andy Grove, the Intel cofounder, once likened it to selling real estate.

But the explosive growth of the personal computer industry in the 1980s had started to raise the stakes, and a handful of VCs began to emerge as important kingmakers, deciding which startups were worth backing. For entrepreneurs, in turn, not everyone's money was equally green: winning the support of a prestigious, top-tier VC was a potent endorsement, helpful for getting publicity and recruiting talent and doing deals.

John Doerr, a St. Louis native with an engineering degree and an MBA from Harvard, emerged as the exemplar of the new-style VC, establishing himself and the venerable firm he'd joined, Kleiner Perkins Caufield & Byers, as the most prestigious of VC brands, uniquely positioned to attract the best startup founders. Doerr had led the Netscape financing and recruited wireless executive Jim Barksdale to be the company's CEO. He'd been an early backer of Amazon too, and then Google.

VCs turned their successful investments into cash with the help of the other big players in the tech-finance ecosystem, the investment banks, who would lead the process of selling shares to the public in an IPO, or perhaps find a buyer for a startup. San Francisco as a money center was tiny compared with New York or London, and Wall Street powerhouses Goldman Sachs, Morgan Stanley, and Credit Suisse First Boston led most of the biggest Internet IPOs. But a quartet of specialty San Francisco investment banks known as the "four horsemen" had for years focused on financing smaller tech companies, and a lot of ships were about to come in. Montgomery Securities, Hambrecht & Quist, Robertson Stephens, and Alex. Brown (the latter based in Baltimore but with historic ties to the city) would all play a major role in the dot-com boom of the late 1990s—and all would be bought up by bigger banks before the decade was out, as Wall Street recognized what was happening in San Francisco.

Once you were able to imagine a world where everyone was connected to the Internet, there was nothing but blue sky ahead. E-commerce, as it was known, was starting to take off, led then as now by Jeff Bezos's Amazon. Financial technology was a hot area too, with a Chicago firm called E-Trade leading the way on Internet stock trading, and established players led by San Francisco–based Charles Schwab & Co. getting into the game.

Chris Larsen, a public school kid from San Francisco, would cofound E-Loan for online mortgages. Elon Musk and his brother Kimbal, immigrants from South Africa, launched Zip2 in Palo Alto, envisioned as an online business directory—the first of Elon Musk's startup ventures. Traditional tech and telecom companies were glimpsing a new order: The Little Garden had sold to a bigger company, Best Internet, in 1996, and now there was a growing industry of Internet service providers, such as Earthlink, that would get you connected. Established Silicon Valley computer and networking companies like Sun and Cisco and HP were offering the picks and shovels of the Internet gold rush.

It wasn't just about the entrepreneurial opportunities in digital commerce and media. The Internet could also provide genuinely new forms of community and human connection, especially for people who were isolated, or discriminated against, or maybe just shy. In the idealistic formulation of EFF cofounder John Perry Barlow's 1996 manifesto, *A Dec-*

laration of the Independence of Cyberspace, it could be a domain of true freedom, not subject to the strictures of the nation-state—though it was never entirely clear what such freedom might really look like.

For Barlow and John Gilmore and the other libertarian champions of the early Internet, making money wasn't incompatible with their vision, but neither was it the necessary or central purpose. A lot of Web pioneers were motivated more by the mission than the financial rewards, though it was often a bit of both. A case in point was PlanetOut, conceived as an online one-stop shop, or "portal," for the gay community. The Internet was an obvious way to connect people who were marginalized, or worse, and gay artists and techies were central to the digital culture revolution that was blowing up in SoMa. Tom Rielly, a peripatetic tech industry marketing executive, had been among the founders of Digital Queers, a nonprofit aimed at building mutual support among LGBTQ people in the business, and he thought the time was right to create a big digital space for everyone. PlanetOut was launched in 1995 on AOL and MSN, an online service from Microsoft, and would debut its website the next year.

Megan Smith, who'd met Rielly through a multimedia project she was leading at Apple, had also been involved with Digital Queers, and would be an early investor and advisor at PlanetOut, and later its CEO. She had quite a pedigree: As an engineering student at MIT in the 1980s, she'd studied under Nicholas Negroponte at the Media Lab and, after a stint with Apple Japan, had landed in Silicon Valley as an early employee of General Magic. Led by Marc Porat, a suave and persuasive big-thinker, General Magic had aimed to build a handheld computer that could communicate—essentially a proto-smartphone—and Porat had convinced Apple's then-CEO, John Sculley, first to hire him and then to let him create a new company in 1990 with key members of the original Macintosh team, including Andy Hertzfeld, Bill Atkinson and Joanna Hoffman. They went on to assemble a formidable group of young engineers, many of whom, like Megan Smith, were destined to play a major role in the Internet revolution.

General Magic was ahead of its time, alas, and folded in 1996. Smith moved to the city and began helping with PlanetOut. Rielly had raised $3 million that year, a little of it from America Online but most of it from Sequoia Capital, and the venture firm's boss, Michael Moritz, a onetime journalist who'd become one of the most successful investors of his generation, had joined the PlanetOut board. With the cash in hand, PlanetOut's small fifth-floor office on Mission Street in SoMa was soon crammed with young people, most of them gay, trying to figure out what an online portal could and should be. There was a veritable movement of "community companies" at the time, Smith recounted later, and they helped each other out. "Remember Net Noir? And Latino Link? And Channel A, for Asian Americans? Blackberry Creek was kids . . . there was iVillage and Women.com," she said. "They got so big they didn't play with us much."

There was trouble at PlanetOut almost from the start though, and Rielly was ousted after the more experienced CEO who'd been brought in by the investors persuaded Moritz to make a change. That didn't solve anything, and a year later Rielly, with Smith's help, worked out a deal where they'd give Moritz back whatever they hadn't spent and buy out his remaining interest for a small sum. It was an unusual agreement, but Rielly got his company back, and Smith then stepped in as CEO. She seemed perfectly suited for the mission, with a cheery optimism that seemed baked into her bones and the technical skills and experience to get her arms around the possible. Yet she'd immediately confront the dilemma that was vexing so many early Internet CEOs: How do you make an online community into a sustainable business, even if you're not trying to get rich?

Like HotWired a few years earlier, a lot of early Internet ventures were discovering that making money online wasn't so easy, and there was a brief blip in investor enthusiasm for all things Internet after the initial euphoria passed. The newly public Internet companies, led by Netscape and Yahoo, were losing a lot of money, and their share prices were falling. It wasn't clear if, or how, people would pay for things online. There seemed to be a lot of hype, but not a lot of profit. And then the Asian financial crisis of 1997 chilled the markets. As the chiding cover-line of my 1997 *LA Times* magazine story said, "Internet, Schminternet. When's This Thing Going to Pay Off?"

Over at 520 Third Street, *Wired* was still in its prime as a magazine, solidly profitable and setting the digital agenda, but HotWired and the other new ventures were a mess: There was a lot of spending, and precious little money coming in. Rossetto had tried to take the parent company public in 1996, positioning it as a cutting-edge digital media enterprise, but it was just after the initial Web frenzy had subsided, and he couldn't hide the fact that most of the revenue came from a print magazine. Nor did he inspire trust on Wall Street with his highfalutin rhetoric and investment documents printed in neon. For its part, Goldman Sachs, which was handling the offering, didn't care much about the comparatively small *Wired* deal and refused to stand behind it when it counted, Rossetto said later. A second effort to sell the IPO also fell short, and suddenly the company found itself over a barrel, even as *Wired* was reaching the peak of its powers as the avatar and tastemaker for what some were now calling the "digerati." Rossetto and Metcalfe were forced to accept financing that stripped them of control, and in 1998 the magazine—though not HotWired—would be sold out from under them to Condé Nast, one of the companies that wouldn't give him the time of day five years earlier. It was a bitter denouement for a venture that more than any other shaped the course of San Francisco's early Internet industry.

Wired would continue to fertilize the new ecosystem in numerous ways though. John Battelle, though he didn't have any management experience when he joined *Wired*, had become the guy who made things happen, and had shown a lot of creative flair too. Figuring he was never going to get the top job at Louis Rossetto's magazine, he'd conceptualized his own publication, a fast-paced weekly newsmagazine that would tell a different story than *Wired*, adjacent to it but less about politics and culture and futurism, and more about the on-the-ground realities of how the Internet was changing the world. That magazine would launch in 1998, and would also be destined for a wild ride.

Just around the corner from *Wired*, on a South Park–adjacent block called Jack London Alley, Fran Maier, a Stanford MBA, was literally sweating it out in the dank office of a company called Electric Classifieds. She'd been brought on to help start a Web business called Match.com aimed at putting the venerable newspaper "personals" online, and it had gone surprisingly well. Even those who were familiar with the Internet were wary about potential dangers in the early days, and thus personal ads seemed like a tough sell. But Maier was finding the opposite. Launched in 1995 as the first online dating site, with rudimentary profile questions and simple text, Match began charging for memberships the next year, and the business was immediately bringing in more money than it was spending—a rare distinction among the dot-coms of the moment.

Maier was hired because the company's founder, Gary Kremen, believed (correctly) that a dating site needed to appeal first and foremost to women. Now it was growing like a weed, but unfortunately the rest of the company was not doing so well, and fundraising efforts were coming up empty amid the brief but steep 1997 downturn. Over the objections of Maier and Kremen, who'd been sidelined by his investors, the company's board of directors decided to sell Match.com. "If I'd had more confidence, more support, I could have done something," Maier said later, with some regret. "Had I been a guy, they would have said, 'Why don't you raise some money and take this thing?' " There were hardly any women in her company, or the VC world, or indeed in much of the startup economy for that matter, a stubborn reality that would never go away. Match.com was sold for less than $8 million in 1997, and then sold again the next year to what would become IAC for $70 million. It would later be worth many multiples of that, but Maier never shared in its success.

Mark Pincus, a wiry and gregarious entrepreneur from Chicago, would have much better luck. He'd moved to San Francisco in 1995 after getting an MBA at Harvard and working for cable television kingpin John Malone, convinced that the city was destined to be the "Motor City" of the incipient Internet revolution. Together with another young entrepreneur, Sunil Paul, he'd built a product where customers could pick websites and topics of interest to them, and the software would download the content automatically, to be read anytime.

The dot-com boom hadn't quite started yet. "We would literally sit in coffee shops in the Mission and talk about tech and no one would understand what we were talking about, no one," Pincus recalled later. "There'd just be people with weird ear piercings and tattoos. I loved it. It was such a different culture. I really believed that San Francisco was the melting pot, the germination dish, the only place in the world that we were going to connect the dots and invent this intersection of tech and creative culture." The company rented an office in the Hamm's building, on the edge of SoMa. "It was really important to me to be a company in San Francisco, to be a part of this shining light that was really a myth in my head."

His timing was excellent: He and Paul sold the company, called Freeloader, after just seven months to a dot-com that had already gone public, and Pincus had his first fortune—$5 million. He rented the most expensive apartment he could find, a $3,500-a-month town house on Geary Street, and shortly thereafter bought a house in Cole Valley, near the

Haight, for about $300,000. Prices were starting to rise, but the city still wasn't terribly expensive.

The 1997 pause in enthusiasm for all things Internet turned out to be very brief. eBay, the online auction company founded by another General Magic alumnus, Pierre Omidyar, was starting to show the unique power of what would come to be called Internet platforms. eBay didn't sell anything, but instead provided infrastructure for others to use for their merchandising, and tried to make it fun. The site would first catch on among people obsessed with obscure collectibles—the company's official founding story, though not entirely true, was about Pez dispensers—and it wasn't taken very seriously in its early days. For one thing, it relied on trust among strangers, which might be fine for low-stakes hobbyists, but seemed questionable for mainstream commerce. Would the person on the other end really deliver the vintage jacket, or rare book, or crate of kitchenware you'd paid for?

Yes, it turned out, much more reliably than one might have expected. The approach could be used for many things, and venture capitalists soon noticed its promise as large numbers of people began building businesses and followings on eBay. Even in the context of buying and selling, there was something to this idea of an online community rooted in trust among people who didn't know each other. eBay showed it could be a big business. A young programmer from New Jersey named Craig Newmark would show that it could be something different: a mostly free public service operated for the benefit of its users.

Newmark was already steeped in the Internet by the time he moved to San Francisco in 1993 for a job at Charles Schwab, and he found himself spending a lot of his time there trying to convince his colleagues in the computer department that the Web was the future. He was on The WELL, and though he was shy and socially awkward, he wanted to meet people, partly because folks on The WELL had been very helpful in giving him tips about moving to San Francisco. He attended some of the virtual community's real-world parties at the funky and low-brow Sausalito Yacht Club, where he got connected to a group exploring virtual reality that met regularly at the Exploratorium science museum. He was tipped to an Anon Salon event that featured a photo exhibit of a fledgling happening called Burning Man, which he thought looked amazing. He was invited to

Joe's Digital Diner, another upstart on the new media circuit, which featured inventive multimedia storytelling and a spaghetti meal. The virtual reality pioneers clustered around Anon Salon were particularly inspiring: "They were about twenty-five years ahead of their time," Newmark said later.

Newmark was grateful to be part of the action, and he wanted to chip in somehow. "I started telling people about stuff I heard about, either arts or technology or a combination. Then I started a mailing list on my WELL account, I would just email people and ask if they'd heard of something. I started with maybe ten to twelve people, and then more wanted to join." By the middle of 1995, with tech and culture events happening all over town, the list had outgrown The WELL, and Newmark needed proper software and telecom connections to send it as an email. He also needed to give it a name. He'd been planning to call it San Francisco Events, but then he learned that the people who were receiving it had already decided on an even simpler moniker.

They just called it Craig's List.

Information about social gatherings and technology happenings was always in demand, but Newmark said later that it was the challenges of San Francisco's rental apartment market that led to Craigslist's surge in popularity toward the end of 1995. "We started to see an apartment shortage here, and I told people that if they see any for-rent signs, they should send them to me," he recounted. Signs on buildings were the primary means of advertising San Francisco apartments in those days, so looking for a place meant walking or biking or driving the streets. It was so much handier to get the information in an email. By the next year, there'd be a website too, kindly supported by a friend who had a high-speed line.

Newmark had just a few basic principles, all based on empathy for the person staring at the screen. He'd become perhaps the only Internet entrepreneur in history to build a substantial and lucrative business on the most commonplace of observations about human nature. People don't like change. Also, free is good. And keep it simple, with minimal rules. It would work extraordinarily well for a very long time.

Cindy Cohn, a young lawyer from Michigan, had been in San Francisco for a few years, living in the Haight, when she and her roommate, who'd gotten to know a crowd of hackers, decided to throw a party. Among the guests was John Gilmore, and Cohn soon fell in with a group of libertarian techies, dating one of them and becoming good friends with Gilmore, while learning about the novel legal and political issues that were starting to churn around the Internet. A couple of years later, Gilmore would call with a request: Would she be interested in working with the Electronic Frontier Foundation and representing a man named Dan Bernstein, who wanted to challenge US government rules banning the export of encryption software? "I thought my boyfriend would think I was cool if I did it, so I said yes," she joked later. (She then fretted about whether her long-ago paramour would mind the reference. "They're privacy people! It's important.")

The regulations at issue in the case essentially treated encryption technology as if it were a weapon, even though scrambling information so it couldn't be read without a special key was a common practice—the capability was already embedded in many computers. Treating it like a gun or missile made little sense as a practical matter, and effectively assured that only the government, and not regular citizens, would have access to truly private communications. Bernstein's was one of several lawsuits against the government that the EFF would take on in its early years, when existing laws often didn't apply very cleanly to the Internet and a lot of efforts to regulate cyberspace looked hopelessly, insultingly ignorant.

A critical test would come in 1996, when Congress passed a law revamping telecom regulations that included a provision called the Communications Decency Act, aimed at blocking children from accessing pornography online. The CDA would have made it a crime to enable anyone under eighteen to view "indecent" material, even though that would have amounted to a ban on words and images that were protected by the First Amendment. The young Internet industry, led by the EFF, rose up in protest, with the rallying cry of "Internet freedom" echoing around SoMa. HotWired was a leader of the "Black Thursday" protest where thousands of websites turned their background colors black as a statement against the CDA, and was a plaintiff in the lawsuit led by the EFF and the American Civil Liberties Union. Louis Rossetto even had signs printed up for a protest in South Park as employees joined the protests against the CDA. "Hands off our Internet" fit nicely with the local left's critique of the capitalist power structure too, though it wasn't a partisan issue in the conventional sense.

Most of the CDA was indeed found unconstitutional, though one important and controversial bit survived, with the strong support of the EFF: Section 230, which said Internet companies were not legally responsible for what third parties post on their platforms. Without that protection, any online service that allowed others to comment or upload content—like PlanetOut, for example, or Craigslist—would face enormous and potentially lethal legal risks. Critics across the political spectrum, including Louis Rossetto, would later attack it for varied reasons, but Section 230 would be a survivor.

The EFF sat in an unusual spot amid the political crosscurrents of the city.

While Gilmore, Barlow and some other EFF backers were true libertarians, Cohn notes that many others involved in the cause were not. "Mitch [Kapor] was never a libertarian," she said later. "He was much more of a more traditional kind of liberal, but there was common cause when all these people were looking at what the government was doing about this new emerging space, and realizing that they didn't have a clue." Kapor, for his part, agrees that the EFF in its early years "was really driven by this libertarian impulse, which I did not have, and it was a constant source of tension."

The group would prove adept at riding the line, and would grow into a major force in the legal and tech policy world, with Cohn taking over as legal director in 2000, and executive director in 2015.

The liberal-libertarian tension, though, would never go away. The EFF was created in the hope that cyberspace could be free from the heavy hand of the state, an open realm of creativity with minimal rules. Yet its mission would grow even more complicated as powerful companies, rather than governments, started to become a big threat to Internet privacy and free speech. It would also have to reckon constantly with the basic dilemma of human governance: If everyone can do and say whatever they want, people get hurt.

The utopian elements of early Internet culture can be seen as efforts to transcend this tension and create new types of communities built in equal part on freedom and trust. Craigslist and community sites like PlanetOut, and even eBay, were showing how it could be done in a limited way with an online service. The adventurers at the San Francisco Cacophony Society would show how a libertarian-themed compact could operate in the physical world. SoMa's new creative class was now conjuring a fantasy twin on a barren salt flat six hours away, one that would suffuse the culture of San Francisco and its tech community, and stand as a metaphor for the city's wonders, and also its woes.

John Law stood forty-five feet above the salt flats of the Black Rock Desert, atop a rickety parody of an office tower that he and his mates at the Cacophony Society had dubbed Helco. Eight thousand people had gathered in this remote corner of northern Nevada for the latest edition of a four-day happening called Burning Man, and Law, a practiced and creative daredevil, could be counted on for a good show. Gasoline-powered flames were soon roaring up the side of the structure, and as it was engulfed, at the last possible moment, Law leaped from the top and sped down a zip line before bursting through a cluster of neon tubes at the bottom in a seeming ball of flame. The fire extinguishers deployed by his friends weren't needed, and the crowd gasped and roared as he dusted himself off.

It was more than just another day at the office, perhaps, but not that out of the ordinary for Law and his crew, who were bent on mocking the pieties of modern life and defying convention in spectacular fashion.

The son of an itinerant left-wing academic and a schoolteacher, Law had landed in San Francisco at seventeen, on the run from the juvenile justice authorities, and he found a place that proved uniquely welcoming to the subversive performance art and urban explorations that would become his calling. He'd fallen in with the secretive Suicide Club, which, after the death of much-loved founder Gary Warne in 1983, had morphed into the less-secret Cacophony Society, committed to the Dadaist ideas of humor, absurdity and sardonic social criticism. An old Victorian on Golden Gate Avenue served as its clubhouse, hosting a nonstop party, but the real organizing principle was an irregular newsletter called *Rough Draft*, where anyone could propose an activity, be it an urban exploration or a stunt or a collective art project, and everyone was welcome to show up. Humor, creativity, bravery and the unexpected were the coin of the realm.

Law and his friend Michael Mikel, who'd been doing the heavy lifting on *Rough Draft*, had become friendly with Larry Harvey, a local landscaper and free spirit who in 1986, along with his buddy Jerry James, had built a small wooden man and burned it on Baker Beach, a secluded strip bounded by steep hillsides out past the Golden Gate Bridge. They did it again the next year and the next, and both the crowd and the Man got steadily bigger, and by 1990 more than three hundred people—including John Law, Michael Mikel and John Gilmore—were there along with a forty-foot-tall man. Then the police showed up, and said the Man could not be burned. Harvey agreed to stand down, despite the sentiment of Law and many others in the crowd, including Gilmore, who'd helped erect the Man. He didn't know Harvey, and was unimpressed. "I think I carried half of one of [the Man's] legs down the path from the road, somebody on the front of the leg, me on the back of the leg. We set the thing up and danced around it, did all the stuff, brought stuff to eat and drink, and then he didn't burn it," recounted Gilmore. "Who is this guy? What kind of urban adventurer are you?"

A few weeks later, Harvey came by the Golden Gate house, which a woman named P. Segal had turned into a Cacophony salon and artists' crash pad. "A lot of Cacophony people were showing up, and then Larry shows up, and somebody was showing a video on the TV of this croquet game that took place in this wide-open desert expanse, only the croquet ball was eight feet in diameter and the mallets were pickup trucks," recalls Mikel. The footage showed a glorious 1987 Black Rock Desert stunt known as Croquet Ex Machina, which had previously caught the attention of Law's friend Kevin Evans. He and P. Segal had subsequently joined a "windsailing" event, where participants cruised the desert in beds and boats and other wheeled contraptions, powered by bedsheets. Those happenings—organized by an artist named John Bogard, who'd built an off-the-grid ceramics studio in the hills above the Black Rock Desert—inspired a Cacophony plan for what they called a Zone Trip, described in the newsletter as a "Dadaist temporary autonomous zone" of open expression, uninhibited by legal or social strictures. *Zone Trip Number 4, A Bad Day at Black Rock* would be taking place in just a couple of months, on Labor Day weekend. Though recollections differ on exactly when and how it was all decided, the Cacophonists invited Larry Harvey to come along, and burn his Man in the desert.

On the Friday ahead of Labor Day weekend in 1990, the crew took off for the Black Rock Desert, two hours north of Reno. "It was daylight when we arrived and came down off the asphalt to the desert, and everybody got out of their cars and trucks," says Mikel. "And I took a stick and I drew a line on the ground and I had everybody line up and I said, 'On the other side of this line everything will be different.' We all stepped across that line together. And then we got into our cars and trucks and we drove way out to the far end of the north end of the Black Rock Desert."

There were eighty-nine people there, by Law's count. "We had a party, a great weekend, we set up the Man and we burned it," Mikel continued. "Me and John and a lot of the people stayed for cleanup. And nobody ever knew we were out there, except the people who were there."

The next year Law got a permit from the Bureau of Land Management, the federal agency responsible for the public lands that included the Black Rock Desert, and decorated the Man with neon lighting, partly to serve as a beacon in the vast, featureless desert. The fire artist Crimson Rose, later the art director for Burning Man, attended for the first time that year, and the group began selling tickets. In 1992, Mikel and Vanessa Kuemmerle, another early participant and Law's girlfriend, created the Black Rock Rangers, both to help the many people who were now getting lost or stuck trying to find Burning Man, and to make sure everyone had paid to get in. There was nothing specific to "do" on the hot, dusty playa, where even a little wind could create a whiteout, only what you and your friends could dream up. Brewster Kahle, the cofounder of Wide Area Information Servers and a committed Deadhead, had become friends with Mikel and Law, and he tied the knot in the Black Rock Desert that year with Mary Austin, a book maven who later started the San Francisco Center for the Book. It was the first Burning Man nuptials.

The Cacophony Society had chapters in Los Angeles and Portland and other cities as well as San Francisco, and many of them began making the journey to the desert. Word kept spreading along the varied tendrils of 1990s West Coast underground culture—the Cacophonists and the ravers and the punk rockers, but also the anarchic robot-builders of the Survival Research Labs, and the sexual experimentalists who'd gathered in an online chat room called Bianca's Smut Shack, and would make a real-world version of it on the playa. An LA Cacophony punk rocker known as Chicken John led the Circus Redickuless, where the performers had no talent. Land artists, fire artists, actors, dancers, musicians and tinkerers from the emerging "maker" movement were all stretching their imaginations in the vast expanse of the Black Rock Desert. Grateful Dead followers were prominent among the old-school hippies drawn to the spirit-moving landscape and the experience that might be had in a community where there was no money. An adjacent happening at several of the remote hot springs in the area, called Desert Siteworks, brought another angle on land-art and environmental performance.

A SoMa DJ named Craig Ellenwood, aka DJ Niles, had stumbled across a Cacophony Society flyer in a laundromat and thought it would be cool to play techno music in the desert, even though Harvey warned him that they weren't all peaceful hippies out there, and some people weren't going to like it. "The acoustics bouncing off the mountains were amazing, the wind swirling the music around," he recounted in a 2021 interview. "Not soon after, boy did we have some angry participants come out." Some local cowboys who'd heard the music and been intrigued ended up being their protectors. Big rave communities known as Wicked and Moon Tribe soon began showing up, camping a couple of miles away and creating the beats that would define the event. It was all still very much on the edge.

"There were these circus troupes and performance art troupes who traveled the world, and they were coming to Burning Man because they could pick up jobs helping clean up the trash afterwards. It was like a meet-up for a lot of these edge communities," recalls Candace Locklear, a Silicon Valley publicist who was involved as a volunteer from the early days. "There were all kinds of gutter punks out there and speed freaks—that's how you stayed up all night to create these big spectacles under immense pressure." Back then, she says, there was nothing in the way of safety precautions; you were on your own. "That element of danger was high. The fact that you could die, which was on my ticket, you really felt that."

Attendance was doubling every year, and by 1996—the year of Law's leap from the Helco tower—it was a $35 ticket for an anything-goes free-for-all. One popular attraction was the "drive-by shooting range," where you could cruise along the edge of the playa and blast away at stuffed animals on the hillside with an automatic rifle. There was only the barest of road grids, with clusters of people and vehicles and makeshift structures scattered across the flat white desert. With twice as many people as the year before, it was a roiling, sometimes-scary affair, with a Dante's *Inferno* theme that would prove a little too on the nose.

The reckoning started with an accident—or more accurately, a tragic incident—before the gates even opened. Michael Furey, a friend of Law's, was drunk as he rode his motorcycle back from the nearby town of Gerlach just before the 1996 event began, and he thought it would be amusing to play chicken with his friends in the accompanying van, riding circles around it as they sped across the dark playa. He misjudged and was killed instantly when his head smashed into the side window of the van as it tried to avoid him. Harvey, arriving at the scene a few minutes later, appeared mainly concerned with the liability, according to multiple accounts of people who were there. That was the final straw for Law, who'd come to detest Harvey, and could already tell that Burning Man that year was going to be dangerously out of control. "Furey killed himself, but it was Larry's response that made me certain I was done after that year," Law said later. As the event was ending, three people in a tent were run over and severely injured—something all but guaranteed to happen with thousands of people, many drunk or tripping, cruising about in the pitch-black desert night.

The truth is, it's a miracle that more people didn't die. Law's leap from the Helco tower, and the fire that then consumed it along with the faux-corporate village around it, were only the beginning of what became a veritable orgy of flame. The elaborate set of the fantastical performance known as Pepe's Opera was the next to go, the actors barely escaping as blazing structures collapsed around them. After the climactic event the next night, the burning of the Man, a mob began burning everything in sight, people with flamethrowers on their backs roaming the playa looking for targets, fireworks exploding heedlessly in every direction. A contraption dubbed the Veg-O-Matic, essentially a flamethrower on wagon wheels, was being directed by Chicken John of the Circus Redickuless when it pulled up to the information board in center camp, the only real means of passing messages among attendees, where hundreds had left notes. Harley Dubois, another early organizer, tried to stand it down to no avail, and it was incinerated in a blast of fire.

Burning Man couldn't go on like this. Nobody wanted people to get badly hurt, much less die, and the legal risk was scary too, even if no one had much money. Law, though tough and brave, was known as the one who kept Cacophony events from tipping over and getting someone killed, and he'd had enough. The guilt and grief were overwhelming for many of the early participants, and bitter resentments boiled over after Law and a few others stayed behind for a month to clean up the playa, only to be accused of taking money. A T-shirt began to circulate: "Burning Man: Woodstock or Altamont? You be the judge."

The Bureau of Land Management didn't care much about the distinction. Flummoxed from the start about what to do with a gathering that they instinctively despised but couldn't simply ban outright from public land, they now had every reason to deny a permit. There were really only two options for Burning Man: develop rules and structure, or call the whole thing off.

For Law, the choice was clear: It was over, or else had to be reimagined as something far smaller. The whole point of the gathering was that there were no rules. Riding out to the desert with a group of friends—even a very big group of friends—to build stuff and burn stuff and get naked and do drugs and conjure an alternate reality was one thing. Organizing an event where ten thousand people could do that without something going terribly wrong was a different matter entirely. Up to that point, infrastructure had consisted of some porta-potties and a temporary orange plastic fence encircling the site; Cacophony events were supposed to be ephemeral, and it was hard to see how Burning Man could survive conversion into a more conventional festival.

The trio, jokingly dubbed "The Temple of the Three Guys"—Law, Harvey and Mikel—agreed to create a company, dubbed Paper Man, which would own the Burning Man trademark. But beyond that Law was out. He'd been a crucial figure, a guy who knew how to talk to people and get things done, and a creative participant who'd shown others the way. But he felt responsible for people's safety, and had an unwavering disdain for commercialism. The day after the inferno, Law told the horrified BLM and county officials not to worry: He wouldn't be back, and Burning Man in all likelihood wouldn't be, either. "It would be insanely irresponsible to do it again," Law said later, recalling his thinking at the time.

At the same moment, though, Harvey, Mikel and about a dozen others were sitting together, in despair, but also thinking about the future. Mocking the system and its values was at the core of Cacophony, and no one was in it for the money, exactly. But for many of the early participants, and the broader collective who'd made the event happen, there was every reason to keep going. They'd created something unique, powerful even, and it wasn't just an accident: They might have been rebels and misfits who lived on the fringe and didn't go to college, but these were some very smart people, with interesting ideas and a lot of tenacity. Their dangerous, off-beat activities weren't easy to pull off, requiring technical ingenuity, people skills, planning and, odd as it sounds, good judgment. They also required a whole lot of work, which a whole lot of people had been willing to do for free. They'd gotten this far. They were all broke. Why give up now?

For the founders of Burning Man, even with the chaos of '96, the event was no longer just a fun adventure. It was the opportunity of a lifetime.

The numbers were hard to ignore, even for anti-capitalists. Getting eight thousand people to buy $35 tickets without doing any marketing, or paying for any entertainment or amenities, or promising anything at all other than outhouses and an opportunity to create a good time for yourself and your friends—bring your own food and water! Leave no trace!—that's actually quite a thing. There was nothing else like it, not even close. And if you wanted to think about it that way, it also had a deeper level of meaning, showing a way that people could live together differently, build and inhabit new worlds, and celebrate creativity and human potential—the same spirit that animated many of the early, counterculture-inflected online communities, and would be part of the ethos of what some were calling the New Economy.

As a venture capitalist would say, Burning Man had shown remarkable viral growth and product-market fit. The *Wired* magazine cover story about the '96 Burn, which would do much to feed the enthusiasm among the Internet crowd in SoMa, hadn't even come out yet. And the nascent Web had only just begun to bring the message to the world.

Marian Goodell, though a newcomer to the scene, was determined to help Burning Man live on.

She'd grown up in Ohio, the daughter of a metal-industry executive whose libertarian politics ran deep enough that he named his private railcar after Dagny Taggart, the heroine of the Ayn Rand novel *Atlas Shrugged*. Goodell became a Rand fan too, and after attending Goucher College, a liberal arts school in Baltimore, she ended up in Boston, working first in advertising and then as a legal assistant. A visit to a friend in the Bay Area in 1988 convinced her to make the move West. "I saw a much more open and less judgmental way of living," she said. "People didn't really care where I went to school. They wanted to know what books I read."

She found a job at a SoMa subsidiary of computer publisher Ziff-Davis that was putting government data onto CD-ROMs, and she was making good money as a sales rep. But she was still searching, and had returned to school for a master's degree in film when she saw pictures from Burning Man in a photography class. She was inspired to go find it—it took a little doing at the time—and eventually tracked down the number you could call to get on the Cacophony mailing list. She went for the first time in 1995 and was transfixed. She went again in 1996 and sought out Harvey, trying to better understand what she was experiencing. "Was it a cult? What was the purpose?" she recalls wondering. Harvey, easy to find in his trademark fedora, gave her disarming answers. "He couldn't burn on the beach so he had to find a place to burn things. I said, 'Why the Black Rock Desert?' And he said, 'Look at it. It's man versus nature.'" Goodell didn't witness the worst of the Dante's *Inferno* mayhem that year, but was spooked enough by the mood. "It wasn't like '95, which felt more civilized. Ninety-six was a crazy shit show."

A few weeks later, back in San Francisco, Goodell went to a party at the Golden Gate House, hoping to meet the Burning Man organizers. "The place was packed, it felt very Bohemian. It was the creative vibe I wanted—the parties were wacky," she recalled. She was formally introduced to Harvey. "I was asking him all these questions and he was drunk as a skunk, couldn't even focus." She gave him a ride home and he asked for her number.

Harvey, who died in 2018, had grown up on a farm near Portland and moved to San Francisco after visiting for the Summer of Love and being "smitten by the San Francisco lifestyle," as his brother would write years later. He lived in the Haight, and was something of a hipster and a layabout, with a low-cost lifestyle that you could get away with in San Francisco in the '70s, '80s and the first few years of the '90s. He had his landscaping jobs, and it would have been a stretch to say that he was an artist. But with his quirky, low-key manner and oracular style of speaking about the big things, he drew people to him. And like the most successful startup founders, he had an ambitious vision of what Burning Man could be and how it might change the world, even if it was sometimes hazy and self-contradictory. His dedication to the Man itself as the beating heart of the event was not shared by everyone, but his instincts proved sound. His admirers were effusive.

"He was just an amazing guy," says Terry Gross, who was Burning Man's lead lawyer for decades and became a good friend of Harvey's. "When you have conversations with him, it could go off in so many different directions. But some dispute would come to Larry, and he'd be like Solomon. He would just come up with some solution that nobody had thought of. It was really amazing."

Harvey and Goodell began dating, and her business savvy—partly absorbed from her father, and partly learned from her work in the nascent tech industry, where she was now helping build the first website for Ford Motor Co.—made her a critical addition to the crew. After the 1996 debacle they couldn't get a permit to return to the Black Rock Desert, and so they'd looked for private land nearby and begun negotiating with a local rancher to hold the event on his property. Reaching an agreement with the half-mad rancher, who was notorious for mistreating his livestock and was motivated primarily by a desire to thumb his nose at the federal government, was arduous. Even worse, since Burning Man would now be on private land, they had to negotiate accords with state and county authorities, whose contempt for the event was matched only by their desire to extract money from it.

The '97 Burn would happen on the Fly Ranch, with the county sheriff seizing the money directly from the ticket booths, convinced the organizers wouldn't pay up for the police and fire services the authorities had required. But the group would fortuitously gain some legal leverage to fight back, in the person of Terry Gross.

A New York native who'd become a lawyer by way of a long stint on a commune in Oregon, Gross attended the Fly Ranch Burn and was entranced—and deeply offended by the approach of the local authorities. He knew more than a little about the government's legal obligations: He worked for a legendary law firm called Rabinowitz & Boudin, which was highly regarded for its constitutional law expertise and defended many left-wing radicals and foreign governments such as Cuba. (Partner Leonard Boudin's daughter Kathy would be part of the violent 1960s radical group known as the Weather Underground and serve more than twenty years in federal prison; her son Chesa would become a political lightning rod as a progressive district attorney of San Francisco.)

"I went to the organizers, and I said, 'Listen, I'm a constitutional lawyer. What happens here, even though there's no speech, is expression that is protected under the Constitution. If this was the Boy Scouts that were burning things or blowing things up, it would be good, clean, American fun. And they're putting all these restrictions on you because of the nature of the expression. We could bring a lawsuit against them for violations of the First Amendment.

"And so then Larry went to the county and basically said, 'Look, I got this constitutional lawyer. They're ready to file a lawsuit. You're going to end up owing us money. It's going to take a lot of time and energy and stuff, but I really don't want to sue you guys.' So he used this to get reasonable rates and things from them," Gross recounted. Burning Man had plenty of need for a skilled lawyer, and for the next twenty-five years, that would be Gross.

The move to private land, as well as the need to avoid a repeat of '96, had forced the freewheeling band of anarchists to grow up. They created a six-person limited liability company that would manage the event, initially planning to dissolve it and create a new one every year, lest the dreaded institutionalization creep in. (That plan only lasted a couple of years.) The authorities in Washoe County, where the Fly Ranch was located, had mandated that they create a street grid, for safety purposes, and a landscape architect and city planner named Rod Garrett designed a two-thirds circle with concentric curving streets crossed by radial boulevards with the Man at the center—an elegant spatial architecture that would stand the test of time.

Goodell quit her day job after the '97 event to devote herself full-time to the cause; it wasn't going to go on without extraordinary effort. "We had to figure out, well, how hard do we work to hold this shit together, to save it?" she said later. The six partners had paid themselves $9,000 each for 1997, but it still wasn't clear if or when Burning Man might be a real job. Goodell saw that public relations, and especially government relations, would need to be handled in a careful and organized way, and she began haunting county supervisor meetings and Bureau of Land Management hearings in Nevada where ranchers, miners, environmentalists and outdoor recreation enthusiasts wrestled over land-use issues. The BLM's mission was to facilitate use of public lands, not block it, and the agency could not legally deny the event a permit without a real reason. Burning Man would now be going back to the Black Rock Desert—and the Internet revolution would be coming along with it.

Brian Behlendorf, now building websites upstairs from *Wired*, had gone to Burning Man for the first time in 1995 and loved it. He got curious about the event and the organizers, and noticed that the "burningman.com" domain name was misconfigured. He sent an email offering to host a Burning Man website on his Hyperreal server. "Next thing I know I'm meeting the early organizers to set up a web site and email addresses," he recalled. By the next year he was working with Goodell on a newsletter called The Jackrabbit Speaks and hosting gatherings of the early Burning Man Web team at his Twin Peaks apartment. "Larry would come by and sleep in the corner," Behlendorf recalls. "Listening to Larry and the founders, I just felt honored to be there."

Scott Beale, who'd been working in the SoMa film-processing and video-copying business when the Web came along, led the creation of the first live webcast from Burning Man in 1996, which involved talking telecom provider Nevada Bell into making a new digital service called ISDN available in Gerlach, the town adjacent to the Black Rock Desert. It was something that was done, as so many things were at Burning Man, simply because it was a complicated challenge, and would be kinda cool. "We had dual ISDN lines coming into a room in Bruno's Motel that would then transmit the Internet back to Black Rock City via a series of microwave antennas mounted on radio towers," Beale wrote later. "It was a pretty impressive setup for 1996."

As the dot-com boom heated up, interest in Burning Man would explode—this one-of-a-kind happening was perfectly tuned to tap the curious, risk-taking, invent-the-future spirit of the early Internet industry. In August 1998, the multicolored logo on the home page of Google, then still a Stanford project, was enlivened for the first time with a doodle—a figure in outline, its arms raised, standing behind the second "o." Larry Page and Sergey Brin had just gotten their first investment to start a company, but had put the $100,000 check (from Sun cofounder Andy Bechtolsheim) in a drawer. They'd be out of the office, at Burning Man.

CHAPTER 3

The Bible of the Dot-Com Boom

John Battelle, the young managing editor of *Wired*, had been fine-tuning his idea for a weekly business magazine about the new Internet economy, and by 1997 he was ready to make it happen. At the same time, Pat McGovern and his International Data Group had their eye on the Internet opportunity too. McGovern, a numbers whiz and MIT graduate, had started *Computerworld* back in the 1960s and built it into a global media conglomerate that included magazines, trade shows, books, market research and even a venture capital arm. Now he was looking for someone to lead a new Internet trade magazine. He found Battelle, and they decided—too quickly, it would later become clear—that their ideas were close enough.

I'd left San Francisco at the end of 1993 for a posting in the *LA Times*' New York office, and after just a year I'd moved to Los Angeles to take a job as the first technology editor for the paper. We'd done some great work, including the creation of a weekly tech section, but the *Times* was big and bureaucratic, and I was ready for a new adventure when I got a call from an IDG recruiter in the summer of 1997. There'd been buzz about Battelle and McGovern's project already and I was definitely interested. Not long after, I flew to San Francisco to meet John.

My first impression left me a little cold. He was clearly a super-smart guy, but with a preppy mien and a whiff of arrogance—a handsome, bike-riding patrician with a perfect tan, quite a contrast to the messy reporters of my professional world. With his sweeping rhetoric he seemed like a bit of a bullshitter, a marketing guy, not someone who'd be interested in the hard-edged, skeptical reporting that I thought of as my calling card. But we kept talking, and I grew ever more impressed with his intelligence and vision. The hype around the Internet was building fast, and I wanted to play against it, positioning as the publication that would cut through all the crazy talk and tell people what was *really* happening. "99.9% Hype-Free," as our T-shirts would later say. John appreciated the approach, and averred that he'd let me be the editor while he built the business—a promise he'd keep through thick and thin.

I soon joined John and a few others in a 1970s-era loft in SoMa, near the Caltrain station, finishing up a prototype issue and working frantically to hire the twenty or so people we'd need to get the magazine started. We decided to call it *The Industry Standard*, which we thought of as a cheeky declaration of our inevitable importance. Like *Wired*, we were a print publication chronicling the rise of the digital economy, an irony lost on no one. But also like *Wired*, we weren't going to be just a magazine. John had expansive plans for what would follow: an online publishing empire anchored at TheStandard.com, alongside a global conference business, data services and research products.

A few months later, we moved into a handsome, century-old, redbrick building on Pacific Avenue in Jackson Square, the historic warehouse district adjacent to downtown that housed a handful of advertising agencies, print shops, TV stations and a few tech startups. My corner office in the third-floor newsroom looked down at the city's oldest bar, the Old Ship Saloon, and the adjacent restaurant, The Globe, known as a late-night oasis for partying food industry workers. I could literally see the neighborhood bubbling with new energy, buoyed by an improving national economy and the nascent Internet boom.

Standard Media International, as we grandiosely dubbed ourselves later, would be a mirror of the industry that we covered. For almost every sector of the global economy, it was possible to imagine a dot-com analog that would be a dominant force. For our part, the goal was clear: We'd be the preeminent business publisher of the Internet era, the Dow Jones of the twenty-first century, proudly representing San Francisco as the new tech capital of the world.

We launched *The Industry Standard* in April 1998, when the first dot-com boom was already well underway, coming out swinging in the first issue with critical stories about a new Intel Internet venture and a dubious effort to create an online investment bank. One of the first reactions was from the enraged subject of one of our stories, screaming at me about how he was going to sue us into the ground. (He never did, though the threats continued for many months.) Two of our charter advertisers canceled right away. That was all fine as far as I was concerned: We wanted to shake things up. Businesspeople don't like aggressive journalism when it's about their own activities, but the business world runs on good information, very much including what journalists dig up. Advertisers would ultimately follow the readers, whatever they might think about the stories.

I thought we were hitting our marks pretty well out of the gate with a straight-talking, no-jargon style, and a distinctive, fun and functional design created by Daniel Carter, who'd come over from *Wired*. There were other titles aiming at the business of the Internet—the *Red Herring*, *Business 2.0*, and *Upside*—but we had a crucial differentiator: They were monthly, and we were weekly. The logistics of getting a national newsmagazine into people's hands within a few days of its rolling off the presses were daunting. But the news cycles were speeding up, even before everything was online, and our timeliness was a huge advantage.

The deadlines were unforgiving: We'd transmit pages to the printer in Minnesota over the course of a few days, and if we were to miss a window, the magazines would miss the air cargo flights that took them to a dozen places around the country for mailing. I appreciated the rhythm of it though. We'd send off the cover and the final pages at two o'clock in the afternoon on Friday, sometimes after a final sleepless rush that included writing the all-important cover headlines in a giddy blur, and then launch into our signature celebratory rituals: smoking pot in the basement and drinking beer on the rooftop. Battelle and I both loved weed, and though it became a sore point with the older and stodgier people who joined the company later, the basement lounge, a cave-like room filled with the detritus of the toy warehouse it had once been, carried on for years. The rooftop parties would become legend.

Pat McGovern had a formula for trade magazines, which involved launching quickly on the cheap and pulling the plug fast if it wasn't working as planned. This was not the best approach for building the big-time brand we had in mind, and we were almost dead before we'd started.

We were doing decently with the journalism, and word of mouth was building. I'd persuaded Michael Wolff, who'd just finished a terrific book about the startup world called *Burn Rate*, to write a column for us, and he was skewering in proper style. The magazine was dense with information, well-designed and well-written, and at its best boasted humor, insight and flair. But with no name recognition and inexperienced reporters, blockbuster stories were hard to come by, and advertisers weren't very convinced by our smoke-and-mirrors circulation strategy, which involved pretending people were paying for copies while mostly giving it away free.

"We're going to miss our $2 million revenue target by $1.75 million," John told me one day with his habitual cool as we sat in his office reviewing the first few months of financial results. I knew things weren't going well financially, but the numbers were bracing. Pat would come to town periodically to review our progress, and though John was a master at making things look good, the next meeting was nasty.

McGovern had an awkward, courtly manner and was never one to yell and scream, but on this day it was clear that he was angry. As John walked through a PowerPoint presentation showing a new plan for business directories that we'd been told to create, Pat interrupted: "This is like a plan from the 1950s," he scoffed. "Why are you showing me this?" The deputy who'd told us to put it together—"Pat loves directories," he'd promised—was sitting right there, but he averted his eyes. Pat's voice rose. "When am I going to get my money back!" he demanded. He was literally pounding the table.

John was a journalist by training and an excellent writer, but his response to this crisis showed why he was a business guy at heart, and a good one. My instinct had been to trim the sails and save money. John's was the opposite. "We need to go big," he said as we smoked joints and drank whiskey in the garden of his bucolic home in Mill Valley, north of the city. "I'm going to double the size of the sales staff." I was baffled. But John walked me through the numbers, showing how much more revenue we could generate, and it made sense, sort of. We also came up with what we thought was a pretty clever trick to get Pat off our back. He'd always disliked the name because it didn't have "Internet" in it, so we agreed to change it, but we'd need to create a plan for doing that, and proposed we aim for the one-year anniversary. That bought us time. We were sure the business was going to work.

And it did. By early 1999, the dot-com boom was in full swing, and the trickle of ad pages in the magazine abruptly swelled to a flood. We had our first big conference, the Internet Summit, which made $2 million in profits and did a lot to put us on the map. Our combination of weekly frequency, edgy but authoritative editorial and excellent marketing proved a winner, and soon I was hiring reporters and editors as fast as I could as the magazine ballooned to a hundred pages, and then two hundred, and eventually three hundred—the most the printer could handle. We were now able to bring on more experienced talent, including *Stanford Magazine* editor Bob Cohn, *Wall Street Journal* reporter Thomas Goetz, *Village Voice* media critic Jim Ledbetter, *Barron's* columnist Eric Savitz and veteran magazine editors Jane Goldman and Amy Bernstein. A shrewd young Brit named Michael Parsons schooled me on my many weaknesses as a manager: I was way out over my skis, as the VCs liked to say. I'd run a team of four at the *LA Times*, and now my staff was on its way to a hundred or more, and we were planning a spin-off magazine, a European edition and a daily online editorial operation, along with a half dozen conferences a year. The company was increasingly chaotic, but we were disciplined, too, and managed to avoid major mistakes while putting out a big, interesting magazine every week, even if the fattest issues sometimes had a quantity-over-quality feel.

At launch, I'd sent a copy to Intel CEO Andy Grove, a discerning media consumer whom I knew from my *LA Time*s days, and he'd politely responded that it was nicely done, but didn't have much he couldn't find elsewhere. He wouldn't be a subscriber. In mid-1999, his assistant called: Andy was very upset that he hadn't gotten his latest edition, could I send one along pronto? Before long, Time Inc., the premier magazine company in the world, was angling to buy us.

We were the bible of the dot-com boom.

In normal times, venture capitalists would carefully vet a startup, put in a little money, see how things went and then maybe put in more. If the company did well, they might join with other VCs to put in larger sums, and if the company continued to grow and started to turn a profit, then it would be time for a sale, or an IPO. It typically took at least four or five years, and often more, to get to the "liquidity event" where investors and employee shareholders could cash in.

These were not normal times.

Netscape had gone public just sixteen months after it was founded, and the bankers and venture capitalists had every reason to feed the suddenly ravenous investor appetite for all things Internet. A ramshackle parade of

IPOs began to take shape, and by 1998 there were multiple Internet IPOs seemingly every week. In the fall of that year a San Francisco outfit called The Globe, a too-early social network created by a couple of Cornell students, established the precedent that a company need not have much of a business at all to convince IPO investors of its value, only a little Internet sizzle. The company had posted a loss of $11.5 million for the first nine months of the year on sales of $2.7 million, but after its IPO it was somehow worth more than $800 million.

Surprisingly, federal regulators who were supposed to protect mom-and-pop investors from being defrauded seemed to be OK with all this, and soon there was drugstore.com, pets.com and eToys.com, to name only a few of the famous startups that raised huge sums long before they'd proven their concept, let alone made a profit. In 1999 alone, nearly three hundred Internet firms went public, a full order of magnitude more than the number of tech IPOs in an average year since then. Most of them didn't have any profits. Some barely had any revenue. But in the moment, nobody cared: Retail investors had caught Internet fever and were ready to buy anything.com, so every IPO was seeing a healthy price spike on the first day. That irregular dynamic meant that the right to buy shares at the offering price, before trading started, was free money for the well-connected. Frank Quattrone, the leading investment banker in the sector, began handing out pre-IPO shares to executives of other startups who then brought their business to him—a scheme that worked nicely when shares went only in one direction.

The IPOs fueled a cash cycle: Venture investors and eventually many others—big corporations, business moguls, celebrities, employees—were making quick returns on Internet IPOs, and oftentimes funneling the winnings back into the Internet economy. Day trading, enabled by online brokerage accounts, became a mania of its own as the soaring stock market made it deceptively easy to make money. Even Mayor Willie Brown, now not so ignorant of the tech world, would get in on the action, enjoying pre-IPO share allocations in seven dot-com companies. *Examiner* reporters Chuck Finnie and Lance Williams noticed the transactions when they were disclosed—a requirement under the city's ethics rules—and they elicited a comical series of responses from the mayor's office on what it was all about. "Like millions of Americans, he is kind of playing around on the Internet and doing trading over the Internet," his spokesman offered at first. Willie Brown was just another day trader! After a follow-up story showed it was implausible that the mayor had received the pre-IPO shares without help, he acknowledged that a Morgan Stanley broker had cut him in. The paper would henceforth refer to the firm as "the mayor's IPO banker."

Silicon Valley, and not San Francisco, was still home to the biggest companies, including Netscape, Yahoo, eBay, PayPal and Napster, the file-sharing service that was sweeping college campuses and giving the music industry fits. If you needed serious technical talent, you were much better off in the Valley, even if you didn't want to live there. Mark Pincus, for one, whose early Web startup Freeloader had netted him $5 million, discovered that reality as he put together his next startup, which would provide technical support for business software over the Internet. He'd started in the city, but just a few months in, colleagues had presented him with a map showing that almost all of their programmers, who constituted most of the company, lived in Silicon Valley. He felt he had no choice but to move the company south and make the reverse commute from the city every day.

"I had this boring-ass enterprise software company and I missed the whole dot-com bubble," he lamented later. "I've been waiting for this moment, for San Francisco to be the Motor City of the Internet, and I can't even go to the launch parties because I'm commuting three hours a day to be with my engineers."

A lot of the new Internet economy wasn't about engineering, though. Design was increasingly central. Digital marketing and advertising had to be invented, and publishing reinvented. Sales and finance professionals were needed. So was something much more ineffable: people with an intuitive feel for the human dynamics of cyberspace, and the creativity to make it alluring and fun. For all of these things, it was better to be in San Francisco.

The city's economic doldrums came to an abrupt end as Web development shops, Internet media and advertising companies, email providers and software developers filled the brick-and-timber spaces of SoMa and the Mission as fast as developers could buff them up. There was spillover to the high-rises of downtown, too, as rents soared. Unemployment in the city had fallen to less than 4 percent by the end of 1996, even though there were almost thirty thousand more residents than there had been at the start of the decade. The city would add eighty-six thousand jobs over the course of the dot-com boom.

You could feel the economic energy on our rooftop, where what had started as Friday beer-and-chips for the staff had metastasized into a lavish party with catered food and fancy booze and careful attention to the VIPs in attendance. The rooftop wasn't that big—there'd be a line to get in, sometimes hundreds of people long—and John had the unlikely idea of creating a party brand, "The Industry Standard Rooftop," and taking it on the road. We had a rooftop in New York and one in London. In San Francisco, we'd host one at the Phoenix Hotel, a former motor lodge in the Tenderloin that an entrepreneur named Chip Conley had converted into a hip rock 'n' roll hotel. We were getting a reputation for big spending, but it was all underwritten by sponsors.

From North Beach to Jackson Square to the Mission, the city sometimes seemed to be in nonstop celebration mode, with companies competing with one another to throw the most memorable events. A young businessman named Gavin Newsom, who owned a bar in the Marina district and was just beginning his climb up the local political ladder, could often be spotted at the venerable Enrico's in North Beach, or nearby Tosca Cafe, sometimes hanging out in the back room with Sean Penn. The musician and DJ Moby, whose *Play* album was vaulting him to worldwide fame, delivered a memorable 1999 concert at Maritime Hall, and his electronica-based music would become the soundtrack for the San Francisco dot-com boom.

People were coming from all over the country, and the world, to get in on the action. Ev Williams, a thoughtful, soft-spoken native of rural Nebraska with an affinity for computers, had arrived in California a few years earlier, after picking up an early copy of *Wired* at a newsstand in Grand Island and deciding it was time. He found a job with Tim O'Reilly's company up in Sebastopol, and after a short stint there he picked up some coding gigs at HP and Intel, and moved to Campbell, near San Jose. That didn't last long, and in 1998 he settled in the Mission. "I just wanted to be in the city," Williams recalled later. "The Web was distinct from tech. The Web was media and publishing and culture."

Together with a woman named Meg Hourihan whom he'd recently begun dating, Williams launched Pyra Labs, with the aim of creating a toolbox for building Web-based software applications, such as calendars and word processors. They raised some seed capital—O'Reilly Media was among the investors—and hired a handful of people and leased a small office in SoMa. Then they discovered that a notes feature they were developing, which made it simple to post writings on a website, could be interesting as a product in its own right—a person could create their own "Web log." They weren't getting much traction with the Pyra product, but the super-simple Web log tool had struck a chord.

At an *Industry Standard* rooftop party in mid-1999, Williams, Hourihan and a programmer named Paul Bausch huddled in a corner, oblivious to the crowd, and decided they'd launch a stand-alone product, called Blogger. It was an immediate hit among the small but energetic cohort of people around the world who were tuned into such things. A few years later, blogging would sweep the Internet, setting off a chain of developments destined to upend the global media industry.

Craig Newmark had been laid off from Schwab in 1995 but was making good money as a freelance programmer, and his passion project, Craigslist, kept growing. People were now offering to pay for job listings, and in 1998 he started sporadically charging for help-wanted advertisements and took on some volunteers to help him keep it all running. He was a one-man show still, writing new software to make things simpler and faster, and paying a lot of attention to customer service. "The lesson I learned is, keep things as simple as possible, and then people won't need much help," he said later. "That worked out pretty well, right?"

By 1999, he realized Craigslist had to be a proper business; he'd been paying himself only occasionally and relying on volunteers, but that wasn't working anymore. He needed to hire people, and he needed a business plan. "At that point, VCs and bankers wanted me to do the usual Silicon Valley thing, to monetize heavily, and then they'd throw billions at me," he said later. But "monetize heavily" meant junking up the site with ads, or charging lots of fees, or selling names to others, and Newmark just didn't want to do it. It was a community service, there to help people, and it didn't cost much to run. "I began to remember a lesson that I internalized from Sunday school back at the Jewish Community Center in Morristown, New Jersey: know when enough is enough," he says. Refusing the VC money was an exceedingly rare choice, but Newmark never wavered, and he'd ultimately do extraordinarily well without it.

Trevor Traina's startup was a more typical case, even if his status as a child of San Francisco's old-money royalty made him an outlier among the city's dot-com entrepreneurs. His father, John Traina, was a shipping executive and a suave, handsome man-about-town, known for his conversational charm and his Fabergé egg collection, and for his marriages. His mother Dede was an heir to the Dow Chemical fortune and an active philanthropist, serving for many years as the chair of the de Young Museum. Now his father was married to the romance novelist Danielle Steel, with whom he had five children, and his mother had tied the knot with real estate investor John Wilsey. It was all as blue-blooded as it gets in San Francisco.

Trevor Traina had an MBA, and in 1996 he raised $350,000 in seed money from friends and family for an Internet venture: a product-lookup website, inspired by Traina's frustration when trying to find an obscure electronic product at the big Circuit City store on Van Ness Avenue. The startup moved into offices at China Basin Landing and seemed to be getting some traction, but by the next year the company needed more money, and it wasn't going well. "We went to the consumer electronics show and we got a bunch of commitments from companies like Sony and Samsung. And I was like, wow, we're really off to the races here," he recounted later. "And for the next few months I followed up diligently with all those big electronics companies, and not one of them followed up."

He'd been working nonstop, desperately trying to raise new financing. "I had called dozens and dozens of VCs, pitched everyone I could think of," he said. "I went to every *Industry Standard* party, kissed every butt." Finally he got a nibble from a hot VC of the moment, Ann Winblad, one of the rare women in the business, who'd built a successful new firm with partner John Hummer. An up-and-coming junior associate, Bill Gurley, was assigned to put together a deal, but it wasn't a pretty one for Traina. "I would have basically had to sell out every one of my friends and family. It was a deal with the devil I couldn't do, and I distinctly remember sitting in my car and crying. I was gonna lose everyone their money, my grandmother, my parents. We were just completely SOL." Then, through a stroke of luck, he ended up on the cover of *Business 2.0* magazine, which led to some introductions that eventually led to the sale of the company to Microsoft, for $100 million in very liquid stock. The fast-inflating bubble could change fortunes in an instant; Traina would go on to do several more startups and eventually be named ambassador to Austria during the first presidency of Donald Trump.

Over at Third and Bryant, *Wired*'s neighbors Organic and Vivid were both in headlong expansion mode. With all the well-funded competition springing up out of nowhere, they had little choice but to go big or go home—a dilemma that many companies would face as the boom roared on. There might be hundreds of companies in, say, the business of building websites, but there would likely be only a handful left down the road, when growth inevitably leveled off and the better firms got bigger and stronger. Vivid sold in 1999 to Modem Media, an East Coast advertising agency. Organic spun out a piece of data software called Accrue that went public in '98, and then began readying its own IPO. CEO Jonathan Nelson was buying out the leases of neighboring companies to accommodate Organic's furious growth, and the company threw a "demo party," where the usual emoluments were accompanied by sledgehammers for breaking down the interior walls that had divided the old factory space. The fire department showed up and shut it down, even though he'd hired off-duty police for security.

A few months later Nelson got a call from the mayor's office: Willie Brown would like to see him. They met, and Nelson vented. "I said, 'We don't ask for very much, I'm basically running my own shuttle bus, I've got my own security to walk people to their cars. And when I want to expand, all I get is shit from the planning department about are you industrial, or are you office, and all this stuff. And I throw a party, and I hire the cops, and they shut us down.'"

Brown was apologetic and moved quickly to reconciliation. "He says, 'Well, two things. First, what do you want?'" Nelson told him he'd always wanted to put up a sign on the building, a permitting nightmare. "He says to the deputy, 'Get him his sign. And the second thing, next time you have a party, invite me.' So I did, and he came. This is crazy, but I'd hired a gamelan ensemble on the other side of the walls. We knocked down the wall, and there are people playing these Indonesian gongs. He loved it."

Nelson, like many people who had direct dealings with Brown, thought he was great. "It was transactional. And it's funny, people say he's corrupt, but what's corrupt? He got things done. I never saw any corruption."

As it happened, installing the signs would fall to John Law and Michael Mikel, who despite their falling-out over Burning Man were still partners in a sign business. Cantilevered beneath the parapet of the old printing factory, they would hang two signs at the corner of the top floor, which read simply "Organic." A prominent sight from the nearby freeway viaduct, the signs would long outlive the company.

One day in 1999, Warren Hellman, a prominent San Francisco investment banker, came by Organic for a visit; he was interested in buying the company, or maybe doing some other sort of deal. Nelson already had Goldman Sachs lined up for the IPO so there wasn't much business to discuss, but he offered Hellman and his colleagues a tour. "We were a bit feral," Nelson recounted later. Hellman and his crew stood out in their banker attire. "We walked by the engineering department, and the engineers all popped up from behind their desks and screamed 'suits,' and then they attacked him with Nerf guns," says Nelson, who was horrified and profusely apologetic. He hadn't yet learned that Hellman was no ordinary banker, and there was little that would delight him more than a Nerf gun attack.

The great-grandson of Isaias Hellman, a pioneering banker who was among the founding fathers of Los Angeles, Warren had graduated from UC Berkeley and then joined Lehman Brothers, where he'd become its youngest-ever partner, and then president. He returned to San Francisco in 1984 to launch Hellman & Friedman, an early mover in the private equity business, and he was now establishing himself as the very likable leader of the downtown business community—and a political power broker in his own right.

"This is the most amazing thing," he told Nelson after the Nerf gun raid, and invited the young entrepreneur to be his guest at Sugar Bowl, the Sierra ski resort where he led the ownership group. There they bonded over a mutual love of music. After discovering that they both adored the Dixie Chicks and Emmylou Harris, Hellman said, "I've always wanted to do a festival."

"I could help you with that," Nelson replied. It was the founding moment of what would become a beloved free music festival, Hardly Strictly Bluegrass, still held every fall in Golden Gate Park.

Dot-com mania was reaching its peak as the millennium approached, and the New Economy was now about more than technology and the Internet. San Francisco was feeding—and shaping—an invigorating new style of business: less hierarchical, more freewheeling, more fun. If the old economy and its hidebound corporations represented everything the city's emerging creative class wanted to leave behind, the new economy now being built in cyberspace represented a way to do it differently, with a lot more power to the people.

As *Wired*'s formulation might have put it:

> **Tired:** Corporate high-rises and wood-paneled offices.
> **Wired:** Brick and timber industrial buildings with big open spaces.
> **Tired:** Sport jackets and wing tips.
> **Wired:** Dogs in the office.
> **Tired:** HR policies and fun-policing.
> **Wired:** Drugs, booze and raucous parties.
> **Tired:** Reagan.
> **Wired:** Anarchy.

The Industry Standard's success made for a heady moment for all of us, and certainly for me. I went to the World Economic Forum in Davos, Switzerland, as a "Global Media Leader" with the coveted white badge, putting me several rungs above the mere "working press" at the status-obsessed gathering of the rich and powerful. I was hiring correspondents around the world, as well as an entire staff for a separate European edition based in London, where two of my top deputies would be in charge. When we finally brought on an HR manager, she was appalled at some of the things people had on their cubicle walls, and nervous about the party culture in general. I brushed her off. One day before a company retreat where we'd be kayaking on Tomales Bay, I suggested to John that maybe open pot-smoking by the two of us, the top executives in the company, wouldn't be a good look with all the new people we'd recently hired.

"Yeah, I thought about that," John said airily. "Nah." It was our company, and we did what we wanted.

There was, however, the annoying wrinkle that it wasn't in fact our company. It was Pat McGovern's company. But we hoped we could solve that one, too. John had fully soured on Pat and IDG, whom he felt (correctly) didn't share his vision, and didn't appreciate what we'd accomplished. McGovern, for his part, greatly resented John's lack of fealty. The only way out would be to spin off the company, which we could do only by promising such a big payday from an IPO that McGovern couldn't refuse. And in the meantime, we could get other people to fund our expansion. We raised $30 million from a group of venture capitalists and the parent company of the *Financial Times* and stepped on the gas—the European edition, a monthly magazine, new conferences and, most importantly and expensively, a big effort to build TheStandard.com.

With the magazine running steady at three hundred pages a week, orders for $10,000 ad pages were coming in unsolicited over the fax machine, and we sometimes had to turn them away—two things that veteran magazine publishers will tell you never, ever happens. The numbers were hard to believe: We went from a standing start to $14 million in revenue in our first year, $40 million in the second and were on our way to $140 million in the third. My back-of-the-envelope math showed us earning a couple of million a month in profits. All around us, companies with little more than a good story were mounting IPOs. We had a real, fast-growing, cash-generating business with a lot of upside, or so we thought. Surely we could go public too.

We'd outgrown our building, so we took over the ground floor of another old brick warehouse up the block, and then a floor in a modern mid-rise building on Kearny Street, and then another farther north, near Levi's Plaza. We leased a crumbling three-story factory building on the edge of Chinatown that needed a full renovation, which would take at least a year. But we didn't have many choices: There just wasn't any space, and what we could find cost four times what we were paying at Pacific and Battery. We put up a billboard across the street mapping all our locations—ostensibly to help people find where they were going, but really to brag about our imperial expansion.

We'd set out to call bullshit on the hype, and now we were among its most famous symbols. We didn't give much thought to what it all meant for San Francisco.

For city residents who weren't interested in or involved with the new economy and the Internet, the dot-com boom was no blessing. Sharky Laguana, who'd arrived as a homeless teenager, had spent much of the 1990s living and working at the grungy Civic Center Motel while trying to make it in the music business, hanging around the indie rock scene at a

Potrero Hill club called Bottom of the Hill. One day in 1998, he found in his mailbox the message of his dreams—a letter from a record label executive praising his music and asking to get together. After a few surreal meetings in Los Angeles he had a record deal for his band, Creeper Lagoon, that was nominally worth more than a million dollars.

But now the band had no place to practice.

Laguana and his bandmates lived and worked in the Mission, a working-class neighborhood flecked with small wood-built factories, workshops and warehouses, and prone to flooding when heavy rains revived the buried remnants of Mission Creek. In more recent decades it had become the hub of the city's Latino community, its commercial streets lined with bodegas, bars, taquerias, secondhand stores selling quinceañera dresses, and much else. The cheap rents and old industrial spaces were also a draw for artists and musicians, as well as young political activists. But suddenly a different sort of tenant was interested in the Mission.

"I'm in a major-label band and I keep losing my practice spaces to tech companies," Laguana recalled. "We went through like six in a year and a half. We'd find a new place, we'd be excited, we'd be there, and then three months later somebody would kick us out so they could put in a dot-com. It was ridiculous."

The downside of the dot-com boom in San Francisco came first in the form of skyrocketing rents, and nowhere was the problem more acute than in the Mission. It was a low-income neighborhood, not the natural home for well-paid tech workers, and there was a lot of crime—not just car break-ins, but violent gang wars that spilled out of the projects along what was then called Army Street. On the other hand, it was more fun than many of the staid residential districts, and in a city known for fog and chilly winds, it had some of the warmest, sunniest weather in town. Crucially, it was also convenient for commuting south, to Silicon Valley.

A wave of prosperous young dot-commers was soon pushing rents far beyond the means of the Mission's incumbent residents. The average city-wide price of a two-bedroom apartment had jumped from around $2,500 a month in 1996 to more than double that by 1999, with even bigger increases in the Mission, and the city's rent-control laws didn't always offer protection. A state law passed in 1985, called the Ellis Act, allowed evictions if the owner was taking a building off the rental market, which they could do by selling units as "tenancies in common," where the residents would jointly own the building. An owner could also kick people out if they planned to move into a home themselves, and could charge whatever they wanted if a tenant left voluntarily.

Evictions swept the city, with more than 2,500 each year in the late 1990s, more than twice the number of a few years earlier. Commercial tenants—including, in the Mission, many nonprofits, arts organizations, small workshops and bands like Creeper Lagoon—were entirely on their own. San Francisco's music scene, though no longer world-changing as it had been in the 1960s, was still buzzing with punk rockers, indie bands, rappers and ravers. Sixteenth Street in the Mission, in fact, was becoming "audio alley," with big startups like Listen.com, the music industry's answer to Napster. But affordable studio space was disappearing fast.

"I felt like tech was very destructive for the arts community," says Laguana. "Very destructive."

It wasn't entirely about the rents either—it was a cultural thing, with an awkward question at the center: Whose neighborhood was it? The khaki-clad financiers just out of business school, the journalists and marketers from New York and Los Angeles, the erudite programmers and math geeks from Berkeley and Boston, the rich kids from Marin—did they have the right to push out the poorer people and transform the culture of the neighborhood? A lot of locals didn't think so, and Mission activists were mobilizing: It was the city's most left-wing district too, steeped in street protest culture and ready to fight.

A development at Twentieth and Bryant was a breaking point. About sixty commercial tenants, including a sex-toy maker, a small furniture manufacturer and several nonprofit art spaces, would be evicted to make way for Internet company offices and so-called live-work lofts, smack in the middle of the neighborhood. Some five hundred protesters, led by a young community organizer named Chris Daly, swarmed a city Planning Commission hearing on the proposal and then did the same when it got to the powerful Board of Supervisors, which had broad authority over development projects.

"We had a five-hour hearing at the Board of Supervisors, and all the local TV stations had reporters there, and it was awful, because the anger towards gentrification came bubbling up," recalls Dan Kingsley, the project's developer. Kingsley, with his partner Paul Stein, had been among the first to see the opportunity in converting industrial buildings into dot-com offices. They'd done a half dozen projects in SoMa that hadn't stirred any controversy, but now he was seeing a new dynamic. "I have to be candid and say I learned a lot from that project," he said, referring to Bryant Square. "It certainly shaped my perspective and my approach."

The Board of Supervisors, including a brand-new member named Gavin Newsom, would ultimately approve Bryant Square, but it nonetheless served as a potent rallying cry for the anti-gentrification cause. The activists would soon turn their attention to an old bank building on Mission Street. It once housed a number of community groups, but they'd been evicted to make way for a dot-com company called Bigstep, and protesters would occupy the offices of the company, the latest symbol of the Mission's controversial makeover.

Chris Daly, who grew up in Maryland, had come to the Mission in the early 1990s after dropping out of Duke University and thrown himself into political organizing, especially advocacy for the homeless. Smart and passionate, and wielding confidence and authority well beyond his years, he was a driving force in the Mission Anti-Displacement Coalition, an alliance of community groups fighting evictions and gentrification. Tall and lanky, and almost too forthright, Daly was the kind of guy that the other side loved to denigrate as a clueless, bleeding-heart trust-funder, and his penchant for angry tirades and drunken screaming matches with opponents didn't help his cause. Moderate lawmakers grew to despise him. The *Chronicle*'s editorial page, and columnists including C.W. Nevius, were almost unhinged in their hatred for Daly, devoting story after story to his alleged sins.

In SoMa and the Mission, though, they loved him.

Daly explains his political philosophy in the simplest of terms. "For me, progressive always meant siding with the underdog," he said later. "Tenants over landlords. Workers over big employers. Small business over big business. I attempted to always speak up for the little guy, and that made part of my job very easy." In a city like San Francisco, he continued, all politics is ultimately about land use, giving it a zero-sum character: A particular parcel could be home to a new condo or office building, or it could be left alone for whoever currently occupied it, or if the money could be found it might be an affordable housing development. It couldn't be all three.

Resistance to development in the Mission, and nearby SoMa, came in many hues. The self-styled "Mission Yuppie Eradication Project"—eventually determined to be just a few people led by a wealthy twentysomething—stenciled slogans including "Die yuppie scum" on sidewalks and streetlight poles, and urged the keying of newer cars suspected of belonging to tech workers. A few vehicles were even burned. Violent resistance was the rare exception, though. Greg Suhr, captain of the police department's Mission Station at the time, viewed the changes in the neighborhood with curiosity. Crime had fallen significantly since the early 1990s, but he was still much more concerned about the gang wars that plagued the district than he was about tech workers getting hassled.

A flash point for anti-development activists was the new wave of so-called live-work lofts, conceived as combined lodging and studio space for artists and craftsmen and exempted from zoning restrictions. Led by the sharp-elbowed Joe O'Donoghue, the city's robust community of small construction companies, many run by immigrants from Ireland, had pushed hard for the special rules, and quickly began putting up small buildings all across SoMa and the Mission. There wasn't a mechanism requiring that they be sold or rented to artists, though, and they were quickly being snapped up by young professionals working in the Internet industry. Many residents were incensed by the workaround.

"Live-work was a representation of the push to gentrify the neighborhood," said Daly. Not only were the new buildings exempt from affordable housing requirements, he noted; in bypassing the zoning laws, they also squeezed out the light industry that had once been an economic pillar of the neighborhood.

The sentiment of the activists fighting the new development wasn't anti-tech, exactly. It was a class conflict. They didn't want upscale young people moving in and driving out the bodegas and secondhand stores in favor of pricey boutiques and high-end restaurants. The Slanted Door and Delfina were beginning to make the Mission a dining destination; Valencia Street was getting crowded with chic boutiques like Aggregate Supply and Therapy. Gentrification was the issue, and there was no obvious compromise.

The free spirits of the cultural underground were certainly on the side of the underdog, and many of them lived or worked in the Mission. They tended to be anarchists more than leftists, though, leading with humor and disruption: Santa Con, which began in 1994 as "Santarchy," with a few dozen Santas pranking Christmas shoppers at Union Square department stores, was more their type of stunt. Parody protests were a popular approach too—a "pigeon roast" outside Macy's, poking fun at animal rights activists, and a rowdy multiday boycott of Disney's *Fantasia*, for a host of fantastical sins; to the delight of the Cacophonists, that one fooled *Time* magazine, which cited it as an example of San Francisco gone mad. A key part of the credo, says Michael Mikel, was "no politics and no religion," but rather provocations that made people think. Even the Billboard Liberation Front, a band led by John Law that altered outdoor advertising signage to make a point (sample work: American Red Cross: '*cuz the government ain't doing squat*'), was as much about the caper as the political message.

Boundaries were fluid in the 1990s underground, though.

Chris Carlsson, a flinty writer and activist who was an important instigator in the city's protest culture, shared the Cacophonist's interest in adventuresome public stunts and the critique of capitalism, but he inverted their priorities.

"Why don't we do something political with bikes?" Carlsson had said one day in 1992, amid the haze of marijuana smoke at the offices of Typesetting Etc., a disheveled two-room suite on the second floor of the old Grant Building, on Market and Sixth Street. Carlsson and his business partner, a talented illustrator named Jim Swanson, provided typesetting services—the main clients were a couple of union newspapers—and made about half their money on the "etc." part of the business—namely, selling weed. There were often people hanging around, bike messengers and temp workers and miscellaneous others who came by to score. It was a clubhouse for the disaffected office workers and pranksters who, along with Carlsson and Swanson, had created the magazine *Processed World*, an arch satire of life in the corporate maze that came out irregularly for almost two decades beginning in the early 1980s.

It was *Processed World* that would first publish the "manifesto" of Law's Billboard Liberation Front, conceived as a parody—"through the Ad and the intent of the Advertiser we form our ideas and learn the myths that make us into what we are as a people." The magazine was often produced on paper stolen from offices, and then distributed in the financial district by costumed anti-corporate warriors. It represented the more traditional left-wing view of technology as a tool of corporate and government repression, rather than personal liberation.

"We dressed up like big computer heads and Liquid Paper bottles to hawk the magazine, it was a spectacle that we hoped would shake people out of their nine-to-five reality and question what they were doing with their increasingly automated lives," recalled Laura Fraser, a writer and longtime city resident who fell in with the group.

The crew took their work seriously—the flaw of *Processed World*, Carlsson would say later, is that it was too pedantic—and they were regulars at protests on causes including US policy in Central America and the Middle East. On this day in 1992, Carlsson and a few others had gotten into a conversation about a recent Gulf War protest where some people had been on bikes. Bike advocacy had a political dimension—cars were the corporate enemy, dominating urban spaces that belonged to everybody, and bikes were transport for the people—and the group arrived at a simple idea: Gather as many people as possible in one place and then bike home together, displacing the cars by filling the street. "We were the traffic," said Carlsson, "if we had enough bikes."

The starting point was the Ferry Building plaza, and the first ride drew about fifty people. But they did it again the next month and there were hundreds, and soon there were a thousand, and then there was a name: Critical Mass. Events began kicking up in cities around the country, and then around the world—a celebration of cycling, community and the outdoors. But in San Francisco it was also a protest against cars, and against the whole idea of a downtown full of corporate office towers. As the rides began to snarl the Friday evening commute, sparring with the police, and with angry drivers, became a part of the ritual, a monthly standoff between those who wanted to keep the city safe for business and those who wanted a revolution, or at least a government that wasn't in bed with monied interests. A police crackdown on a ride in the summer of 1997 resulted in more than a hundred arrests, with Willie Brown denouncing the ringleaders as "lawless, insurrectionist types."

Carlsson, though no tech enthusiast, had taken an early interest in the tools and had worked at Community Memory, an ambitious 1970s project conceived by PC pioneer Lee Felsenstein that put computer terminals in public locations for people to share their thoughts. Carlsson would later devote years to a San Francisco history project based on interactive CD-ROMs.

But if he wasn't always anti-tech, Carlsson was most definitely an anticapitalist. He'd grown up in North Oakland, and a high school gig caddying at the exclusive Claremont Country Club was a formative experience: His fellow caddies were a motley crew of mostly African Americans, many of them addicts of one sort or another. But he found that unlike the club members, they had a lot to offer.

"They were really smart and funny, and I learned a lot from them," he recalls. "And the people we were working for were the richest people in the Bay Area like [Bechtel Corporation scion] Stephen Bechtel Jr. and a couple of 'Bankers,' of Coldwell Banker, and Dick Landis, who was the CEO of Del Monte, and all these kinds of people. And they were idiots! Later, in retrospect, I realized oh my god, what a class lesson that was, to learn about that disparity of polite but really stupid people who own everything, versus the very bright and sarcastic and in some ways, very cynical, poor guys who were the caddies."

Carlsson's first political cause was the movement calling for the shutdown of the Diablo Canyon nuclear power plant, on the Central Coast. "The anti-nuclear power and weapons movement was very much a part of the left in San Francisco," recounts Tim Redmond, the *Bay Guardian* editor. So was the Central America solidarity movement and, in 1991, opposition to the first Gulf War. Those involved in the protests over foreign policy issues "intersected with the folks working for rent control and trying to prevent too much development," Redmond recalled. There seemed to be marches about one thing or another almost daily.

The protesters didn't win very often as the dot-com economy roared, but they often shaped the political dynamics. On Market Street, the Grant Building was sold and the new owner moved to impose massive rent increases on everyone, including Typesetting Etc. Carlsson was part of a campaign to resist it, which succeeded in delaying any evictions long enough for the market to turn and provide a reprieve.

On a far larger scale, Willie Brown's big development projects, which initially had wide support, were being considered with suspicion just a few years later. The City Hall renovations were complete, and the results were spectacular, though very expensive. The new ballpark at China Basin, approved by voters in 1996, would see its first pitch in the spring of 2000, and by general acclaim was a stellar venue, even if some lamented the loss of the funky warehouses and greasy-spoon cafés that dotted the old China Basin waterfront. Less popular were all the nearby condo developments, the fancy residential towers rising on Rincon Hill and the emerging plans for Mission Bay, Hunters Point and Treasure Island. Brown never seemed to see a development project he didn't like, and his imperial style was wearing thin.

Brown was a master at bringing his rivals close, and back in 1995 he'd shocked the local political world by persuading Calvin Welch, a longtime leader of the anti-growth movement, to support his first mayoral campaign. Welch had been fighting redevelopment and office construction for decades, but he believed the city needed investment and fresh energy, and Brown, with his civil rights credentials and deal-making skills, could be the ticket.

Welch and his allies would sour on Brown in short order, though. The rush of new projects, and workarounds to the rules like the live-work lofts, were a betrayal; one person's economic development was another's gentrification. "Landlords were evicting long-term tenants left and right, rents were absolutely soaring, and Willie didn't do anything about it," said Tim Redmond. "We call it the economic cleansing of San Francisco."

The mounting resistance to Brown wasn't just about protecting tenants. It was also about preserving the charming old buildings and small businesses that gave the city so much of its flavor. It wasn't downtown high-rises that made San Francisco the Greatest City in the World, after all, and letting developers have their way would be its ruination.

North Beach, a symbol of San Francisco's bohemian soul since the Beat poets made it their home in the 1950s, was already an expensive neighborhood in the 1990s, and though there had always been a lot of tourists, many locals wanted to keep things from changing, as locals often do. The young creatives and Latino activists of SoMa and the Mission had found their champion in Chris Daly. The white and Asian middle classes of North Beach, and some of the other northern and the western neighborhoods, would find theirs in Aaron Peskin.

Barely five feet tall, with a healthy salt-and-pepper beard and an impressive head of hair, Peskin moved with the ramrod posture of a military man, but spoke in the seamless paragraphs of someone who had made his way in life with his intelligence. He'd grown up in Berkeley, the son of two psychologists, and not far from his kindergarten classmate Kamala Harris. After attending UC Santa Cruz, where he previewed his political approach by leading protests against the university's expansion, he settled in North Beach with his wife, buying a condo and leading the residents' association in the building. He got involved with a neighborhood group, the Telegraph Hill Dwellers, known for battling projects that might block its residents' spectacular Bay views, and his first task was a tree-planting initiative. A local nonprofit, Friends of the Urban Forest, had challenged the group to plant four hundred trees around the neighborhood, and legwork would be needed, Peskin recalled years later.

"My job was to go and talk to these old Italian ladies and old Chinese landlords and convince them that having a street tree in front of their house at no cost to them was acceptable, and get them to sign the forms," he recounted. But Telegraph Hill Dwellers—where Peskin's wife, Nancy Shanahan, served on the board for decades—was also involved in far more contentious anti-development fights, and would ultimately prove a powerful political springboard for Peskin.

The cause that would vault the young Peskin into the top ranks of local politics came in 1997, when a twenty-four-hour Rite Aid drugstore was proposed for a spot adjacent to Washington Square, the bucolic centerpiece of the neighborhood. The burghers of North Beach and adjacent Russian Hill, though often at odds, were united in horror at the prospect of a round-the-clock chain store in their midst. As residents debated their options, Peskin, who had met Willie Brown a few years earlier, agreed to seek an audience with the mayor.

It was a cold and rainy winter afternoon, Peskin said later, when the neighborhood leaders filed into Brown's sprawling office at City Hall. "I said, 'Mayor Brown, do you remember me?' And he did. And I said, 'Here's why this is a terrible idea and yada yada yada.' And he said to me, I kid you not, 'Peskin, if you don't like the way I run this town, why don't you run for office.'" Peskin took the message to heart.

Brown has his own story about his early dealings with Peskin, an anecdote about outsmarting him in a fight over a proposed parking garage in North Beach. According to Brown, he pulled strings to get the lot cleared over a weekend, immediately following a favorable legal ruling, so it would be too late when Peskin went to court on Monday morning to stop it. It was classic Brown: He lived for outmaneuvering his political foes and then boasting about his victories. Years later, at a birthday roast for Peskin, Brown paid him his highest compliment: "This guy was bright," he said of his initial reaction upon meeting him. "I'm not used to bright elected officials." The two struck up an unlikely camaraderie over the decades, having coffee every week at the venerable Cafe Greco in North Beach, and the affection was genuine. Brown playfully called Peskin a "pain in the ass" and ribbed him about being cheap; Peskin said Brown ran the city "like a company store."

Yet none of this could paper over the fact that it was Peskin, more than anyone, who had brought Brown's reign as an imperial mayor to a premature end, and would be an obstacle to his agenda for twenty-five years.

Peskin would eventually find himself cast as enemy number one of the tech industry too. His detractors would accuse him of trying to freeze the city in amber and doom it to a future as a tourist theme park—a critique that would grow more pointed over the years as the city, and much of the state, prioritized preservation over development and failed to build enough housing.

It's surprising that a political operator as perspicacious as Willie Brown didn't get ahead of the shifting mood as the dot-com boom and all its dislocations washed across the city. His 1999 reelection campaign had provided an urgent warning that the benefits of the Internet frenzy weren't obvious to everyone, and that the Willie Show was getting old: Tom Ammiano, a popular gay politician and onetime comedian, gained enough votes as a write-in candidate to force a runoff election, and though Brown ultimately prevailed by a solid margin, it was an embarrassment—and a sign of what was to come.

The next year there'd be supervisor elections, and voters in 1998 had approved a measure mandating that supervisors be elected by district, rather than citywide. This was intended to give neighborhoods—and, by extension, the various racial and ethnic groups who inhabited them—more power over civic governance. Citywide elections had been put in place after the 1978 assassinations, in the hopes that they'd produce a less polarized Board of Supervisors, but twenty years on, with the city changing fast, a lot of people were feeling unrepresented. Confident in his power and popularity, Brown opposed the change but didn't do much to fight it. The result was that in 2000, all eleven seats would be up for grabs, some for two-year terms and some for four, with staggered four-year terms beginning in 2002.

When the election came, Brown had his handpicked candidates. "I'm not the issue," he insisted ahead of the vote. But he was very much the issue. Chris Daly, running for office for the first time in a district that included SoMa and the Civic Center, won an astounding 81 percent of the votes in a runoff against the Brown-backed candidate. Pastor Amos Brown, a leader of the Black community and a good friend of the mayor, lost unexpectedly, as did Brown's pick for the Bayview. Peskin easily won the North Beach seat. A handsome young Stanford Law graduate, Matt Gonzalez, ran on the Green Party ticket and prevailed, representing the Haight and the Western Addition.

It was a progressive sweep—a rebuke to Brown personally, and also a signal of the discontent the Internet economy had brought along with its riches. The ideological split between the Board majority and the mayor's office didn't do much to hinder Brown in his final years, but his successors would prove far less adept at managing it. In what had become a one-party town, the fault line separating "moderate" Democrats from their "progressive" counterparts would persist in city politics for decades—though it was hardly the source of all problems, as partisans often wanted to believe.

The dot-com economy, meanwhile, was about to catch city leaders across the political spectrum off guard yet again.

On a Friday evening in early 2000, I was taking the short, steep walk down from my apartment on Telegraph Hill to our latest "rooftop" party, at Bimbo's in North Beach. I was weary of the routine, truth be told: I'd been working seventy to eighty hours a week for almost three years, and the Friday afternoon closing of the magazine was still very stressful, in part because it had gotten so massive. When we were done, I was usually exhausted and just wanted to chill.

My presence was mandatory though, and as I crossed Washington Square, I saw a line of mostly white youngsters in jeans and shirtsleeves snaking up Columbus Avenue and around the corner. Surely this couldn't be a line for the party, I thought; Bimbo's was a big place, and there was no way it could be overflowing. But it was. I made my way inside and forged through the raucous, shoulder-to-shoulder crowd, eventually reaching a terrace where I could observe the scene below. The buffet table was piled high with shrimp and champagne, the open bar was flowing and I didn't recognize a single face in the young crowd as they scarfed the delicacies. It was all being paid for by Hewlett-Packard.

It felt unnatural, in more ways than one. North Beach was composed of old, low-rise wooden buildings, intimate in scale, with cozy cafés, family-owned restaurants and shops selling Italian sausage and pastries on the ground floors, and modest apartments above. It held treasured landmarks like the City Lights bookstore, still owned and operated by Lawrence Ferlinghetti of Beat poet fame, and Tosca Cafe, where booze was allegedly served throughout Prohibition and generations of celebrities were known to enjoy back-room cocaine. The bohemians of the 1960s had long since been priced out, but this mob of young people on the make seemed strangely out of place.

This "rooftop" also struck me as unserious, from a business standpoint, an expensive event crowded with twentysomethings, some of whom were committed entrepreneurs and major talents of one sort or another, but many of whom were recent college grads who'd come to San Francisco for the party, or the chance to get rich, or simply the opportunity to be themselves—the reason so many before them had come. Nothing against any of that. But my gut said most in this crowd were tourists, here for the boom today, but gone tomorrow if and when things went south.

"This can't possibly be worth it for HP," I muttered to a colleague who'd joined me as I estimated the catering bill at about a hundred grand. There was a late-empire feel to it all, partying too hard for reasons everyone had forgotten, a premonition in the air that it was all about to come tumbling down.

Just a couple of months later, it did just that.

Most dot-com stocks traded on the NASDAQ exchange, and the index tracking share prices peaked on March 10, 2000, after more than doubling in eighteen months. Suddenly it was heading fast in the other direction, with individual Internet stocks in a freefall, and within a few weeks the index was down by 25 percent. Our cover story called it "The End of the Beginning," which was a good take and turned out to be true, though it wasn't very helpful in the moment.

We'd been warning in the magazine from the very start that we were in a bubble that would eventually pop, but when it actually happened, I didn't really see it for what it was. The stock market was a leading indicator: If public investors lost confidence, then the venture capitalists and other financiers counting on IPOs to cash out would pull back, and then companies that weren't yet profitable wouldn't be able to raise more money and would fail. There would be a contagion effect, with investors fleeing even healthy companies. I'd covered business long enough to know that it was all but certain to play out this way.

What I didn't absorb was that this process would take a while, and in the interim I managed to persuade myself that it wouldn't be *so* bad. We had our third-anniversary party the following month at the magnificently restored City Hall, with hundreds of guests and a toast from the mayor. We tried not to think about the stock market.

Our advertising stayed very strong throughout 2000, and we'd even set the all-time record for most ad pages ever sold by a magazine in a single year—7,558, a record likely to stand for all time. I interpreted this as evidence that we were going to be okay. Some of the startup money would go away, sure, but we had plenty of big companies too, not just in tech but in luxury consumer goods and business services, and they weren't going bankrupt.

What was really happening, though, was that big companies were spending money that had been budgeted the year before, and startups were spending their last dollars in a desperate effort to find customers, or investors, or better yet a buyer before closing their doors. And honestly, much like the companies we covered, we just didn't want to believe we were going down. *The Industry Standard* might have staked its claim on clear-thinking analysis and telling it like it was, but we were fully implicated now, with our own glamorous lives and paper fortunes (my shares were nominally worth $2 million at one point, though I never really believed it). We mostly did a more-than-respectable job of covering the bust, but there was a plaintive tone to the still-fat magazines of late 2000, insisting on reasons for hope, even though most of the stories suggested there wasn't much. There was still a very big magazine to fill, and I foolishly kept hiring people through the summer.

John called one day around Thanksgiving of 2000, and I was abruptly awakened to the devastating reality. "I'm looking at the thirty-sixty-ninety," he said, referring to a sales-forecasting tool, "and after December, there is nothing." We were doing about $12 million a month in revenue through the end of 2000. In January 2001, we did $4 million.

Well, January is always the slowest month, we told ourselves. But it was downhill from there. We were losing $1 million a week.

I was in serious denial. In January 2001, it was obvious that major cost-cutting was essential, and we shut down some ancillary products. But I resisted big cuts to the core editorial staff, arguing that most of the bloat in the company wasn't in editorial (true, but beside the point) and that we had to maintain product quality to get out of our predicament (a clueless-editor thing to say in the circumstances). I ended up laying off about a dozen people, which only assured further rounds of layoffs in the months to come, each more painful than the last.

We still had north of $10 million in the bank, but our chances of survival ended with a game of chicken among our investors: IDG, which still owned 85 percent, and the venture firms, who had only 15 percent but two of the six board seats. IDG wanted to slash and burn, and fold the remnants into a different business unit, *PC World*—and, though I figured this out only later, avoid paying the $10 million it owed to Standard Media under a "tax treaty." The VCs, led by Jerry Colonna of Chase Capital Partners and strongly backed by John, had no interest in IDG's plan, which would wipe them out, and they were working on a deal to inject another $20 million into the business. But neither plan could happen without the others' consent. And the bad blood between John and IDG had poisoned the discussions, with McGovern determined to punish John, who was fed up with the fighting and mostly wanted out anyway.

We had a final round of negotiations in a basement conference room of a luxury hotel in San Diego while our last big conference was underway, a dozen or so grim-faced bankers, lawyers and John talking tensely in technicalities about the possible form of a new financing. I was a board "observer" so tended to keep my mouth shut, and I thought about the similar versions of this drama that were undoubtedly playing out among many of the companies in attendance upstairs—desperate efforts to maintain a facade of normalcy, accompanied by desperate efforts to find a way to fight another day. For most of us, there would be no way out.

About ten days later, in late July 2001, I was at a secluded beach south of the city on a Sunday morning, hoping to enjoy the magic of California for a few moments amid the stress, when my phone rang. I'd need to dial into an urgent conference call.

"We've asked bankruptcy counsel to join us today," said our lawyer, kicking off the call, and you could feel the mood shift crackling through the phone. "Just to be prepared in case, of course," he added, but no one had any illusions: We were going down. And the end was going to be ugly.

We filed for Chapter 11 bankruptcy a few weeks later, and I found myself standing in front of a hundred people on the first floor of 325 Pacific and telling them they were all laid off. And unlike those who'd been let go earlier, they'd receive no severance or health-care benefits.

People were pissed, naturally. A lot of them had become personal friends, and all were comrades in arms, and I'd been assuring everyone for months that the worst wouldn't happen. Given the $10 million in the bank, and the seeming folly of folding a business that had built a global brand in just a few years and was still generating about $3 million a month, I'd been confident there'd be a deal. But no.

Dazed, I retreated to my office and stared at the original *Dilbert* cartoon that was supposed to front the next issue—a tech worker in chains. Our final cover story the week before, about the federal investigations into the financial shenanigans of the dot-com IPOs, featured a fitting epitaph: "The Party's Over. Let the Blame Game Begin."

Our bankruptcy was national news, a perfect peg for a story everyone had been waiting to write: The dot-com bubble had finally and truly burst. A *New York Times* editorial, headlined "Economic Tombstones: The Industry Standard Unplugged," intoned on how we were a symbol of the moment, the *Rolling Stone* of our era. "It is a dizzying reversal of fortune for a publication whose meteoric rise and fall mirrors the euphoric, then harrowing times lived by the dot-com entrepreneurs it covered."

Most accounts of our downfall would be far less kind, of course. On the way up, we'd been feted as a publishing sensation, with glowing stories in the media trades and frequent mentions in the national press. My one-eyed elkhound Kaela, who came to work with me every day, had become practiced in posing for the magazine photographers who came through, a charismatic canine mascot for the liberating workplace style of the Internet economy. But now it was our turn in the barrel. Stories about us went negative after the bust began, questioning our prospects and mocking our hubris, often with barely concealed glee. The snarky corners of the Internet, led by a site called FuckedCompany, showered us in ridicule, down to making fun of my haircut.

I tried not to be thin-skinned—we dished it out, after all, so we'd have to take it too—though not always successfully. It was certainly an excellent if unpleasant education for me, as a journalist, to be in the media spotlight for a bad situation. *The Wall Street Journal* ran a long, well-reported story about our demise, by media beat reporter Matthew Rose, which was pretty tough on John, and me by extension, though it was mostly accurate. *The Washington Post* assigned a Style section reporter, Sharon Waxman, later founder and CEO of *The Wrap*, who by her own admission knew nothing about the magazine when she started in and, to my astonishment, wrote a long piece full of anonymous, self-serving quotes from IDG executives blaming John and me for the failure. There were also a number of factual errors.

I never got a response to the letter I wrote to the *Post*'s Style editor asking for corrections, though I don't blame him for that: I sent it on September 10, 2001.

The terror attacks of 9/11 put a definitive coda on the dot-com era. The old federal bankruptcy court on Pine Street, where the *Industry Standard*'s assets would be unceremoniously auctioned for a pittance just a few weeks later, was a busy place that fall; by one estimate, some five thousand US Internet companies went out of business in 2001. Even before 9/11, all the cities that had been lifted by the Internet economy were suffering. But no place was more exposed than San Francisco, with its comparatively small size, exceptionally large number of dot-com startups and an economy with few other pillars besides tourism, which was also crushed by 9/11.

Some thirty thousand people moved out of San Francisco in the early 2000s, with downtown and SoMa offering a preview of what would come in far more dramatic fashion with the Covid pandemic twenty years later: deserted streets, shuttered stores and restaurants and an eerie mood of apprehension. Unemployment spiked back over 7 percent, while commercial rents, which had more than doubled in a few years and briefly approached $100 a square foot for the best space, dropped by more than half. To the relief of those who didn't leave town, residential rents plummeted too, from an average of about $2,300 in early 2001 to $1,700 a couple of years later. Ominously, though, that was still far above the 1994 average of $1,010. And while condo prices slipped a little bit, housing prices never fell at all, and in fact continued a steady climb that began in 1994, took a breather in 2008 when the housing bubble burst and then marched quickly to new heights. The median price of a house in the city was less than $300,000 in 1994, and hit $500,000 in 2000. (By 2019, it would be $1.6 million.)

Winners and losers in the bubble era were often separated by nothing more than random quirks of timing. Jonathan Steuer, after bouncing from HotWired to CNET to The WELL—a veritable grand tour of first-generation Internet media startups—ended up as a vice president at Scient, a company that combined building websites for big corporations with expensive digital-strategy consulting. It had gone public in 1999, and Steuer's options were worth millions. But he sat in the wrong meeting one day in 2001; a division manager shared a disastrous sales forecast, and the company's general counsel immediately declared that everyone present was "locked up," meaning they couldn't sell their stock lest they be illegally capitalizing on inside information. By the time he was able to sell, the proceeds didn't even cover the taxes. Fran Maier, the Match.com cofounder, and many others were similarly burned by a quirk of the tax code where paper profits on stock options were counted as income as soon as they were converted to shares, even when those shares lost most or all of their value before they could be sold.

The *Wired* founders, though devastated by the loss of their company, at least had some luck on that front. Lycos, a Boston startup that was trying to rival Yahoo as a Web portal, had gone on a buying spree in 1998 and agreed to acquire Wired Digital—left for dead by Condé Nast after it bought the print magazine—in exchange for stock. By the time the deal closed in July 1998, that stock was worth an estimated $84 million. It went quickly downhill from there, but Rossetto and Metcalfe got out in time and made about $15 million apiece, according to one-time employee Gary Wolf's book about the company. Battelle earned enough to buy a house and set up a college fund for his kids.

Over at Organic, Jonathan Nelson acted sooner than most. He'd been boarding a plane for Singapore in April 2000, and the markets, after a brief respite, were falling like a stone. Organic had gone public just two months earlier, a superstar of the new economy, with international advertising giant Omnicom in its corner. Now it was all suddenly in doubt. By the time Nelson landed at Singapore's Changi Airport, the markets had reopened again—and were still falling like a stone. That fall he'd lay off seven hundred of Organic's twelve hundred employees, and himself as CEO, and that was just the start: By the time he was done, more than 90 percent of the company was gone. "The only way to stop a falling knife is to lie down underneath it and hope it doesn't take out your heart," he said ruefully years later.

Big chunks of the new economy were absorbed into the old economy. Organic was ultimately taken over by Omnicom altogether. (Nelson would continue to run the digital division, which eventually accounted for the bulk of the company's revenues.) CNET was bought by CBS. Wired Digital was reunited with its print sibling at Condé Nast. Time Warner, whose merger with AOL (which had already bought Netscape) in January 2000 is still regarded as one of the worst deals of all time, slowly marginalized its online division and resumed its old-school Hollywood dealmaking. It would be half a decade before the media industry realized that the dot-com era was only a preview of the upheaval that was to come.

On Wall Street, the IPO rules were tightened and the highest-profile touts were punished. Securities analyst Henry Blodget, who'd become famous for an over-the-top bullish call on Amazon and would later start the news outlet *Insider*, was banned from the industry for dishonest stock promotion. Investment banker Frank Quattrone, the architect of many dot-com IPOs, was convicted of obstructing justice, though the verdict would be overturned on appeal. A few executives, notably Bernie Ebbers of the telecom firm WorldCom, and several perpetrators of the Enron fraud, would be sentenced to long prison terms.

Yet the sanctions for bad behavior were the exception more than the rule. And the Internet, and its transformational potential, hadn't gone anywhere, with a few corners of the new industry still going strong. Google, down in Mountain View, had proven itself a superior search engine and was beginning its long march to dominance. In Palo Alto, Elon Musk had already sold Zip2, the classifieds site he'd founded with his brother, and launched X.com, the predecessor to PayPal, which would take off in the early 2000s. PayPal later became famous for its alumni, the PayPal Mafia—Musk, Peter Thiel, Reid Hoffman, Max Levchin, David Sacks, Jeremy Stoppelman and Keith Rabois, to name only the most prominent. They'd go on to launch some of the most important companies of the next Internet era.

Brewster Kahle, after selling WAIS to AOL, started a Web navigation service called Alexa Internet, which he sold to Amazon after just three years for $250 million in shares. After working for Bezos for a few years, he'd used the proceeds to buy an old church on Funston Ave. in the Richmond district, not far from the Presidio, to house the Internet Archive. Kahle had realized that a lot of what was being created on the Internet would simply vanish absent proactive efforts to preserve it, and the Archive would become a lifelong passion project and an important utility for many millions of people, though legal fights over copyright would never go away.

The list of startup obituaries was long, but the idealistic vision that animated early Internet pioneers would survive the bust. The notion that the Internet could be a means of personal empowerment and community connection—and maybe even, as John Perry Barlow had advocated, a domain of true freedom, not subject to the rules of nation-states—might have been overshadowed by the flash and glamour of sudden riches. But a lot of people were still dedicated to the idea that the Internet would make the world a better place. The venture capitalists hadn't gone anywhere either. Now there'd be another chance.

PART TWO

The City Family (2002–2011)

CHAPTER 4

Gavin Newsom Meets Web 2.0

Gavin Newsom, the handsome young scion of an influential San Francisco political clan, had moved to the city in 1992, and though he wasn't a tech maven, he quickly established himself as an entrepreneur, cofounding a successful wine and hospitality business called PlumpJack. It would prove a good stepping stone to his natural calling.

Tall and charming, with an impish grin, Newsom began his political career as a volunteer for Willie Brown's first mayoral campaign, in 1995. He squired the candidate around the bars of the Marina and Cow Hollow, buying drinks for the house and generally proving himself "invaluable," Brown said later, though his talents didn't come as a total surprise. Brown had known Gavin Newsom since he was a kid.

Gavin's grandfather, Bill Newsom II, was a close confidant and campaign manager for Edmund Brown, the towering two-term Democratic governor credited with building much of the infrastructure of modern California in the late 1950s and early 1960s. Gavin's father, Bill Newsom III, was a highly regarded attorney and state court judge, and was close to Edmund Brown's son Jerry, also a powerful multiterm governor. Bill Newsom III was also fast friends with a classmate he'd met at St. Ignatius: Gordon Getty, the oil heir who was among the richest men in the country.

Bill Newsom had split with his wife, Tessa, when Gavin was still a toddler, and she raised Gavin and his younger sister in Corte Madera, in Marin County, sometimes working multiple jobs to make ends meet. Bill Newsom had been appointed to the bench in Auburn, a couple of hours away, and young Gavin went to public high school, the son of a single working mother. Newsom and his associates would later cite the comparatively straitened circumstances of his upbringing as formative influences that shaped his later ambition. In 2003, Newsom said he suffered from "pretty severe" dyslexia that had made schoolwork difficult.

But Corte Madera was still a very upscale suburb, and Newsom had attended Redwood High, a feeder for elite universities that was considered among the best in the state. Gavin Newsom also had another, far more glamorous family, the one that revolved around his father's best friend.

The fourth son of the oil baron John Paul Getty, Gordon Getty was enamored with music and anthropology from a young age and never got much respect as a businessman from his father. But he was a central player in a convoluted series of events that led to the sale of Getty Oil in the mid-1980s, which yielded a family fortune that in today's dollars would be worth more than $10 billion. Gordon had eloped with a striking redhead and Berkeley grad from the Central Valley named Ann Gilbert back in 1964, and she'd ensconced herself as the undisputed queen of San Francisco high society, a small but robust clique of families who'd made serious money building California into an economic colossus, and later in marketing its style to the world. Gordon and Ann had four boys, and Gordon settled into an exotic life as a composer and art collector (and, as it was revealed only in 1999, the paramour of a woman in Los Angeles with whom he had three daughters).

Dynastic fortunes had a lot of throw weight in San Francisco. There were the Stanfords, the Crockers, the Huntingtons and the Hopkins, all of the Southern Pacific Railroad, and then the Hearsts (mining, media and real estate), the Floods (silver), the Fishers (The Gap), the Haases (Levi's), the Swigs (hotels) and the Bechtels (construction), to name only the most prominent. All wielded great influence at various junctures. But none in quite the style of the Gettys.

Ann and Gordon bought a Pacific Heights mansion and eventually the one next door, and then another, creating a compound where Ann would throw lavish parties, ranging from fundraisers for politicians (mostly Democrats, with Barack Obama and Kamala Harris eventually among them) to intimate salons for the wealthy and high-minded. Among their circle were Ron and Barbara Pelosi, as well as Ron's brother Paul and his wife, Nancy.

The Pelosi brothers were both wealthy real estate developers, and Barbara had been reared in the political world as the daughter of Bill Newsom II and the sister of Bill Newsom III. Gavin Newsom was her nephew. Barbara's sister-in-law Nancy Pelosi, born into an influential political family in Baltimore, was still raising her five young children, and it would be years before she ran for office. But those early bonds among the Gettys, the Newsoms and the Pelosis would shape the politics of San Francisco—and the country—for many decades.

Young Gavin came of age straddling proximate but very different worlds: bucolic Marin County, full of erstwhile hippies who'd found the pleasures of money, and the rarefied corner of San Francisco society where vast fortunes and political power were all but a birthright. Gavin became close to Gordon and Ann's son Billy, and the Gettys treated him as one of their own, even taking him on a safari to Africa.

Newsom went to college at Santa Clara University, and though his hopes of baseball stardom didn't come to fruition, he graduated with a degree in political science and arrived in the city at an opportune moment. Smart and ambitious, with a gift for patter, Newsom liked to party, but he had more conventional tastes than the ecstasy-loving ravers and anarchic performance artists who were then lighting up SoMa. Newsom preferred drinks, and felt at home at the Balboa Cafe, a bustling bar with a colorful history that sat in the Marina district, just down the hill from the gaudy mansions of the Pacific Heights Gold Coast. Those extraordinary houses, standing tall on the ridges of Broadway and Vallejo and Pacific Avenues, often with dozens of rooms and terraces sprawling across four or five floors, were a looming reminder of the generations of business moguls who'd shaped the city, and usually done their best to keep the lower classes down, or at least down the hill.

The Balboa Cafe, indeed, was once a working-class dive. But the Marina, with its peerless location on the city's northern waterfront, had been moving upscale for years, and extensive rebuilding after the 1989 quake—which destroyed or seriously damaged about seventy buildings in the neighborhood—solidified it as a prime destination for moneyed young professionals, many in finance and law. Animated young women in short summer dresses, seemingly straight out of the magazine ads for Banana Republic or Esprit, crowded the Balboa Cafe night after night, along with their boaty, Polo-wearing male counterparts, a running frat party for the established upper crust.

Newsom was just out of college, working as an assistant for politically connected real estate developer and family friend Walter Shorenstein, when he and Billy Getty, with financial backing from Gordon Getty, opened a Marina wine store, which they named PlumpJack in honor of a fun-loving Shakespeare character from one of Gordon's operas. The idea was to make wine less pretentious and more accessible, and the business was well-done and well-timed, with interest in wine on the rise and the new Internet economy just starting to hum. The wine shop, and then a nearby PlumpJack café, did well, and Gordon kept investing: They bought the Balboa Cafe in 1995, opened an elegant ski lodge in Tahoe that same year and acquired a Napa winery a few years after that. Newsom had a flair for marketing, a taste for risk and was very much at home in the well-lubricated hospitality industry. He certainly had a lot of help with PlumpJack. His sister Hilary and a cousin, Jeremy Scherer, joined the company to run operations, and Gordon Getty would remain the financial lynchpin of the growing empire. It was never very clear how much Gavin Newsom had to do with the company's success. But PlumpJack would provide him with a healthy paycheck and solid résumé entry as a successful small businessman and entrepreneur.

He was already eyeing his next move.

Though just twenty-eight years old, the extraordinarily well-connected Gavin Newsom was a very good fit for Willie Brown's mayoral campaign, and a natural part of what Brown would come to call the City Family. The young entrepreneur shored up some important constituencies, namely the young and flush Marina crowd, and the adjacent ultra-rich in Pacific Heights. Brown needed the help: He represented a San Francisco district in Sacramento and worked closely with the Burton brothers, but for decades his focus was the expansive, big-spending state government. City politics was a different zone, and he didn't have it wired up in quite the same way.

Newsom was a good talker, and seemed to have the political gene. There was also that Getty money sitting behind him, a formidable asset by itself, with the potential for a lot more from the Gettys' social circle. Susie Tompkins, who with her ex-husband had founded North Face and Esprit, and later married a well-liked real estate developer and former city official named Mark Buell, was a big fan of Newsom. The style moguls Don and Doris Fisher, and the local members of the Pritzker hotel clan, would be big supporters too.

Newsom's efforts in helping Brown win the mayoral race were rewarded: He was named to the Parking Commission, which may not sound like much, but in San Francisco such oversight bodies often wielded real power. There, too, he impressed Brown with his work ethic. "I wanted to do something about cleaning up the taxicab industry," Brown recounted. "I called a taxi summit. The only person that spent the same number of hours at that summit as I did was Gavin Newsom. He was a real student of public activities." The next year, when a seat on the Board of Supervisors came open, Brown, with the encouragement of John Burton, gave Newsom the nod, appointing him to the vacant office and supporting his run for a full term the next year.

Tom O'Connor, a friendly, strapping young fireman who'd always been interested in politics, was tapped by the powerful firefighters' union to work the Newsom campaign. "I was a junior man in the union and they said, 'Hey, this pretty boy from parking and traffic, he's making a run for Supervisor. He's going nowhere. We'll put O'Connor on it.' So I got assigned to Gavin, walking precincts, building-to-building. Generally the firefighters always went with incumbents. But he had a strong pedigree and he was close friends with John Burton, so as a favor to Willie we endorsed him. It was part of the Burton/Brown machine." It was the last year of citywide supervisor races, and Newsom was elected easily.

It only seemed fitting that Newsom would soon begin dating the daughter of another political power broker, Anthony Guilfoyle, who'd worked his way up through the construction trades to become a successful real estate investor, and was now known locally as the Godfather. As a feisty young assistant district attorney, Kimberly Guilfoyle made a name for herself in 2001 prosecuting a notorious dog-mauling case, with a glowing profile in the *Chronicle* describing her as "a bespectacled former underwear model who loves animals and prides herself on hard work and her track record in the courtroom." She and Gavin Newsom would marry the next year. He was already eyeing the mayor's office.

Willie Brown tended to people—that was one of his superpowers, and the foundation of his political success—and it was a long game, involving a wide circle. His commitment to diversifying government ran deep, and he thought women were an untapped talent pool. He and the Burtons backed Dianne Feinstein in her failed bid for governor after she left the mayor's office, and then again when she ran for Senate, even though she was considerably more conservative than they were. They also supported the much more liberal Barbara Boxer, a former Burton aide and Marin County supervisor, who won a seat in the state assembly, and then the US Senate. Brown developed a tight relationship with Nancy Pelosi who'd emerge as the other samurai of contemporary San Francisco politics. She now held the House seat once occupied by the master of them all, Phil Burton, who'd died young from a heart attack in 1987.

High-level alliances were only the tip of the political iceberg for Brown. He was often peering far below the surface, figuring out obstacles and opportunities long before his rivals were paying attention. With numerous boards and commissions overseeing city departments, and various powers reserved for the Board of Supervisors and the state, getting anything done required exceptional skills of orchestration and persuasion. Brown had his methods. "We hired a person whose job was to figure out every committee or organization in a state agency or whatever that would have to approve something, and then list all the people, and who appointed them," recalled Ed Harrington, who served as city controller—a quasi-independent fiscal overseer—during Brown's mayoralty. "He knew that in six years we would need their votes, and he was starting to figure out getting his people appointed to those places, so when he needed them, they would be in the right place for him. It was all very long-range."

Brown enhanced his power by pushing through a charter reform measure—one initially intended, ironically, to strengthen his hapless predecessor, Frank Jordan. He packed his office with well-paid deputy mayors and special assistants, some 350 in all, according to a *Chronicle* investigation in 2003. He steered City Hall sinecures to whomever he liked, including, at the end of his second term, an aide whom he'd impregnated. Brown was married but had long lived separately from his wife, and the relationship was no secret around the office: When the baby came, Brown happily shared the news.

Brown was never shy about mixing the personal and the political. In 1994, he'd begun dating an attractive lawyer half his age named Kamala Harris, then working in the Oakland District Attorney's office. He appointed Harris to two plum state commission posts and squired her around, from the celebrity-studded sixtieth birthday party of his billionaire friend Ron Burkle to a New York jaunt where Brown met with Donald Trump. The relationship brought Harris invaluable connections and a whiff of celebrity, and they were together for about eighteen months. But Brown wasn't in it for the long term and they split in late 1995, shortly after Brown was elected mayor. Harris moved to the San Francisco District Attorney's office in 1998, and then the city attorney's office a couple of years later, before challenging incumbent DA Terry Hallinan in the 2003 election.

The favorite son of a family of left-wing lawyers, Hallinan was a notorious figure in town, famous for settling disputes with his fists; he'd managed to get admitted to the state bar despite numerous arrests for assault only after Willie Brown and Phil Burton testified to his character. But his name and heritage—his father, Vincent Hallinan, was once a presidential candidate on the Progressive Party ticket—along with a strong civil rights record and a lot of swagger, had won him an enthusiastic following. He was appreciated for defending the musicians and hippies of the Haight-Ashbury from drug charges in the 1960s. He'd even represented Anton LaVey, of the Church of Satan, over alleged crimes relating to a pet lion.

Despite the fierce opposition of the police union and much of the city establishment, Hallinan was elected DA in 1995 and then reelected in 1999, boosted by his willingness to take on the police department in a scandal known as "Fajitagate."

The next election would be different. David Chiu, then a deputy in Hallinan's office, said it was clear from early in his second term that he was vulnerable. "At the time we had the lowest felony conviction rate in the state, and we all talked about how at some point someone was going to have to challenge the boss," Chiu recalled. He told his friend and colleague Kamala Harris that she ought to do it. And she eventually did, tapping the power circles that Brown helped introduce her to and looking to capitalize on weariness with the incumbent. She reached out to Mark Buell, and they went to lunch at the Balboa Cafe. "I'm thinking, she's pretty smart," Buell said later of their meeting. Buell had worked at City Hall under Mayor Joe Alioto in the 1970s before making his career in real estate, and he and his wife Susie, the Esprit founder, occupied an opulent penthouse at 2500 Steiner Street—one of the city's best addresses—where they'd throw soirees for the Democratic elite. He was the kind of guy whose support could mean everything for an ambitious San Francisco politician.

Buell had only intended to make a token contribution to Harris as a favor, but was impressed enough by the end of the lunch that he'd agreed to chair her campaign finance committee. He had another motive too. "It wasn't as much that I liked her, it was that I *hated* Terry Hallinan," Buell half joked years later. When Hallinan was supervisor, he explained, he'd agreed to support Buell's appointment to a city commission, but then didn't do so when the vote came. Buell was livid and hadn't forgotten. He would put his shoulder into Harris's campaign, serving as chair and sitting with her while she dialed for dollars alongside other supporters, including David Chiu, who'd later win election to the state senate and serve as city attorney. A formidable fundraising haul would help Harris win the DA race, and launch her meteoric career.

City leaders, including Brown and Newsom, had been caught off guard by the dot-com boom, and it had come and gone so fast that it was easy to see it as an anomaly, a one-off gold rush moment. For the city, there were a couple of tight budget years, though even at the peak of the mania the dot-coms still accounted for only a small part of the city's economy. In much of government, and many parts of the business world too, there was a distinct if misguided sense of relief that the Internet frenzy was over, even if the economy was suffering for it. A "nuclear winter" had descended on the local Internet industry, and there didn't seem to be much life in the rubble.

But anyone deeply involved with the Internet knew that it was certainly coming back, and the early 2000s would be a charmed moment of possibility for those who'd stuck around. A number of San Francisco entrepreneurs and creative talents, many of whom had made some money in the first dot-com boom, were ramping up new enterprises, or in some cases reviving ideas they'd worked on earlier. They were laying the groundwork for a new burst of Internet innovation, soon to be dubbed Web 2.0.

Ev Williams and Pyra Labs, in the wake of the bust, had found themselves with a hot product, Blogger, but not a dollar in the bank. After a bitter split with cofounder Meg Hourihan, Williams ended up running Blogger by himself out of his Mission district apartment. The product was popular, but it was also free, with Williams at one point forced to mount a donation campaign to buy a new server. But with a well-timed investment he was able to hang on until he caught the attention of Google, not yet a public company, which bought Blogger for stock in early 2003.

Williams and a couple of people he'd hired in the final stages of Pyra Labs, Biz Stone and Jason Goldman, began carpooling together from the city to the Google offices in Mountain View every day, and not very happily. Blogger kept growing, but Google didn't seem to have much of a plan for it. Williams didn't love it at Google, and found Mountain View terribly dull. "I just didn't really click that much," he said later. In 2005 he quit, and before long he had a new company, and a small office in South Park. The terms were never disclosed, but he'd made good money on the Google deal, and was ready for the next thing.

The blog, in the sense of personal publishing on the Internet, had begun to emerge in the '90s along with the Web itself. Justin Hall, a teenager from Pennsylvania, had caught the early Web moment with a site called "Justin's Links from the Underground," where he surfaced interesting things from the brand-new medium alongside intimate personal commentary. He was a fixture at the *Wired* offices on Third Street and Bryant. Suck.com, created surreptitiously by two HotWired employees, Carl Steadman and Joey Anuff, offered cultural commentary on the emerging world of the Web that was by turns hilarious, erudite and cynical. It was a viral hit for many months in the mid-1990s before anyone knew who they were. Dave Winer, an insightful and argumentative software entrepreneur, was publishing an influential proto-blog via email, and along with several others had developed new software for personal web publishing.

Mark Frauenfelder and Carla Sinclair's print zine *Boing Boing* had gone dormant in the mid-1990s, but after Frauenfelder interviewed Ev Williams for a story it was reborn. "I was writing about blogging and I got interested in blogging," Frauenfelder recounted later; as he'd done with the original print zine, where he pursued a passion for comics and sci-fi, he started writing about things he found interesting on the Web, and publishing via Blogger. A few folks he knew started reading and got in touch to see if they might contribute, and it settled into a four-piece band: Cory Doctorow, a prolific writer of science fiction and hard-left political commentator in equal measure; futurist David Pescovitz; culture critic Xeni Jardin; and Frauenfelder. It proved so popular that the cost of the "bandwidth" needed to accommodate all the readers was becoming significant, and Frauenfelder got in touch with his friend John Battelle to see if he might have ideas on making it a business. That conversation would ultimately lead Battelle to start a new company, called Federated Media.

New tools and new ideas about Internet publishing were suddenly springing up everywhere. Matt Mullenweg, a teenager from Houston, Texas, had created an open-source blogging platform called WordPress, building off a popular piece of software called b2. By the time he made a trip to San Francisco in the spring of 2003, he was already Internet-famous: Thanks to the quick uptake of WordPress, he'd achieved a personal goal of being the first link that came up in a Google search for "Matt."

His personal blog had serious reach. "I posted a message that I was going to be in San Francisco, and a lot of people got in touch," he recounted later. Mullenweg visited Yahoo, and then Google, where he met Ev Williams and the Blogger team, and then CNET, which was very interested in WordPress for its own operations. "I really fell in love with the city," he said. He returned to Houston, where he was supposed to start his junior year in college, when the job offers started to come in, and he worked out a deal with CNET where he'd join as an employee while continuing to develop WordPress. "So I dropped out of school and drove my Chevy to San Francisco with my mom." He crashed with a friend in a shared house in the Richmond district, where his roommates included a young Korean American lawyer and aspiring politician named Jane Kim.

"There was a lot of integration between all of our communities—the artists, the activists, the tech folks . . . San Francisco at that time just really felt like a creative hub for all of this," Kim said later. "Matt was a couch surfer, I'd wake up and he'd be on our couch." It wasn't until a year later, when she was running for school board, that she realized Mullenweg wasn't just any couch surfer: A brief blog post he'd written about her candidacy was now the number one result for a Google search on her name. "I'm like, who is this guy?" recalled Kim. Not long after that, they went to an event together at a nearby bar. "There was a huge crowd of young folks waiting for Matt—it was an event they were hosting for him. We walked in and everyone just erupted, they were so excited to finally meet this person. I was very struck by the community that existed that I wasn't really aware of, the blogging community that really admired what Matt had built."

Blogging hadn't yet broken through to the mainstream, but enthusiasts were consumed with a number of seemingly simple technologies and business concepts that built off early ideas about what the Web could be. Brian Behlendorf's Apache project was all about a free and open Internet, where corporate behemoths like Microsoft wouldn't be in charge and the user would come first. Craigslist, the Electronic Frontier Foundation and the Mozilla Foundation, which was ramping up an open-source browser called Firefox, were similarly committed to that philosophy. "WordPress was part of this wave where things were open by default," Mullenweg said later. He cited Flickr, which had recently pioneered photo-sharing, del.icio.us, which developed the idea of "tagging," Technorati, which built a leaderboard of blogs, and the news-reading tool called RSS. "They all had APIs [application programming interfaces], there was tagging, there was linking, there were a lot of integrations, everything would work together. It was just a very generative time." It didn't have a name yet, but later on, when big companies were taking over more and more of cyberspace, what Mullenweg and Behlendorf and Newmark and Battelle were championing would become known as Web 2.0, and then the "independent Web." It would be a font of innovation and a source of inspiration for a generation of San Francisco entrepreneurs, and a major force in the transformation of media. It would also lay the groundwork for social media—destined to prove a far less beneficent innovation.

Mark Pincus was also appreciating the relative calm, and creative possibilities, of the early 2000s. Remarkably, he'd managed to score with two dot-com era companies, Freeloader and Support.com, taking the latter public just as the bottom fell out entirely. Now he was enjoying his neighborhood, a particularly pleasant enclave off Haight Street called Cole Valley, lined with colorful Victorians in varying states of repair. Strolling the streets with his bulldog mix Zynga, Pincus was outgoing and charming when he was in the mood, and could talk a blue streak about the possibilities of the Internet, or the future of politics. Yet he also carried the mien of a hippie dog-lover, and was anything but out of place in what was sometimes called the Upper Haight.

Craig Newmark, who lived in Cole Valley too, took a liking to Zynga, and eventually to its owner. "I learned so much from Craig," says Pincus. "I remember him telling me that after two years he was going to turn on pictures with listings, and he was worried about what the community would think. And I'm like, 'Dude, of course I want to see a picture of the couch before I drive across San Francisco to buy it. You're that worried?' And then I realized that he empathizes with his consumers on a level that most of us don't get."

The people-driven Internet was coming, Pincus was convinced, and it would take control away from the big corporations that dominated media, finance, retailing and everything else. A young intern at Freeloader, Sean Parker, had called Pincus one day about a company he'd cofounded, Napster, which allowed people to share music freely on the Internet. Pincus wrote a check for $100,000, becoming the company's first investor. Napster was shut down after a couple of years following a brutal legal assault from the music industry, but it was a taste of what was to come. The "Revolution of the Ants," as Pincus summed it up in an essay, was at hand.

People power was coming to politics too, and Pincus spun up something called eParty.org, a "Napster meets eBay for politics," as he put it later, striving to build a "Web-based coalition" that might give a "voice and choice to the people." He'd begun throwing dinner parties at his home and meeting more fellow travelers interested in the nexus of politics and the Internet. He'd become friends with Brewster Kahle, who'd just begun building the Internet Archive, and they'd work together on something called SF LAN, an early effort to provide free Internet connectivity in the city. Newmark pitched in on that too. Another friend, Reid Hoffman, was also fixated on the idea of social connections on the Internet; Pincus and Hoffman would invest in a pioneering social network called Friendster, which Jonathan Abrams, a former Netscape engineer from Canada, was running out of his San Francisco apartment. A genial Englishman and dot-com entrepreneur named Michael Birch, who'd moved to the Bay Area in 2002, was similarly inspired by Abrams, and would spend thirteen straight days coding up his own social network. Ringo.com, as he called it, didn't prove a winner, but his next attempt would bring an improbable fortune.

Hoffman, who would prove to have an amazing eye for startup investing, had launched a very early social network back in 1997 called Socialnet.com. That didn't work out, but after his stint at PayPal he returned to the idea and in 2003 would create LinkedIn. Pincus rented a space in a low-rise commercial building on Eighteenth Street in the Mission, where the resident entrepreneurs included Kevin and Julia Hartz, who were ramping up what would become the listings platform Eventbrite. Pincus decided to build his own social network, which he called Tribe. It got some early traction, but was ultimately a bust.

"It was trying to be Craigslist meets Friendster," Pincus said later. "It turns out you want to see a picture of the couch you're buying, but not the person who's selling the couch, because they're weird. I got that wrong with Tribe." A young Harvard dropout named Mark Zuckerberg had taken a different approach, trying to bring order to the chaotic early online world by requiring real names on his social network—in contrast to the break-out success MySpace—initially limiting it to college students and keeping the design understated and rigid. He'd moved the company from Cambridge to Palo Alto in 2004 and hired Pincus's former intern and Napster cofounder Sean Parker as president. Pincus, along with Hoffman, was an early investor in Facebook—taking some of the sting out of the failure of Tribe.

Pincus was also immersing himself in the city. With his friend Emily Morse, later a popular sex blogger, he prowled the streets in a satirical search for sex clubs, interviewing people and filming it for the local cable access channel. He went to Burning Man for the first time in 2003 and would become a regular. He was meeting lots of people around town, including Matt Gonzalez, a dashing young left-wing lawyer who was now president of the Board of Supervisors and had been among the first politicians to make use of Tribe. Pincus was impressed.

"He was just cool as fuck. I didn't agree with any of his politics, I just liked him. He was going to make San Francisco cool again," he recalled. "I liked Gavin, but, you know, he's not that cool."

For Willie Brown, ever-attuned to what it would take to win, Gavin Newsom was the obvious choice as his successor, even if he wasn't that cool.

The rookie supervisor was living up to his promise, working hard and building up to a run for mayor with a ballot initiative called Care Not Cash, aimed at ending homelessness. He was clever and articulate, and the Pacific Heights crowd was ready to go all-in on his bid for City Hall. Tom Ammiano, the supervisor and gay rights activist who'd almost beaten Brown as a write-in candidate four years earlier, would take another swing, but he'd had a falling-out with Matt Gonzalez, who joined the race on the Green Party ticket. Newsom easily defeated the field but fell short of 50 percent, and would have to face Gonzalez in a runoff.

The Newsom-Gonzalez matchup was a pure expression of the city's main political divide. Newsom, tall and angular, with mousse in his hair, expensive clothes and the natural confidence that comes from growing up around powerful people, was the centrist, pro-business candidate. He was the clear choice of the wealthy, as well as small-business owners and others worried about homelessness and crime. Gonzalez, who was from a wealthy family in Texas and landed in San Francisco after graduating from Stanford Law, cut the figure of a Mission hipster, also smart and handsome, and advocating for the poor and vulnerable. Gonzalez was often righteous and doctrinaire in his approach, as the local left tended to be, inveighing against developers and the police and money in politics. As president of the Board of Supervisors he refused even to meet with Brown, lest he be tainted by corruption.

Newsom would prevail, though at 53–47 it was no landslide, despite his spending some $4 million, a full order of magnitude more than Gonzalez.

Newsom was now the mayor, but the surprising enthusiasm for Gonzalez had put him on the back foot. He quickly began shoring up his left flank by supporting striking hotel workers, and the gay community was also a key swing constituency. Mike Farrah, an aide to Newsom, recalls traveling to Washington, DC, with the new mayor to attend George Bush's State of the Union address, where the president proposed a constitutional amendment to ban gay marriage. "I witnessed his anger" over that, says Farrah, and they began plotting a response. The gay marriage question was in the air: Eric Jaye, Newsom's top political advisor at the time, recalls seeing a flyer about it in the Castro, and seizing on it as a promising way to win over the gay community. There was already a protest tradition of gay couples going to City Hall on Valentine's Day to get married only to be turned away, and after some strategizing, on February 12, 2004, barely a month into his term, Newsom directed the city clerk to begin issuing marriage licenses to same-sex couples, in defiance of California law.

It was a shocking move, and most mainstream Democrats, including Nancy Pelosi and Dianne Feinstein, declined to endorse it, worried it would hurt the national party's presidential nominee, John Kerry, in the fall election. Even in the LGBTQ community many were wary; Farrah recalls a conversation with Barney Frank, a Massachusetts senator and the country's most prominent gay elected official, who thought it reckless and damaging to the cause. Surely the courts would shut it down before it started.

Word had spread fast in the gay community the evening before and there was a crowd on the steps by morning, as there would be the next day and the next day and the day after that. The marriages began right away. To the surprise of Newsom's political team, it would be a month before the state supreme court stepped in—a month that would propel Gavin Newsom to the top ranks of American politics.

"When it started to happen, I'll never forget that," recalled Sean Elsbernd, who at the time was Newsom's liaison to the Board of Supervisors and would later serve as a supervisor himself, as well as an aide to Dianne Feinstein and then chief of staff for Mayor London Breed. "It was just excitement, lines of people around the building. It didn't matter if you had to wait two hours or fifteen hours, everyone was so happy, it was unbelievable."

The LGBTQ community quickly came around.

"We were strong feminists. We were very ambivalent about marriage," Toni Broaddus, then head of a statewide LGBTQ rights group, said in a magazine interview years later. "Suddenly, my relationship was recognized in a way that it never had been, and it really did feel different. It really felt profound to me in a way I had not expected."

Newsom's approach to politics is all about the bold stroke. Gay marriage was one for the ages.

Locally, the political benefits were immediate and game changing. "It was the best time to be board liaison," says Elsbernd. "No one could touch whatever the mayor needed. 'The mayor would like to vote this down.' 'OK, whatever you need.' I had the easiest job in the building."

Newsom now enjoyed nearly unanimous support in the gay community, which had typically been on the progressive end of the local divide. He'd locked in his civil rights credentials and shown himself willing to take political risks in supporting the underdog, and the city greatly appreciated it.

Nationally, Newsom had pushed gay marriage, and gay rights more broadly, to the top of the agenda, and new state laws sanctioning the practice came fast and furious. Despite no-holds-barred legal battles in California and across the country—and continued criticism from centrist Democrats who feared it was hurting the party—it would be little more than a decade before the US Supreme Court would enshrine gay marriage as a constitutional right. There are precious few cases where the actions of a mayor have had such profound national consequences.

Newsom was now a global symbol of San Francisco values. The downside of his historic success, according to several former aides, was that it reinforced the idea that only the big things mattered.

"I bet it was like a drug," says Elsbernd. "'Look what I can do.'"

It would be a hard act to follow.

Newsom had hired a press secretary from New York named Peter Ragone, and he and Eric Jaye would help hone what became a media-first governing style. Where Brown was pragmatic and transactional, Newsom was "an idea volcano," as Tom O'Connor of the firefighters' union put it, constantly coming up with new policy proposals that he'd announce at press conferences before they were even fully formed, in the hopes of generating momentum. Jaye, who later broke with Newsom, called him a "generational talent" as a political communicator, and they'd built the strategy to play to his strengths. Newsom would lean into policies that were bubbling in liberal Democratic circles both locally and nationally, especially on the environment and health care.

Phil Ginsburg, who'd serve as Newsom's chief of staff and then as head of the parks department, compared him to former Vice President Al Gore for his forward-thinking approach. "On the environment, on tech, on food sustainability, there were some places where he really was seeing around the bend," Ginsburg said.

Follow-through would be a very different matter.

Newsom's personal style was also the polar opposite of Brown's. Jesse Blout, whom Newsom had tapped to lead a revamped Office of Economic Development, pointed to their lunch habits. "When Willie was Mayor, every Friday he'd have the corner table at Le Central, and it would be a place where people would come and see him if they had business with Willie. Gavin was much happier reading a twenty-page policy memo with a turkey sandwich at his desk." Brown was a quick study and a master of the off-the-cuff speech; Newsom worked long hours to be prepared on policy, and enjoyed getting into the weeds. Brown wasn't afraid to call department heads on the carpet; Newsom hated such confrontations.

Those differences aside, though, Newsom's administration was very much a continuation of Brown's. Steve Kawa, Brown's chief of staff at the end of this term, would assume the same post for Newsom. Rudy Nothenberg, who'd been a top Brown aide in his Assembly days and held a series of powerful city jobs under Feinstein, would come out of retirement to get the Mission Bay project, stalled by slow-growth opposition, back on track. Newsom tapped Angela Alioto, daughter of the 1970s mayor Joe Alioto and a perennial candidate for City Hall herself, to develop a ten-year plan for homelessness. He cultivated the influential public safety unions, later raising a lot of eyebrows when he granted the police (and therefore the firemen too) an exceptionally large raise of almost 25 percent in their 2007 contract negotiations. Newsom had settled in comfortably with the City Family.

In the wake of the dot-com bust, San Francisco's economic future was again front and center. In most American cities, the downtown business district accounts for 10 percent to 15 percent of overall economic output; in San Francisco, where the financial district serves as the downtown for a much larger region, that number was closer to 30 percent. The business taxes paid by those downtown companies, and the lodging levies from the nearby hotels, were critical to the enterprise of San Francisco, so the city had to do what it could to keep its corporate citizens happy. And they were among the many constituents who were very unhappy about the growing problem of homelessness.

"Care Not Cash," initiated by Newsom while still a supervisor and approved by voters the year before he was elected, loomed as an important test.

Homelessness had begun to be a major problem in the city in the 1980s, and by the time Newsom was elected mayor, residents had become accustomed to the specter of thousands of people living on the streets. Progressives pinned the blame on rising rents combined with the federal government's retreat from public housing and mental health services under Ronald Reagan, while conservatives pointed to San Francisco's generosity in giving about $400 a month in public assistance to almost anyone with the wherewithal to fill out the forms. Both were contributors to the problem, and whoever was to blame, the reality was obvious enough: There were about eight thousand people living outside, and the number hadn't changed much in a decade. The problem had helped sink the career of Art Agnos—he'd allowed the homeless to live on the lawn of the Civic Center, across from City Hall, in what was quickly dubbed "Camp Agnos"—and residents were demanding an answer.

Brown was accused of eliding the issue after calling it a national problem that the city couldn't solve, though Jennifer Friedenbach, longtime head of the nonprofit Coalition on Homelessness, says he got a bad rap. "I would have taken Willie Brown over any of the people who came after him," Friedenbach said later, crediting him for supporting low-income housing and having genuine empathy around the problem, even if he eventually embraced some hard-line tactics. But Brown didn't have any solutions.

Newsom's Care Not Cash initiative was based on a simple idea: Instead of $400 a month in so-called general assistance, indigent people would be offered shelter and meals, along with $59 in cash. Critics on the left attacked the program for falling short on the "care" part and depriving people of the help they needed. But it was obvious from overdose and crime statistics, if nothing else, that a lot of general assistance payments were being used to buy drugs. And though Friedenbach and other advocates insisted that homelessness was mostly a result of San Franciscans being priced out of the housing market, that wasn't the picture that emerged if you walked the streets of the Tenderloin and Civic Center and asked people where they were from. Certainly displaced families and elderly people without means made up a significant portion of the homeless, but it was also true that the city had always been a refuge for people running from something, or seeking a fresh start—a big part of why homelessness was tolerated—and from its very beginnings had been shaped by people who came from somewhere else. That made the politics of welfare especially complicated.

With its commonsense appeal and snappy name, Care Not Cash was very popular, and Newsom boasted a couple of years into it that the number of people on the streets had dropped by 40 percent. Randy Shaw, a journalist turned housing activist, pioneered a "master-lease" model in which run-down SROs in the Tenderloin and SoMa, which developers had been eager to tear down or convert to tourist hotels, would instead be leased to nonprofits and operated as so-called supportive housing, with services for people who needed them. Shaw's Tenderloin Housing Clinic would eventually run twenty-three such buildings. But the services were often sorely lacking, and with tiny rooms and no-visitor policies, residents often passed their time on the streets.

Shaw considers the master-lease idea a great success. But the approach reinforced the Tenderloin's status as the city's "containment zone," where drug use and crime and extreme poverty were tolerated—even though the neighborhood was home to many immigrant families and had the highest density of children in the whole city. Del Seymour, who was himself once a homeless drug dealer in the neighborhood and now leads a community group called Code Tenderloin, says most of the SROs should have been torn down a long time ago.

"We have buildings where people have to go to the next floor to use the bathroom—a hundred people using one bathroom," he says. "Unbelievable living conditions. No one should live like this." This was the Achilles heel of anti-gentrification politics: Its logic led to preserving old, substandard housing, and keeping poor neighborhoods poor. The supportive housing system was also flawed, with widespread drug use and disorder in some buildings and not enough properly trained staff—problems that would become a debilitating crisis a decade later when the fentanyl epidemic hit.

While Newsom celebrated falling homelessness numbers, the annual count wasn't very precise, and homeless advocates including Friedenbach said Care Not Cash did little more than cut benefits and shuffle people around to make the numbers look good. Later on in his term, Newsom would launch another major initiative called Project Homeless Connect, designed to cut red tape and use technology to offer needy people better access to social services.

Whatever the programs' merits, though, there is one thing they definitely did not do: solve the city's homelessness problem. There was no major effort to spur housing construction under Newsom, and drug-treatment and mental health services were still wanting. Angela Alioto says her ten-year plan for homelessness, which included a lot more supportive and subsidized housing, would have worked were it not for the new tech boom that began driving up prices and squeezing affordability at the end of the decade.

In truth, nobody had achievable solutions. Poverty, mental illness and drug addiction were at the root of homelessness, and the scope of those problems was daunting and stretched far beyond the city.

Over the long run, though, one key policy choice made San Francisco different. Unlike other major cities including New York, San Francisco favored spending money on permanent housing, rather than overnight shelters; the weather was pretty nice most of the year, after all, so there was no life-and-death concern about forcing people inside. Everyone hated the dirty and dangerous shelters that housed people only at night. Why not spend the money instead on more permanent solutions? The evidence also showed that only after people had a reliable place to live could they kick drugs or deal with their mental health problems, and the city embraced the approach known as Housing First. A major consequence of that policy was troubled people living on the streets becoming a symbol of San Francisco, a routine part of everyday life for residents and visitors alike. It would be nearly two decades before it brought a political crisis.

Gavin Newsom wasn't a techie by vocation, but just as he had a policy-wonk side, the burgeoning Internet tapped his inner geek: How might digital technologies make San Francisco a better place? He was already pushing to upgrade creaky city computer systems, and an online system called SFGov would be a leader in providing easy access to government data. Still, the city's dense bureaucracy and thicket of rules made innovation on the tech side tough.

But Newsom was looking for opportunities. In 2005, he was at an event at the home of Ron Conway, a prominent tech investor, when he struck up a conversation with a couple of Google employees, Chris Sacca and Minnie Ingersoll. The company was trying to figure out how to get closer to the customers who were using its search engine, and one means would be to provide Internet access. Google engineers had gotten excited about the concept of "mesh" networks, where a whole building or neighborhood might be served by a handful of interconnected wireless access points. Sacca and Ingersoll were part of a small team that had been assembled to see what could be done. Ron Conway's son, Ronny Jr., was their intern.

"We got to talking with Gavin about the power of this technology to bring the Internet to a wide area, like potentially an area as big as San Francisco," Ingersoll recalled later. "At the time we had a pretty technical team that was looking at signal propagation and trying to understand what sort of area it could cover. We hadn't gotten to, 'Where would we do this?' or 'What would be the business model?' But Gavin was like, 'Wouldn't it be amazing to bring this free to San Francisco? Wouldn't it be amazing if you could sit at a public park or the car wash and get Internet? Wouldn't it be amazing to offer this to underserved communities?' He was just so fired up."

Before Ingersoll knew what was happening, Newsom held a press conference to announce free citywide wireless service in San Francisco. Then he got back in touch with the Google team. "He said, 'Hey, remember that thing we talked about? I just announced we're going to do it, so are you guys on board?'" Google was indeed on board, and Ingersoll's group secured a $10 million commitment from the company's philanthropic arm. "We were like, great, everybody's going to love this," Ingersoll recalled.

Wi-Fi technology was just becoming a standard means for computers to connect to the Internet, and there hadn't been much experimentation with new public services that the Internet might bring. Here was a chance to do something big.

It didn't take long for Ingersoll to realize she'd been a bit naive. The first broadside came from an unexpected quarter: the Electronic Frontier Foundation, which was always nervous about what the government might be up to and had a long list of questions about privacy policies. The ACLU had concerns too. Ingersoll was blindsided; Google hadn't yet defined how it would address any of those issues.

That was just the beginning of her education. "We ran tons of community meetings, we did a listening tour of every supervisor district," she recalled. Residents would air their grievances, though most had little to do with Google, or even technology. After a while she began to see a pattern. The supervisors, who had to approve Newsom's initiative, were willing to support it, but only if they got something in return, such as money for an unrelated initiative. Ingersoll remembers trying to explain that she had no authority over Google contributing a million dollars to pet projects like improving the local branch library.

An even bigger problem was Aaron Peskin, now president of the Board of Supervisors. "He said to us, 'I'm opposing this because it's likely a political win for Gavin,'" Ingersoll recalls. "I'm like, what the fuck? I can't even believe you're allowed to do this."

The city process ground along for a while, with local telecom companies throwing sand in the gears too. Google wasn't prepared for all that would be required for being an Internet service provider, and a new plan was devised calling for Google and Earthlink, an established ISP, to collaborate on the network. But Google lost interest, and instead did a deal with the city of Mountain View to try out its mesh network. Earthlink, facing its own troubles, walked away too. Newsom, already eyeing the governor's office, did little to fight Peskin and persuade the Board. (A mesh network operated by a company called MonkeyBrains now provides comparatively inexpensive Internet in some parts of town, but free citywide wireless never made it off the drawing board.)

It was a great example of how politics and process could sabotage even comparatively uncontroversial efforts at innovation. Who could be against free wireless? Plenty of people, it turned out, and not always for very good reasons.

The wireless project was far from the only Newsom initiative to stall in the face of board opposition. He was a policy wonk but not a coalition-builder, focused on the ideas far more than the execution. "He was very interested in shiny things, and he was not very interested in managing things," recalled Ed Harrington, the city controller. Newsom never had the long-serving team of can-do aides that were behind so many successful executives. He seemed to lose interest once the headlines had faded. He never cultivated relationships with the supervisors, which made it hard to get things done. "He wasn't into small talk, and building relationships with some of the folks on the Board," says Phil Ginsburg. "It was pretty adversarial."

As his first term was ending, Newsom was distracted with his personal life too. Newsom and Kimberly Guilfoyle had their moment as a glam power couple, with an infamous spread in *Harper's Bazaar* showing them sprawled across an Oriental rug in the Getty mansion. But not long into Newsom's tenure at City Hall, Guilfoyle moved to New York for a job with Fox News, and the couple soon separated, their ambitions and beliefs moving in opposite directions. Newsom was now the city's most eligible bachelor, and he was having a little too much fun.

In 2006, after dating an actress whose commitment to Scientology had gotten the media's attention, Newsom, then thirty-eight, was seen around town with a nineteen-year-old, and there was a small scandal when she was spotted drinking wine. Rumors began circulating that Gavin himself was drinking too much—and worse—after he showed up inebriated at a hospital vigil for a police officer. The dam would break early the next year, when Alex Tourk, who'd been his campaign manager, stormed into his office and confronted him with the confession of his wife, Newsom's appointment secretary, that she'd had an affair with the mayor. Newsom quickly acknowledged the sordid story at a press conference, where he preempted questions by allowing that "it's all true." In what's now a familiar ritual for public officials facing humiliation, he said he had a drinking problem and would seek treatment, though he wouldn't step aside as mayor.

The treatment—outpatient visits that ended after a few months—never amounted to much. The episodes might even have helped him politically, keeping him in the headlines and painting a picture of a game if reckless divorcée who was having some luck with the ladies—not the worst public image for a San Francisco politician. He was reelected by a wide margin in the fall. But he'd permanently lost some trust. Eric Jaye, though accustomed to duplicity among politicians, split with the mayor over the episode. That kind of behavior with people in the inner circle was very unbecoming of the City Family. Newsom had shown that he wasn't loyal to people in the appropriate, familial way. And they weren't loyal back.

The Palace Hotel, alone among the city's grand old hotels, sits on Market Street in the financial district, its plain exterior belying the soaring courtyard dining room and elegantly restored fixtures and stonework of the nearly century-old building. Elite tech conferences were typically held at luxury resorts in Southern California or Arizona, requiring a plane ride and a commitment of a couple of days, but the Palace was just a few minutes' walk from SoMa. It was a fitting venue for a new conference, conjured up by Tim O'Reilly, the Sonoma computer-book publisher and Web pioneer who was emerging as an industry intellectual. His colleague Dale Dougherty had coined the moniker Web 2.0, and John Battelle signed on as a partner and onstage impresario for a conference that would do much to define the city's next tech wave.

Web 2.0 was a big idea: The Internet was no longer about individual websites, but rather had become a platform upon which all kinds of new products and services could be built. O'Reilly's core insight was that the companies that survived the bust had "harnessed the power of collective intelligence" in some fashion, and that understanding the data that was being generated by people's actions on the Web was the key to the next stage. Blogging and Craigslist were among the first to show the potential of a medium that wasn't just about what was published, but how people acted, and interacted, and what could be learned from that.

The tech was improving fast: Websites and Web browsers were becoming much more capable. Software innovations based on code known as Javascript were enabling much more interactive experiences, while open-source tools including the Apache Web server and the Linux operating system (a variant of Unix) were making it possible to build high-performance websites for very little money. The more websites out there, the more collective intelligence they could generate: "Network effects" and "big data" would become the terms of art.

O'Reilly, a native of the city, combined a deep understanding of technology with a gift for bringing people together and an instinct for what was important. His company published a range of computer titles and early websites, and in 2003 he'd debuted Foo Camp (for Friends of O'Reilly), an "un-conference" where people would show up, camp in tents and invent for themselves how they would share ideas. It quickly became a hot ticket: Scott Beale, who'd brought webcasting to Burning Man back in 1996, had a blog called *Laughing Squid* that was chronicling the rise of the Web, and he recalled standing on the broad lawn at the O'Reilly campus in Sebastopol when a helicopter descended, to drop off Google's Larry Page. "Our whole idea with Foo Camp is to bring together people from different disciplines," O'Reilly said later. "I love helping people make connections."

It was an auspicious moment. Web 2.0 was starting to produce a mini-boom of its own as first-generation winners like Yahoo and eBay tried to consolidate their positions, and a wave of startups destined to define the next phase of the Internet began to emerge. Garrett Camp, a college student from Calgary, had created an unlikely early Web success called StumbleUpon, which surfaced semi-random Web pages when people clicked the "stumble" button (it was more fun than it sounds). In 2005, he raised money from a group of angel investors, including EFF cofounder Mitch Kapor, and moved the company to SoMa. A little more than a year later he sold it to eBay for $75 million. A couple of years after that, he'd dream up the idea for Uber.

Michael Birch, the Brit who'd been inspired by Friendster, and his wife, Xochi, had gotten traction with Bebo, which they'd launched in 2005 as a way to share contact updates and then turned into a social network. "We were briefly the largest website in the UK, Ireland and New Zealand," Birch related. "I was very proud of that." They sold the company to AOL in the spring of 2008, just a few months before the financial crisis, for $800 million in cash. Bebo never went anywhere and would prove a disastrous deal for AOL. But the Birches pocketed $590 million and would quickly begin spending it in the city, starting with a Gold Coast mansion and then creating what would become the city's hottest social club, the Battery.

Stewart Butterfield, a hippie from Vancouver, and Caterina Fake, a former art director at Salon, all but invented photo sharing in 2004 with Flickr, which was quickly snapped up by Yahoo and was now bringing images to every corner of the Internet. The early Web entrepreneur Paul Graham, who'd done a lot to establish the foundations of e-commerce with a 1990s company called Viaweb, had cofounded a tech incubator called Y Combinator in 2005, devoted to helping the most promising entrepreneurs get their startups off the ground. The founders it nurtured would soon be descending on the city: Brian Chesky, who'd graduated from the prestigious Rhode Island School of Design before moving to San Francisco and starting Airbnb; Steve Huffman and Alexis Ohanian, two University of Virginia students, who along with a much-admired hacker named Aaron Swartz would launch the social network Reddit; and Drew Houston, an MIT grad from Massachusetts, who created Dropbox, an easy-to-use online file storage service. Tony Xu and three Stanford classmates invented DoorDash, destined to be the most successful of a wave of app-based delivery services.

Another Y Combinator alum, Justin Kan, was live-streaming his life 24/7 from the Crystal Tower, an anomalous short-term rental building that loomed over the North Beach side of Russian Hill, and would soon start a company called Twitch. A few other early Y Combinator companies were in that building too, and it would earn the nickname the "Y-Scraper." Y Combinator would be a major force in the startup world, and the city, for many years to come.

Chris Larsen, the San Francisco native who'd started the dot-com-era online mortgage company E-Loan, was now on to his next venture, Prosper Marketplace. The idea was "peer-to-peer" lending, with individuals offering to lend money to other individuals at a mutually agreed rate. Larsen, with one success under his belt, got a fleet of VCs to back the effort, and though his first startup had been launched in Palo Alto before moving to the East Bay, Larsen wasn't doing that again. "With Prosper I was like, 'We're starting the damn thing in the city. And we're staying there,'" he recalled. Larsen hadn't yet gotten involved in local politics, but later on, after becoming a billionaire with the crypto firm Ripple, Larsen would emerge as a major Democratic donor and one of the most influential tech executives in the city.

Craig Newmark, meanwhile, was solidifying his place in the rarefied club of entrepreneurs for whom fast growth and big profits were not a priority. He was again ignoring the waves of venture capital dollars that had started to sweep across San Francisco. He'd promoted a programmer from Michigan named Jim Buckmaster to CEO of Craigslist, and they resisted constant entreaties to sell, even fighting a nasty legal battle with eBay after the auction site, which pined for the classifieds business, acquired shares of the privately owned Craigslist from a former employee. The two of them—Buckmaster very tall and thin, Newmark short and round—gave a memorable comic performance at an early Web 2.0 conference, riffing off each other as they mocked the pieties of the startup world and voiced their commitment to "unbranding" and "demonetization."

Newmark was a brilliant representative of the city's early Internet spirit, and the idea that the Web was a medium for the people. At the same moment, though, Web 2.0 was seeding a generation of San Francisco companies that would raise more money than any startup ventures had ever done before. Their extraordinary growth would bring both prosperity and political strife to the city—and ultimately challenge the idea that the rise of the Internet was good for San Francisco, and for humanity.

The enthusiasm—and the dollars—of Web 2.0 were also lifting Burning Man to a new level, transforming the offbeat happening in the desert into a touchstone experience of the burgeoning Internet economy.

The event had survived the near-death experience of 1996, the departure of cofounder John Law, the difficult 1997 event on private land and the inherent challenge of wrangling and managing an army of volunteers and deciding who gets paid. After the 1998 Burn, the one commemorated with the Google Doodle, the founders and other key participants had met at Marian Goodell's house to figure out some basics. "If we're going to do this, what's the budget we need?" Goodell recalls saying. "What do we need to survive as individuals, and how are we going to do it? Who's going to quit their job? What's the income?" They needed to commit on whether Burning Man could be a career.

The answer was yes. The group first rented a tiny office at Third and Cargo Way, an industrial area in the south of the city, before moving to Mission Bay, which was still in the early throes of redevelopment. Goodell had very much wanted to be in SoMa, but this was during the dot-com boom, and it was too expensive. "I would look at these places and they had glass block in their reception areas and soundproof carpeting and we couldn't afford that," she recalled. The Mission Bay space was gritty and a bit out of the way, but the big second-floor room was where the organization now known as the Burning Man Project would take shape, and formalize how the event would be run. "We'd have these public meetings and people would show up to help the different departments," Goodell recalled. "You'd stand in front of the room and you'd say what it was you do. And then people would come over to you and say, 'I can do this, I can do that.' You'd have to figure out really quickly how to organize people. If someone said, 'I'll do it, I'll write the names down,' you knew you had a leader."

"Those were the years of creating leadership and figuring out," she added. "Even amongst us—learning how to trust each other, learning how to communicate. We had shitty communication skills. We would just raise our voices and walk out of the room."

There was always a lot to argue about. It wasn't just the sensitive distinction between volunteers and employees that created friction, or the related issues that arose when the organization began giving grants for art projects. There was a very meta question of what, exactly, Burning Man was supposed to be. Goodell said later that after she began dating Larry Harvey in late 1996, she undertook her own inquiry into the split with John Law, suspecting that Harvey might be some sort of cult leader. With a literal burning man as its centerpiece, there was an idol-worship dimension to the event that some people found off-putting, at the least. "John was very clear that Larry was a messianic leader and he, John, wanted anarchy, and Larry wanted civilized society. I got that off both of them. But John said that Larry wasn't a manager, and it was never going to work."

Law, for his part, agrees with that assessment. He says he wasn't wrong about Harvey—Goodell turned out to be the manager he needed.

"We were working all the time," she recalled of the late 1990s and early 2000s. "The city was hopping, there were a lot of wacky art events, machine art out in front of buildings, performance stuff taking over." The Mission Bay building had a gravel backyard, and they put in a fireplace and developed a regular Saturday ritual: a volunteer meeting in the afternoon, and then a party in the backyard in the evening. The event in the desert was growing fast—fifteen thousand people in 1998, twenty-three thousand in 1999, thirty thousand in 2003. Everything needed to scale up with it.

Art provided one answer to the existential question. Christine Kristen, the first art curator for Burning Man, credits the late Argentinian artist Pepe Ozan and his Pepe's Opera—which burned so spectacularly in the mayhem of 1996—for setting the aesthetic tone, with elaborate sets and costumes created from playa mud. The event was even briefly named the Black Rock Arts Festival, to highlight and encourage such pursuits, and though that label wouldn't last, there were more and more art projects every year, many of them land-art installations that played off the simple angles and light effects of the endless white playa. In 1999, the organization began offering grants to underwrite some of the considerable costs of building in such a remote location, with Harvey and Kristen picking among various proposals.

"The thing that was wonderful about it was that the grant program was pretty much based on trust," Kristen said later. "I came out of the New York art world, where if you had a show you'd need slides, a résumé, gallery and exhibition history, letters of recommendation—you were heavily scrutinized before you got anywhere near a grant. We didn't require any of that. We wanted to see a good, interesting idea, and that you had the ability to pull it off—we didn't care about your past artwork or anything like that, it was more about the idea and whether you could actually get a crew of people together to build it."

As central as the art was to become, with four hundred works eventually gracing the desert each year, it was only one aspect of Burning Man. Theme camps were becoming more and more elaborate, with every sort of bar and various clever games—an ersatz ski slope over here, a mobile steam room over there. The "orgy dome," to the chagrin of Kristen and others, became a symbol of the project's enthusiasm for sexual experimentation and nudity, and the flip side was what might be called the spirituality track, where meditation and yoga, and feeling the force of unmediated nature, were at the center of the experience. The Berkeley-based filmmaker and artist Harrod Blank had spearheaded an "art car" movement that began in the 1980s, and Burning Man, a perfect venue for every type of "mutant vehicle," would become an epicenter. The rave camps, with towering sound stages glistening in neon while propane flames and fireworks lit up the skies, drew the many people mainly looking for a dance party. Electronic dance music blasted across Black Rock City 24/7, even though a lot of people didn't like it.

The experience was different for everyone. Yet there was something that cut across all these varied endeavors, an idea that would be expressed in the "Ten Principles" developed mostly by Harvey in the early 2000s: Radical Inclusion, Gifting, Decommodification, Radical Self-Reliance, Radical Self-Expression, Communal Effort, Civic Responsibility, Leaving No Trace, Participation and Immediacy. The ban on money was especially powerful.

The principles had some practical utility for guiding behavior at the event itself. But how deep was it all meant to be? Were they quasi-religious beliefs, a manifesto for a new way of organizing society? Or did what happened at Burning Man stay at Burning Man?

Stuart Mangrum, who holds the title of "philosophical director" of the Burning Man Project, is quick to wave away the more elaborate interpretations. "The ten principles are easily mistaken for a philosophy, which they are absolutely not. They're actions in the world. They're not beliefs." He traced their roots to the preamble of the Cacophony Society's *Rough Draft* newsletter: "The Cacophony Society is a randomly gathered network of Free Spirits united in the pursuit of experiences beyond the pale of mainstream society."

Burning Man certainly had plenty of detractors, many of whom found the whole thing ridiculous. Back in 1999, journalist Matt Taibbi, in *Rolling Stone*, had set the standard for derision, concluding that "it ultimately came down to just a few basic truths: it is a bunch of hairy art snobs running around with no pants in the middle of the remotest desert, calling themselves agents for transforming social change." It was an easy critique to make, but like so much else that was said about Burning Man, it was more revealing about the writer than it was about the event.

By 2006, Black Rock City had grown to nearly forty thousand people, and the fees and public safety precautions demanded by the Bureau of Land Management and state authorities—which involved hiring a lot of police and fire marshals, a lucrative bit of overtime for public safety officers around the region—were increasing steadily too. Tickets were now in the $200 range, and even beyond that it was an expensive endeavor, what with the need to bring all your own supplies and procure camping gear sturdy enough for the weather. But the costs were not a huge problem for the San Francisco tech professionals who were now a big part of the event and were underwriting increasingly exotic attractions. They sometimes arrived in luxury motor-home convoys, or flew into the airstrip that had been set up, though not everyone thought that was in the spirit.

But building an interesting Burning Man camp with your comrades turned out to work perfectly as a creative expression of New Economy ideas about collaborative work and personal empowerment. For San Francisco's tech community, the collective spirit of the desert endeavors was an idealized version of the startup ethos, but with the only objective being to have some fun.

San Francisco would empty out for ten days before Labor Day, the normally bustling restaurants of downtown and SoMa and the Mission suddenly quiet, the busy offices now dark. As in the late '90s, when the dot-com bubble was inflating and Burning Man was coming into its own, a large part of San Francisco's Internet community, and creative class, was relocating to Black Rock City.

The continued growth and popularity of Burning Man, with spin-offs popping up in many other places and a steady drumbeat of media coverage, was again ramping up tensions over a central contradiction. Burning Man was dedicated to the principle of "decommodification," and a complete ban on any kind of buying, selling, advertising or marketing at the event was strictly enforced. The organization was militant about its trademarks. Anyone using the iconic Burning Man symbol for almost any purpose could expect an immediate cease-and-desist letter from Terry Gross and his team. Yet Black Rock City LLC, which had ticket revenue of almost $9 million in 2006, was a private company owned by the six original partners, and it now employed a number of people in a normal, capitalist way, alongside the volunteer army.

The company did not, however, actually own the trademarks; they were held by Paper Man LLC, the triumvirate of Larry Harvey, Michael Mikel and John Law. In 2006, Harvey attempted to grant Black Rock City LLC a multiyear license to the trademarks for a nominal fee, prompting legal action by Mikel against Harvey for breaching his duties to the LLC. "Larry was effectively stealing the trademark from both me and John," Mikel said later.

Though he wasn't a target of Mikel's initial action—technically a "demand for arbitration"—Law received a formal notice as the third Paper Man partner, and he was offended and angry. He hadn't spoken to either Harvey or Mikel in years.

"I get a fucking notice in the mail," he recounted later. "I hadn't tried to get any money from them, I hadn't tried to take them. They were making money by '99, and by 2007 they were making bank. I'm doing my own thing, I don't care, and get the notice of this fucking suit." He got a lawyer and sued both Harvey and Mikel, demanding the trademarks be put in the public domain, in keeping with his distaste for the institutionalization of Burning Man. The legal proceedings were lengthy and convoluted, and Law lost on the public domain argument, but he ultimately got a financial settlement, which totaled $400,000, plus $100,000 to be distributed to Burning Man artists that Law would designate. After the lawyers and the taxes were paid, Law says he ended up with about $120,000. Even if it wasn't the point at first, the money would go a long way toward soothing old wounds.

"I put my whole life into that thing for seven years. It bankrupted me, literally. I couldn't rent a car for seven years. And with that whole experience [the lawsuit and settlement], any residual anger and anguish were largely dissipated. I was able to buy a house in Detroit and pay some of my friends some money, literally struggling starving artists. I paid for some stuff for my son, set him up a little. I burned through it right away, and had a fucking blast. So it went a long way."

Law got some resolution, but the contradictions in running a multimillion-dollar event and building a temporary city in the desert while advocating ideas like "decommodification" and "leave no trace" would burble to the surface in various ways as Burning Man continued to grow.

Similar contradictions would plague an Internet industry that was doing very well, but also wanted to do good.

CHAPTER 5

The Twitter Tax Break

Ev Williams had quit Google in 2005 and, "lacking anything better to do," as he put it later, joined a fledgling podcasting company called Odeo, started by his San Francisco neighbor Noah Glass. They soon moved to a small office in South Park, and his Blogger colleagues Biz Stone and Jason Goldman joined them there too. A quiet young man from St. Louis named Jack Dorsey, who had some programming skills and a gift for design, talked his way into a job. After Apple came out with its own podcasting platform, Odeo switched tracks to focus on a new idea.

In March 2006, Dorsey sent out the first tweet: "Just setting up my Twttr." Together with Glass, Stone and a few others, he'd been working an idea for sharing short missives via text, building off the idea of the "status updates"—"I'm on the phone" or "I'm out of the office"—that had become popular in online chat rooms. Williams had raised $5 million for Odeo, and with the company dumping its main product and casting about for a new direction, he offered the investors their money back. (Some, including EFF cofounder and software mogul Mitch Kapor, took him up on it, to their great regret later.) He then created a new firm called Obvious Corp., with key Odeo employees owning some shares, and fired Noah Glass, whom he wasn't getting along with anymore. Infighting among the founders would later devolve into bitter feuds, with momentous consequences for the company (and eventually the country).

After a modest but promising start, Twitter became a sensation almost literally overnight, in the spring of 2007, when it stole the show at the South by Southwest (SXSW) tech festival in Austin, an emerging hub for the Web set. The ability to trade observations and news and real-time location with a large cohort of friends proved instantly addictive for the entrepreneurs and creative techies in the sprawling Austin conference center. "This year's rockstar service at SXSW Interactive is Twitter," wrote Scott Beale on the *Laughing Squid* blog. "Super-active here with people posting updates on where they are and what they are doing." Traffic on the nascent service tripled during the week of SXSW.

Twitter worked with almost any cell phone—the first iPhone launched in 2007, but the smartphone era was still a few years away, and the 140-character limit on Tweets derived from the maximum length of a text message. Twitter quickly caught on in many parts of the world, partly as a means of avoiding government censorship and repression. Iranian dissidents protesting that country's 2009 election were among the first to show its powers of subversion, and the platform would play a major role in the Arab Spring protests that began in late 2010.

With all the desire to see the Internet as a force for good, here was something that might help deliver on the dream. It was a gift from San Francisco to the world: a medium for self-expression and a tool for building community, egalitarian by design, open to everyone without judgment and free from the corrupting power of governments or megacorps. "One could change the world with one hundred and forty characters," Dorsey had tweeted portentously before SXSW, and there seemed to be something to that promise. Dorsey and the rest of the company founders were unmistakably liberal idealists, fervent in their belief that they were doing right by humanity. "I believe fundamentally that the next Gloria Steinem, the next Gandhi, the next Martin Luther King—they're out there and they're actually using Twitter today," Dorsey told *The New Yorker* a few years later. "And our job is to insure that people find them."

Twitter was also a tantalizing opportunity for venture capitalists, even though it had no business to speak of. Facebook was proving that you could indeed make a lot of money with social media advertising, and Mark Zuckerberg made a bid to buy Twitter for $500 million in 2008. But investors thought it could be worth far more than that; it had the chance to be one of a handful of companies that would dominate a huge new global industry. The founders would just need to get over their distaste for selling ads.

Even with all the Web 2.0 ferment in SoMa, the late 2000s were a precarious moment for the San Francisco economy. Downtown was still heavily dependent on financial services, and the nationwide real estate meltdown and resulting Great Recession had hit hard. Unemployment had jumped back up to almost 10 percent in 2009 and the office vacancy rate leaped to 15 percent. Newsom was slashing the budget as tax revenues dipped.

Yet the city's second-generation Internet companies were rapidly laying the groundwork for a new boom, even if it wasn't yet visible beneath the rubble left by the financial crisis.

Mark Pincus's Tribe, with its optimistic cross of Craigslist and Friendster, had faded, partly because of frictions over explicit photos uploaded by users; The wars over what should and shouldn't be allowed on a website were being joined in earnest now. Facebook was offering a clean and well-lighted place, where people used their real names (or were supposed to anyway). It had also begun inviting others to build complementary products, and Pincus saw another chance. A committed poker player, he and a group of cofounders launched "Texas Hold 'Em Poker" on Facebook in 2007, the very first game on the social network. He incorporated a company named after his dog, Zynga, and it took off, with discrete teams each working on their own games, including the smash hit "Farmville." "It felt like this federation of startups, because they all had different personalities," Pincus said later. "It was startup culture. I got my wish, and somehow Zynga seemed to be a turning point in San Francisco, being a consumer Internet company. We were hiring so many people." Zynga eventually moved to an imposing, brick-clad building near the Caltrain tracks in SoMa that had previously been occupied by the Japanese game company Sega.

Pincus's old business partner, Sunil Paul, had begun working on an idea for summoning taxicabs via a mobile phone: The San Francisco–based ride-hailing industry, soon to be led by Uber and Lyft, was just starting to take shape. Airbnb was getting its feet under it, becoming one of the early participants in Paul Graham's Y Combinator incubator in Mountain View and raising money from Sequoia Capital in 2009. Reddit, Dropbox and Twitch were growing fast, and so were the juggernauts down south—namely, Google, Facebook and Apple—which were becoming significant employers for San Francisco residents. The companies were now running shuttle buses from the city, destined to become an issue of its own.

It wasn't all about the consumer Internet either. Salesforce, founded in 1999 by Marc Benioff, a star salesman at Oracle, was early to the idea of offering business software that was hosted on machines sitting in remote data centers, otherwise known as the cloud. The business proved much more robust than that of most bubble-era dot-coms, and Salesforce had gone public in 2004. By the late 2000s it was the biggest tech company in the city, occupying several floors of the elegant Landmark Building at the foot of Market Street, the historic headquarters of the Southern Pacific Railroad.

Twitter, though, was San Francisco's breakout Internet star, the company that more than any other symbolized the new tech era in the city.

Fittingly for the flagship company of a town whose ideals often outran its capabilities, Twitter was committed to noble principles of free expression and power to the people, but it wasn't exactly a smooth-running machine. A certain amount of chaos was typical at a fast-growing startup, but Twitter was a problem child even by those standards; Mark Zuckerberg mocked it as "a clown car that ran into a gold mine." Dorsey, the first CEO, was inexperienced and indecisive, and was ousted by Williams and the Board in 2008. But Dorsey was a big shareholder and still on the Board himself, and he was immediately plotting his revenge.

Twitter was falling down on the basics, what business-school types liked to call "execution." Technical weaknesses led to frequent outages, making the "fail whale" that people would see when the service was down an Internet meme in its own right. Nobody anticipated the abusive behavior that would plague the service, and even after the company hired a young woman named Del Harvey to head a new "trust and safety" function, it didn't take the issue very seriously. Harvey, who'd cut her teeth on the controversial vigilante television show *To Catch a Predator*, was hardheaded about the problems, but didn't get a lot of support, with the bosses equating the abuse to annoying junk emails. "I met with Biz [Stone] and Jason Goldman for the first time and they told me they didn't think spam was going to be a big problem, because you could choose who you followed," she recounted later. "I was like, 'Oh, that's really cute. That's sweet that you guys think like that.' "

Rapid growth covers over a lot of sins in the tech world, though. Twitter closed its third round of funding in early 2009, bringing the total to almost $60 million, and then in the fall raised another $102 million, giving the enterprise a value of about $1.2 billion, four times what it had been worth at the start of the year. Office space was a constant issue as hiring accelerated. The company had moved from its original office on Bryant Street to a six-story building at 845 Folsom Street in the heart of SoMa, which was bursting at the seams almost immediately. Now Williams wanted to find a permanent home for Twitter somewhere in San Francisco.

It was a good time to be looking. There'd been a flurry of office construction in the 2000s, with gleaming new towers along lower Mission Street, but with the Great Recession, many buildings in the financial district were wanting for tenants, and new projects had come to a halt. At the same time, with Web 2.0 sprouting, there wasn't all that much empty space in SoMa and the Mission, and there were very few buildings that could offer the large floor plates that Twitter and other tech companies loved.

Walking down Market Street with his wife and son after the Pride Parade in mid-2010, Ev Williams was struck by the decline of Civic Center and Mid-Market. "I'm thinking, why is the middle of San Francisco so dead and dilapidated? These big wide sidewalks and these huge buildings, and they're literally boarded up. This makes no sense. And I was thinking about what we can do about this and then shortly after that, we got the presentation at Twitter for the new office spaces."

The company's real estate brokers had collected various possibilities. Among them was the cavernous Art Deco structure at 1355 Market that had been a furniture showroom, but it wasn't on the agents' A-list. Nobody wanted to be in that part of town.

But Williams did. "We were like yes, we *do* want to be in that part of town. We want to bring thousands of people into that part of town and breathe life into the city. That would be awesome and it's an awesome building. Let's go check it out."

First built in the 1930s, with new floors added in the 1950s and 1960s, 1355 Market was an odd but alluring structure, a sprawling layer cake of stone and concrete, twelve stories tall, taking up the entire block between Ninth and Tenth Streets. It was oversize and drab from a distance, but more appealing the closer you looked, with soaring interior spaces and intricate, orange-hued terra-cotta stonework featuring Mayan themes. A large annex was built in the 1970s, connected to the main building by a footbridge.

For its first few decades, the Western Furniture Exchange and Merchandise Mart, as it was formally known, fit naturally as an anchor of the booming commercial and entertainment corridor that was Market

Street. In the mid-twentieth century, the thoroughfare was lined with gaudy, neon-clad movie palaces, grand theaters, elegant cocktail bars and rowdy street life, a veritable Great White Way on the Pacific. Huge department stores like the fabled Emporium were destinations for all classes, while moneyed tourists reveled in the luxuries of the Palace and St. Francis hotels. The city's storied history in the vice trades lived on, if in slightly diminished fashion, in the semiofficial red-light district of the nearby Tenderloin.

But the theaters and shopping meccas began losing patrons in the 1950s as television and suburbanization took hold, and by the late 1960s the district known as Mid-Market, which extends for six long blocks from Fifth Street to Van Ness Avenue, had fallen into steep decline. The Fox Theatre, commissioned by motion picture pioneer William Fox and considered among the most beautiful movie palaces in the world, had been torn down in 1963, a grim harbinger of redevelopment crimes to come. Construction of the BART rail system in the mid-1960s, with a tunnel beneath Market Street as its centerpiece, had left the streets torn up for three years—a devastating disruption that the city made little effort to mitigate. The flight of businesses and shoppers from city centers in the post–World War II period was a national phenomenon, driven by "white flight" and government policies promoting highways and automobiles. But the decline of San Francisco's Mid-Market proved especially dramatic.

Discussions about reviving the area bubbled up from time to time, but there were always other priorities. Dianne Feinstein had been fixated on the creation of the Moscone Center and the Yerba Buena Center for the Arts in central SoMa, which displaced thousands of poorer residents and small manufacturers but solidified the financial district as a top-tier business destination. Willie Brown had his long list of infrastructure priorities that didn't involve the messy business of navigating a neighborhood full of poor people.

Beginning in the mid-1990s, local businesses and residents spent a decade working with the city on a redevelopment plan, which involved creating a special improvement district financed by the increased tax revenues that would come from upgraded property. But redevelopment remained controversial, viewed by many as a tool of gentrification. Chris Daly, who'd been elected to represent the district in 2000, would support it only if it included a lot of affordable housing, like the nearby Trinity apartment development that he'd helped make happen. Lacking support from Daly or a push from Newsom, the effort faded. (At the end of the decade Governor Jerry Brown would abolish redevelopment in California altogether.)

There were a number of crumbling architectural treasures along the Mid-Market corridor, including several grand theaters and a couple of good-sized office buildings. The sheer scale and central location of the Furniture Mart, though, made it the key to any effort at revitalizing the neighborhood. Fortunately for the city, its owner was now motivated to make something happen.

Alvin Dworman was a real estate operator of the old school, a talkative charmer who relied on people skills as much as spreadsheets. Growing up in Coney Island, in Brooklyn, he was an entrepreneur from his earliest days, parlaying high school sports stardom into his first big business venture: running an offseason barnstorming tour for baseball pioneer Jackie Robinson, whom he'd met through a teammate. After finishing law school and determining that the courtroom was not for him, he worked first in the mortgage business, and then with the help of mentors began buying development properties in New York and Los Angeles. By the late 1950s he'd built multiple apartment buildings on the east side of Manhattan, as well as an early residential high-rise on Wilshire Boulevard in Los Angeles.

He arrived in San Francisco at a transitional moment. A friend had tipped him off in 1960 that the city's extensive redevelopment program—which would all but destroy the Fillmore district, the heart of Black San Francisco, as well as neighboring Japantown and parts of SoMa—was opening up a lot of land in the center of the city, a rare opportunity for an ambitious apartment builder like Dworman. He soon started accumulating property on a hilltop at the edge of the redevelopment district, in what's now known as Cathedral Hill, not far from City Hall.

When a fire destroyed the grand St. Mary's Cathedral on Van Ness in 1962, Dworman had a chance to win some favors. He got a call from Justin Herman, the city's redevelopment honcho, asking for help in rebuilding the church, according to a rare profile of Dworman by then–*San Francisco Business Times* reporter J. K. Dineen. Dworman agreed to help assemble the land, and even donated some of his own property, earning political capital that would only grow over the years. Willie Brown, who met Dworman in the 1980s, later called him "one of my best friends." Dworman was an expert at working local politics, even winning over Aaron Peskin, who was hardly a friend of high-rise apartments. Peskin recalls picking up his phone one day to hear a stranger's voice: "I'm Alvin Dworman and I want to meet you." They met, though Dworman had no specific agenda. Later on, Peskin continued, "I was back East visiting relatives and he took me to his favorite lunch club. He introduced me to Henry Kissinger." Dworman was a regular at the Four Seasons, then the premier power-lunch spot in New York.

Dworman would do well on Cathedral Hill, where he'd build two big residential towers (Willie Brown would live in one of them for a time). He would also stand up a condo project in SoMa, called Museum Parc, as well as a twelve-story office building across from the Moscone Center.

The Furniture Mart, which Dworman bought in 1968, would be his other big San Francisco play. Its furniture and design showroom operations and twice-yearly trade fairs had carried on profitably even as the neighborhood deteriorated over the decades, though by the 1990s tenants started to drift away. The dot-com bubble had only briefly grazed Mid-Market. Dworman managed to rent some space to a couple of dot-coms, including the online version of Ticketmaster, but they wouldn't last long. There were special events too, including one that few like to talk about to this day: a 1997 birthday party for Jack Davis, a flamboyant campaign strategist who'd helped get Willie Brown elected and was known for outrageous antics. Held in the building's soaring ninth-floor ballroom, the festivities featured a gory onstage satanic ritual involving blood and urine and the drinking of said-same, all while the city's power elite mingled in their evening finery. It was freaky even by San Francisco standards.

By the early 2000s, the jig was finally up for the Furniture Mart. "The furniture industry was changing, because so much was being done online, and some of the manufacturers were actually taking people to their factories," recalls Linda Corso, who was Dworman's right hand in the city, working alternately at the Furniture Mart and the residential buildings (Dworman died in 2022). "So it started shrinking." Worse, a competing facility to serve the furniture trade was being built in Las Vegas, and most of Dworman's remaining tenants decamped, leaving only a few lonely design showrooms at 1355 Market.

Now the building was a tombstone of sorts looming over the civic embarrassment that was Mid-Market. The neighborhood was going nowhere; across the street, Fox Plaza was as grim as ever, and down the block only a couple of the big theaters were alive amid the liquor stores, smoke shops and homeless people. A handful of long-standing businesses were still hanging on. But anyone who spent even a short time in Mid-Market could see that the neighborhood needed help, badly.

The plight of Mid-Market had never quite gotten to the top of Newsom's frenetic agenda. He'd tried to revive the redevelopment effort after he was elected mayor, but hadn't made it happen. He'd launch a new Mid-Market initiative in early 2010, the Central Market Partnership, but it was more aspirational than anything else.

Dworman knew he couldn't wait for the city. And he had some ideas of his own.

"We developed a whole repositioning plan," recalls Eric Grossberg, one of Dworman's lieutenants, which would transform the building into offices and condos and retail space. It would be hugely expensive, and the only way to pull it off would be with a big, high-profile anchor tenant, one that might bring thousands of office workers to the neighborhood and give other companies the courage to follow. The building had one rare asset: those coveted, extra-large floor plates. There were good transit connections too.

Still, there weren't a lot of candidates. Discussions with Levi's, which showed some early interest, didn't go anywhere. Neither did negotiations with the FBI. With the possible exception of Salesforce, tech companies big enough to lease 1355 Market were still all down in Silicon Valley and didn't yet see much reason to consider the city at all.

Twitter would turn out to be the special tenant the building needed.

"Twitter in the early days wanted to save the world, they were really all about doing good," recalls Chris Roeder, the broker who represented Dworman. It was Williams's then-wife, Sara, who first toured 1355 Market on the company's behalf in mid-2010, and Roeder sold her hard on the vision. On Dworman's suggestion, Roeder had recently visited a massive Art Deco building in Manhattan's Chelsea district that had been acquired and remade by Google. Twitter could have the same transformative impact, Roeder told her. "This building could be like Google at 111 Eighth," he said. "You can change this neighborhood."

Soon Sara came back with Ev, and they fell in love with both the building and the idea of leading the rebirth of the area. "The next thing you know they have eighteen people on a tour including all the founders," Roeder recalled. The company had a lot of cash in the bank with all the fresh VC funding and brought in a celebrated architect, Olle Lundberg, to help envision what the Furniture Mart could be. Williams would later write that even though the brokers were waving him off Mid-Market, he considered the location "a pro rather than a con."

Lease negotiations got underway in earnest, with Twitter demanding a cut-rate price for being the anchor tenant. The renovations were going to cost hundreds of millions and it was a tough time to get financing with the Great Recession still weighing, so in August 2010 Eric Grossberg reached out to the Shorenstein Company about a possible partnership. Doug Shorenstein, the CEO and son of the company's famed patriarch Walter Shorenstein, was a shrewd political operator in his own right, and knew how to work the levers of the city. But nothing immediately came of the discussions.

Lawyers and brokers representing Dworman and Twitter traded paper for months through the fall of 2010 as Twitter continued to grow, at one point cramming people so tightly into its Folsom Street building that employees were starting to worry about the health risks.

There was another issue on the table, though, one that neither the company nor its prospective landlord could resolve on its own: The city at the time had a payroll tax, which cost all but the smallest companies 1.5 percent of their wage bill. Twitter was moving toward a public stock offering that would rain money on many top execs and employees, and the payroll tax, which no other city in Northern California imposed, would put the company on the hook for tens of millions of dollars in taxes for the IPO gains alone.

If it was going to move into 1355 Market, Twitter wanted tax relief. Otherwise, company executives said, they would have to leave town. The cities of South San Francisco and Brisbane were only a few minutes away on the 101 freeway.

It would be a very bad look for the city to lose one of the hottest companies of the moment, especially one whose philosophy and purpose seemed to fit San Francisco so well. Even more urgently, the Great Recession had again spotlighted the city's vulnerabilities. "It was a really scary time in terms of the economy, it felt bottomless," recounted Jason Elliott, a top aide to Newsom at the time. "The only thing anyone cared about was jobs, jobs, jobs. When you're at double-digit unemployment, you can't be sitting there talking about mandatory composting programs or plastic bag bans. People need jobs."

There was a precedent for how to solve Twitter's problem, which various city officials, commercial real estate brokers and tax attorneys were well aware of: Earlier in the decade, the city had offered a time-limited payroll tax break to attract biotech companies to Mission Bay, and that had proven quite successful. Dworman, having learned a thing or two about how things worked in San Francisco, reached out to his friend Willie Brown.

They'd met decades earlier and hit it off, the self-made developer and the political entrepreneur nonpareil, both fond of pressing the flesh, both showmen, both dealmakers to the core. Brown wasn't in office anymore, but that hardly mattered: He was the most influential lobbyist in town, with protégés throughout city government and dozens of powerful clients. He technically served only as a legal counsel so he wouldn't have to register as a lobbyist, but everyone knew he was the man to see if you needed something from the city.

In the fall of 2010, in a conference room at the Furniture Mart, Brown convened a gathering that included Mayor Gavin Newsom, City Administrator Ed Lee, the city's head of economic development, Jennifer Matz, Twitter CFO Ali Rowghani and several other aides and city officials, as well as representatives of Dworman.

And so the Twitter tax break was born.

A series of meetings to hammer out the details ensued, with Matz, a sharp and well-respected attorney, playing a central role. Brown was involved too, according to multiple people familiar with the discussions, though he wasn't being paid to represent anyone and would later deny having anything to do with the policy. The outlines of a "community benefit agreement," under which companies receiving the tax break would invest money and resources to help the neighborhood, began to take shape. City officials tried to devise plans on how the district could be made appealing for upscale tech workers without evicting the low-income residents who occupied the SROs and subsidized housing complexes scattered across the area. For the business-minded, this sort of tax incentive was a no-brainer. It wouldn't cost the city much money, and the potential payoff was substantial.

Politically, though, it wasn't going to be easy. Chris Daly was dead set against tax breaks as a matter of principle—a widely shared view among the city's left-wing factions. Keeping Twitter in the city wasn't an especially important priority from his point of view, not when so many of the people in his district were suffering from poverty and related ills that the city didn't seem very interested in addressing. When Matz proposed the idea to him, "I think I laughed," he recalled later. Daly was an outcast among the city's power players, shunned by Willie Brown and his omnipresent machine, and it didn't look like there was a deal to be had that would bring him around. It would be very hard, maybe impossible, to do something big in his district over his opposition. But time was on the side of the tax break supporters. Daly was due to be termed out at the end of the year.

The fall, though, would bring fresh complications.

In October 2010, Ev Williams was pushed out as CEO of Twitter as its venture capital investors grew impatient with the "fail whale" and what some considered Williams's indecisive management. He was replaced by Dick Costolo, a dot-com entrepreneur who'd also sold his company to Google and then been recruited by Williams as Twitter's number two. Dorsey would soon return to the fold as chairman (a prelude to his eventual return as CEO). The culture of a startup is usually a direct reflection of its founders, and the soap-opera quality of Twitter's leadership struggles proved a constant.

Dick Costolo, too, liked the Furniture Mart, but he wasn't a fan of Alvin Dworman, whom he later described as "prickly." Dworman didn't want to make the investments that the dilapidated building needed, in Costolo's view, and was driving a very hard bargain in the lease negotiations. Costolo, a Michigan native, didn't feel the New Yorker's charm.

At the same time, another momentous development was unfolding in the city's political ranks. Mayor Gavin Newsom had had his sights set on Sacramento ever since his success with gay marriage, and against the advice of his father, as well as Willie Brown and many others, he'd announced a bid for the governor's office in 2009. It was too much, too soon: Trailing far behind front-runner Jerry Brown in the polls, he dropped out in the fall of that year and instead ran for lieutenant governor and won that office in November 2010. Because San Francisco elected its mayors in off years, that meant he'd be leaving City Hall a year before his term ended, and the Board of Supervisors would name an interim replacement. Straightforward enough, or so it seemed. But the drama that ensued would showcase San Francisco's peculiar brand of political dysfunction in vivid style—and result in the tech industry winning far more influence over city policy than it could have dreamed of just a few years earlier. It was a very local political story, but like so much else that happened in San Francisco, its effects would ripple far beyond the city.

San Francisco City Hall, with its gilded dome, grand central staircase, and stately, marbled chambers, could readily be taken for the capitol building of a small country, or a large US state. Willie Brown had prioritized a post-earthquake renovation that cost $300 million, and while critics sniped at the price tag, it was hard to argue with the results. The steps out front offered a perfect perch for protests or press conferences, and the majestic central atrium was often crowded with wedding parties, the photographers gaming out the many intriguing angles. It was a fittingly extravagant centerpiece for a city of expansive dreams.

In the first days of 2011, the elegant building was all but vibrating with intrigue as the eleven-member Board of Supervisors considered candidates for interim mayor. Whoever was chosen would have a big leg up in the race for a full term in the fall, which complicated matters immensely, since the Board was both divided on ideological lines and full of aspiring mayors.

The left included Daly, in his final act as supervisor; John Avalos, a progressive who dreamed of becoming the city's first Latino mayor; and David Campos, an up-and-coming leader in the Mission. The Board president, Bevan Dufty, who himself sought to be the city's first gay mayor, leaned left but tried to walk the center, alongside David Chiu, a rising star in Chinatown. On the right side of the spectrum—the "moderates," in local parlance—were Sean Elsbernd, Carmen Chu and Michela Alioto-Pier, all aligned with Newsom and Brown. The supervisors informally interviewed candidates, struggling to find someone who could get six of their votes.

Early on it looked like Willie Brown himself could be the pick, but he was ultimately considered "too toxic," according to one of the supervisors; he hadn't yet risen to the status of beloved elder statesmen that he'd later enjoy. Ed Harrington, the respected controller who would have been an obvious choice as a caretaker, pitched a plan to bring some much-needed structure to the mayor's office, which could then be handed off to whoever was elected in the fall. But he was a bit too independent for some of the supervisors' tastes. After weeks of maneuvering, former Sheriff Michael Hennessey, a well-liked progressive, emerged as the candidate who could secure a majority, and when the Board convened for the formal decision, he was in the room, ready to be anointed.

But Willie Brown and his allies had other plans.

It wasn't that Brown didn't like Hennessey, exactly. But he couldn't necessarily be counted on to support policies—like, say, the Twitter tax break—that were favored by Brown and Newsom. They wanted someone loyal and pliable in the mayor's office, someone who would tend to and extend the reign of the Brown machine. Rose Pak, a powerful political operative who held down the Chinatown corner of the City Family, was also unenthused about Hennessey, and saw an opening to push one of her biggest causes: putting a Chinese American in the mayor's office.

Together, Brown and Pak decided that San Francisco's next mayor would be Ed Lee, a little-known civil servant who held the post of city administrator, a powerful but behind-the-scenes role that involved overseeing big chunks of the bureaucracy. He was experienced and popular among his charges, but nearing retirement and, importantly, not very ambitious—a critical quality in the view of those who were.

There was only one problem: Ed Lee did not want to be mayor. He was very clear about that.

As the Board haggled and cast about for candidates, Lee wouldn't even talk about it; he took to crossing his arms over his face when anyone came to his office and raised the subject. He was looking forward to retirement, he told people, with a fat pension, and being interim mayor would delay that and could ultimately reduce his payout. It wasn't just that. Ed Lee was a bureaucrat's bureaucrat, low-key and self-effacing, with none of the bravado and ego that animates almost everyone who runs for public office. He was proud of his career, and being mayor was a tough job. He didn't need the stress. Electoral politics was never his lane.

What Ed Lee wanted, though, was less important than what other people wanted.

He was in Hong Kong for the Christmas holidays in 2010 when Rose Pak called. Lee had known her for decades, ever since his days as a tenants-rights lawyer, and the two were close, with his kids calling her Auntie Rose. And now she was asking him if he'd like to be mayor. Or perhaps telling him that he was going to be mayor. Neither of them ever gave a full readout of the conversation, but the outcome was clear enough.

Lee had grown up in Seattle, the son of poor immigrants from southern China, and worked his way through college, and then Hastings law school, before beginning his career advocating for tenants in Chinatown, where immigrant families often lived in seedy, dangerous SROs. After punching his ticket with both progressive housing advocates and Pak—whose formal role as head of the Chinatown Chamber of Commerce belied her broad influence—he'd caught the eye of Willie Brown, who brought him into city government. He then climbed to the top of the bureaucracy under Newsom. Lee grew into a perfectly chiseled piece of City Hall machinery: capable, affable, loyal and not very greedy.

Lee was a good soldier, with lots of debts to his patrons and no elaborate vision of his own as to what the city could or should be, and from a certain point of view that made him the perfect interim mayor. A competent loyalist who would do what he was told: That's what more than a dozen elected officials and senior political operatives said privately in explaining what Brown saw in Ed Lee. The optics were extra-compelling too. It would be hard for the Board's progressives to oppose the first Chinese American mayor. And the key to it all was that Lee would promise to be *only* an interim mayor; he would not stand for election in the fall. That was enough to persuade David Chiu, who was planning a bid himself. Chiu liked Lee, but agreed to back him only after Lee swore he wasn't going to run for a full term, a promise Lee would repeat again, months later, while encouraging Chiu to get in the race.

The Board seemed ready to vote for Hennessey when it convened on that January day—including Board President Bevan Dufty, who'd thought Lee was a good choice early on but had accepted the city administrator's insistence that he didn't want the job. But as the meeting ground on the Board was still deadlocked, and Dufty grew wary of what the progressive faction was up to. He finally said he wanted to speak to the mayor and called a recess. He retreated to Newsom's office, where the mayor's chief of staff, Steve Kawa, was waiting for him, with Newsom dialed in on the phone. He brought board colleague Sophie Maxwell along too. "Newsom told me the sheriff had a list of twenty people that would be terminated on day one," Dufty said later, a list that included most of the department heads. "I thought about it, these were people I'd worked with for years, and I just did not want to risk doing that." Then Ed Lee called—he was in China—and Dufty asked for assurances on a couple of issues, including that Lee wouldn't run in the fall.

Dufty heard what he needed to hear. He returned to the chambers and said he'd be voting for Lee.

Pandemonium ensued as Daly and his allies erupted in rage, but they'd run out of options. "To present an opportunity to a person of color, well-credentialed and well-qualified, ought to be one of the tenets of the progressive movement," Brown declared piously ahead of the Board decision.

Lee was duly installed—a monumental defeat for the progressives, who'd thought they were finally in reach of the mayor's ring. "We got played," griped Avalos. Chris Daly vowed in a widely circulated video to "haunt" Chiu for his betrayal of the progressive side.

For her part, Pak cackled in delight over her victory, telling reporter Gerry Shih at *The New York Times*: "Now you know why they say I play politics like a blood sport."

With Lee in place, the path was clear for the Twitter tax break. He introduced the proposal as one of his first official acts as mayor, and soon the Board of Supervisors was negotiating the details, though the outcome was never in doubt. In the meantime, Alvin Dworman decided he wanted to sell, what with the daunting renovation costs and Twitter's aggressive stance in negotiations, and he had another big project going on that needed capital, a resort in Santa Barbara. He called Doug Shorenstein directly, and in January abruptly sold the building to Shorenstein Company for $120 million, plus a piece of the profits if it was later sold again. The sale blindsided Twitter and the city, and the price seemed low, but it would ultimately be an excellent deal for all involved.

There was horse-trading to be done on the tax-break details, and negotiations were being led by Jane Kim, who'd served on the school board before succeeding the termed-out Chris Daly as supervisor for the district. She didn't believe in tax breaks but had agreed to support the Twitter deal, to the fury of her predecessor and others on the left. "It was going to pass anyway, so why not try to influence the outcome?" she explained later, adding that she didn't want her term as supervisor to be defined by losing a battle on something that was already a done deal.

The boundaries of the tax-break district moved around as negotiations proceeded. Randy Shaw pushed for parts of the adjacent Tenderloin district to be included. Buildings that already had tenants were drawn out of the map. Kim wanted the payroll tax exclusion to apply only to new hires and also wanted the time frame reduced from eight years to six, but Twitter pushed back.

"They're like, 'We will not accept this under any circumstances,'" Kim related later. "And we said, 'That's our best offer.' I thought it was over and then they called back a few days later and they're like, 'Actually, we're willing to take it.' I don't know what happened."

Twitter was still threatening to move to South San Francisco or Brisbane right up to the end, with chief operating officer Ali Rowghani writing a letter to the city in March 2011 promising that Twitter would leave if the tax break wasn't approved. Rowghani didn't think Twitter should have been in the city in the first place, he made clear to others, and certainly didn't think it should be paying tens of millions for the privilege of staying there. That wasn't a consensus view in the company's executive ranks though, let alone among the rank and file.

All of which makes it impossible to know what the company would have done *without* the tax break. Costolo would say later that it was never a bluff: The payroll tax hit for the upcoming IPO alone would have been upward of $30 million, and the company just wasn't going to do it. "There was zero chance we were staying in the city," Costolo said. The truth is likely a bit subtler. The search for office space south of the city was a "ruse," according to a person familiar with internal operations, deliberately concocted to pressure the city.

Ev Williams says there was a lively internal debate at the time. "It was a pretty complicated discussion over some number of months as we looked at possibilities and negotiated different options," he said. "There were many, many proponents on multiple sides." Employees clearly favored staying in San Francisco—and what employees wanted mattered quite a lot as Web 2.0 heated up.

The city was beginning to buzz with a new wave of Internet companies that were competing for talent, and trading a rejuvenated landmark location on Market Street for the dull confines of the peninsula would carry some risk. Williams and Dorsey were both on board with the Furniture Mart plans, and Twitter was emerging as a glamorous international brand. Why would it alienate its employees and dim its luster by moving to the suburbs, especially when it had all that VC cash in the bank?

If it was a bluff, it worked beautifully. The Board approved the tax break in April 2011. Twitter signed its lease. A handful of other companies began to show interest in Mid-Market. Officially the "Central Market/ Tenderloin Payroll Tax Exclusion," it would be universally known as the "Twitter tax break"—to the great irritation of city officials—and become an indelible symbol of San Francisco's tense relationship with its tech companies.

There was still some unfinished business though.

Mark Pincus's game company, Zynga, was blowing up, and it was on a path to an IPO too. It had just moved into its new home in the SoMa neighborhood known as Showplace Square, but it was outside the tax-break district, and Zynga leadership hadn't been aware of how the city's payroll tax would hit an IPO. Costolo says it was Twitter's finance team that alerted Zynga to the issue, and though Pincus doesn't recall that detail, he does remember his indignation in learning about it. Between the IPO and the regular payroll tax levy, Zynga looked to be on the hook for $80 million in city taxes, he says, and the company's CFO was soon raising the relocation issue with the Board. "He's like, 'Here are some headquarters options, because we're definitely not going to pay the $80 million,'" Pincus said later. He thought the payroll tax was outrageous. And like many tech CEOs, he was offended by the attitude behind it. "There's a cultural thing that's getting lost here," Pincus said. "Politicians should *want* you here." Certainly Pincus's own vision for San Francisco was a business-friendly one, with entrepreneurship and economic growth at the heart of it.

Pincus then did what you have to do when you have a problem with San Francisco's government: He hired a lobbyist. It turned out a lot of people in the business community were unhappy that the Twitter tax break offered relief only for companies in a specific part of town. Soon the Board of Supervisors was considering another piece of legislation that would eliminate the payroll tax on IPO proceeds for *all* companies. Pincus's lobbyist, Chris Gruwell, told him the swing vote on the Board was the idiosyncratic Green Party supervisor Ross Mirkarimi, and the way to get to him was through the hotel workers' union. So Pincus took some meetings, including one with the service employees' union boss, a famously tough character named Olga Miranda.

Mirkarimi came around, and just a couple of months after approving the Twitter tax break, and with far less fanfare, the Board of Supervisors passed the IPO tax-relief bill. The Twitter tax break was attacked for years as a giveaway to the tech industry, but this second step would be a much bigger hit to city tax revenues.

There were many more battles to come. This new wave of Internet companies needed things from the city, and they were about to find their champion.

Ron Conway was not an obvious candidate for the role of San Francisco power broker. With snow-white hair, a stocky frame and a distracted but unfailingly polite manner, he was too old to be a true "tech bro," and lacked the studied, moneyed, faux-casual persona of so many of his fellow venture capitalists. He wasn't a scion of a dynastic family nor the founder of a tech company, and he didn't sport the charm or charisma of a political leader.

It wasn't even clear if he was a particularly shrewd venture investor, at least in the traditional sense. After beginning his career in the marketing department of National Semiconductor, an old-line chip company, he eventually got rich as CEO of a long-forgotten computer firm called Altos, and then, during the first dot-com boom, made a name for himself by building an impressive network of angel investors among the growing Silicon Valley tech set.

Angel investing would later become a common pastime for the rich, but back in the 1990s Conway was a pioneer, rounding up wealthy individuals and successful entrepreneurs who contributed five- or six-figure sums to a fund called SV Angels, which in turn made hundreds of investments in fledgling startups. The vast majority of these dot-com era investments were a total loss, but with his vast network and dogged insistence, Conway managed to get into a very early Google funding round in 1999. That single deal resulted in SV Angels investors earning three times their money.

Conway's approach was innovative, and he was of course correct that the Internet industry was going to be huge, maybe bigger than anything the business world had ever seen. But if the partners at top-tier VC firms like Sequoia Capital won their stripes by analytically picking winners among the scads of wannabe tech stars, Conway had a different sort of edge: He was a consummate salesman and networker, a person who always knew someone who could help, and who expected help in return. He wouldn't take no for an answer. He cultivated anyone he thought could help his network—with money, sure, but even more importantly with connections that might help one of his companies close a deal, or recruit a star programmer, or find a buyer, or even just throw a good party.

He was perfectly suited to prosper as a member of the City Family.

Conway had lived in the city as a kid before his father enjoyed success in the shipping business and moved the big Irish Catholic family to tony Atherton, in Silicon Valley. Conway would later buy a sprawling property of his own there. But as the industry's center of gravity shifted north, Conway was right there with it, buying a condo in one of the best buildings in Pacific Heights and standing ready to serve the needs of his companies and his network.

By Conway's own account, it was Gavin Newsom's first campaign for mayor, in 2003, that animated his interest in city politics—partly because he liked Newsom, but mostly because he hated his opponent, Matt Gonzalez. Later on, in 2010, Conway would be a major donor to a contentious, Newsom-backed ballot initiative that banned sitting or lying on city streets. He'd known Jason Elliott, the Newsom advisor, for many years. Elliott was the son of Marc Porat, the founder of General Magic, and the nephew of Ruth Porat, later the CFO of Google parent Alphabet.

Conway says his friendship with Ed Lee didn't begin until 2011. The extraordinarily close relationship they would develop had led many to assume that the Twitter tax break was a Lee-Conway project, but Conway says it was in fact the passage of the tax break, in the first months of Lee's tenure as interim mayor, that got his attention.

"A bunch of us got together with Ed Lee while he was the interim mayor and we said, 'You have to fix these two issues,'" Conway said, referring to the Mid-Market tax break and the stock option tax. "And he went and fixed those two issues. And I said, 'Oh, my God, this guy's awesome.'"

Conway's single-mindedness when he wants something is his greatest strength, and now he really, really wanted Ed Lee to be the mayor for the long run. Once again, Lee resisted.

Conway says it was he and Warren Hellman, the investment banker and Hardly Strictly Bluegrass founder, who ultimately persuaded Lee to run for a full term—partly by abandoning their support for the political project of another uber-rich venture capitalist, Michael Moritz. As the managing partner of Sequoia Capital, Moritz far outranked Conway in the VC pecking order, and he'd begun getting involved in city politics himself, pouring money into a ballot measure to overhaul the city's generous and very costly pension plan. Union bosses were beside themselves and launched a scorched-earth campaign against the initiative, with official tactics that included picketing Moritz's house in Pacific Heights and unofficial ones like having uniformed policemen harass signature gatherers on the streets. The political establishment, including Lee, fell in behind the unions.

Conway supported Moritz on the pension changes, which would have cut retiree health benefits and upped the retirement age for some workers. But he was prepared to switch sides if it meant keeping Lee in office. "It was just [Lee] and me and Warren," Conway said of their meeting about the proposal. "I quickly caved on the pension thing and said, 'Not only that, I'm going to donate a hundred grand and be active [in helping defeat it].'" Lee was pleased, since he'd only been hoping that Conway would be neutral.

"And then I said, 'But you have to run for mayor.'"

Lee said he couldn't do that; he had made a deal. Conway chided him for making a "backroom deal," and then Lee got angry, the "most furious I have ever seen him," Conway recounted.

Conway vowed that if Lee did run, he would do everything he could to help get him elected. Hellman, who'd also been inclined to support the pension measure, agreed to stay neutral, and later joined Conway in opposing it, after extracting a commitment from the unions to negotiate.

Conway and Hellman were soon part of a veritable army bent on strong-arming Lee into running for the full term. Brown and Pak had made it clear from the start that they didn't view him as an interim mayor, and soon there was a full-on marketing campaign—"Run, Ed, Run!"—featuring TV ads and signature gatherers all over town, animated by acolytes of the City Family. A priceless video called "2 Legit to 2 Quit," featuring the Oakland rapper MC Hammer, former Mayor Willie Brown and local celebrities including football legend Ronnie Lott, Giants pitcher Brian Wilson and Google exec Marissa Mayer, implored in truly hilarious fashion that Lee was just too good not to stand for a full term. It was pretty clear where all this was going.

At the time I was editor of *The Bay Citizen*, a nonprofit news organization funded by Warren Hellman. When Lee visited our offices near Union Square in May 2011, we grilled him at length on whether he was really, truly not going to run in the fall. He tried his best not to answer before finally mumbling, almost inaudibly, that yes, he was not going to run. We didn't believe it for a second.

Sure enough, by the end of August, Lee had joined the race. And despite his protestations, by then he was clearly having fun as mayor, gladhanding his way through ribbon cuttings and community events with an elfin smile and a spring in his step. Conway helped mobilize support, hiring consultants for the campaign and rounding up millions in contributions. Lee cruised to an easy victory in November over Chiu, Avalos and two other experienced politicians—Dennis Herrera and Leland Yee—who were once considered up-and-comers.

"We did get Ed Lee elected," Conway later boasted.

He had reason to be pleased. Lee's installation was a crucial development in allying City Hall with the needs of the rising Internet industry. It was also the key to extending the reign of the Willie Brown machine. All in all, it was truly a Family affair.

CHAPTER 6

The Media Shift

Warren Hellman had been easing into retirement after a good run with Hellman & Friedman, which was prosperous and well respected, even if his most high-profile deal, a 1995 buyout of Levi Strauss & Co., had yielded mixed results. He was passionate about music, extreme sports, the outdoors and civic affairs, and now he was busier than ever, from leading the fundraising for an underground parking garage in Golden Gate Park to helping Ron Conway convince Ed Lee to run for the full term as mayor. He constantly cajoled the local moguls and high-society doyens to support the city's arts institutions, including the symphony and the ballet. At the same time he cultivated warm relations with the unions—especially the firefighters, whom he adored. He'd won endless affection around town by creating and underwriting Hardly Strictly Bluegrass, which over three days every fall offered dozens of top acts across six stages set in the meadows of Golden Gate Park, all for free. A registered Republican in a town run by Democrats, he was perhaps the only man trusted by all sides.

Many progressives loathed the idea of "downtown" having a big say on policy, but the plain fact was that politicians, city agencies and labor unions had come to appreciate—and depend upon—the money and support it could bring. Back in the day, the big companies "were all deeply involved in helping the city, as long as they felt they were being paid attention to," recalls Rudy Nothenberg, a fixture in San Francisco government for many decades who served as Dianne Feinstein's deputy mayor and a close aide to Willie Brown. Transamerica Corp., whose pyramid skyscraper was a distinguishing feature of the San Francisco skyline, once ran a full-scale management training program for city employees. Wells Fargo might pitch in on a customer-service problem, recounted Ed Harrington, who'd meet with the big companies regularly in his capacity as city controller.

They, too, were part of the City Family.

Hellman sometimes operated like a shadow mayor, with his own capacity to make things happen. Relentless in his pursuits even well past retirement age, he'd led the negotiations to reform the city's pension plan after helping defeat the ballot measure on the topic that was anathema to the unions. He'd also financed what he hoped would be a solution to what he viewed as another major civic problem: the decline of the city's news media.

The *Chronicle* was supposed to have been on solid footing. A morning paper, it had been owned since its inception in 1865 by the de Young family, and had been the dominant daily paper in town for much of the twentieth century. The Hearsts' afternoon *Examiner* was its only major rival for much of that time, though since 1965 the two papers had run a combined business-side operation under a so-called joint operating agreement, which enjoyed a special antitrust exemption. In 1999, Hearst agreed to acquire the *Chronicle*, but a federal judge forbade Hearst from shutting down the money-losing and unsaleable *Examiner*. It was then offloaded—along with $66 million that was supposed to be used to keep it healthy—to the wealthy and well-connected Fang family, who published a local paper aimed at the Chinese American community.

The end result was the Hearst-owned *Chronicle* effectively enjoying a daily newspaper monopoly. It was early to the digital game too: *SF Gate*, launched during a newspaper strike in the mid-1990s as the joint home of the *Chronicle* and the *Examiner*, would long remain one of the most highly trafficked news destinations on the Internet.

But the Great Recession had taken a hammer to an already wounded industry, with advertising revenues plummeting as the Internet shattered newspapers' stranglehold on local information and advertising. The *Chronicle* no longer had Herb Caen, who died in 1997, and in what could only be interpreted as a desperate effort to get some of Caen's sizzle back, it had agreed to run a column by none other than Willie Brown. He'd go on to use the "Willie's World" column—widely understood by those in the know to be ghostwritten by veteran *Chronicle* city hall reporter Phil Matier—to puff his friends and pet causes in predictable fashion. It did prove to be a popular read, even if it was seen in the newsroom as a journalistic disgrace, but it didn't provide a palpable lift for the business. Even with the early success of SF Gate and the Internet revolution happening all around it, the paper didn't have much of a digital strategy, and it lost $50 million in 2008. Hearst said that without labor concessions it might even close.

The worst-case scenario seemed unlikely. Hearst Corp., based in New York, was a major national media company with interests in a variety of newspapers, magazines and business-information ventures, and it was doing better than most of its peers, even if that wasn't saying much in 2009. The family had huge landholdings on California's Central Coast, near the famous Hearst Castle—built by their patriarch and later donated to the state—and also owned a lot of real estate in San Francisco, where the old *Chronicle* building on Fifth Street and Mission would later anchor a major development project. For all its troubles, the paper still had a lot of clout in the city and the state, and the Hearsts had plenty of reasons not to give that up.

Still, both the quantity and quality of local journalism in the city was clearly in decline. The *Bay Guardian* and the *SF Weekly*, both once weighing in at over a hundred pages every week—and at their best delivering stylish, irreverent and well-reported stories alongside the invaluable entertainment listings—were being decimated, with Craigslist eating the classified advertising business and a host of online competitors joining the entertainment-advertising fray. They were also beating each other bloody. The *Guardian*, now based in a three-story industrial building in Potrero Hill that owners Bruce Brugmann and Jean Dibble had the foresight to buy in better days, hadn't lost much of its political clout with the prolific and articulate Tim Redmond at the helm. But the *SF Weekly*, launched by Phoenix-based *New Times* in 1995 and now led by its bombastic co-founder, Mike Lacey, was a stiff competitor, with contrarian politics that took on progressive pieties in slashing style. It was also very aggressive on the business side, ultimately prompting a highly unusual lawsuit by Brugmann, who accused his rival of violating state law by selling ads below cost in an effort to drive the *Bay Guardian* out of business. In 2010, Brugmann won a $15 million verdict that would be upheld on appeal and lead to the closure of the *Weekly*.

That wouldn't be enough to save the *Bay Guardian* though. As the business collapsed, it was sold and finally shuttered, though Brugmann at least still had the building. Tim Redmond had succeeded in getting a clause in the sale agreement giving him the right to buy the assets for $1 if the paper was shut down, and he managed to save the archives on his own. Lacey's legal woes were just beginning: Kamala Harris, now state attorney general, would make a crusade of criminally prosecuting Backpage.com, a lucrative online bulletin board for prostitution listings that Lacey and his partner had spun out of *New Times* in 2012. He would eventually go to prison.

For all the carnage in the media business, there was still hope that the explosion of blogs in every shape and form, appealing to audiences no one even knew existed, could be harnessed for advertising and reinvigorate journalism.

John Battelle had seen the promise back in 2003, when he was working on a book about Google and started his own blog to write about search and the Internet. He was blown away by the numbers. "I'm like, 'Hey, wait a minute, one dude writing on a blog can get a couple of hundred thousand people to pay attention to him?' " Battelle recounted later. "And I was linking to all these sites and they were linking back to me, and I was thinking, I want to make some money from this thing."

In 2005, he raised $8 million and launched Federated Media to serve as the advertising and infrastructure arm of blogs and small publishers. By selling ads across many individual blogs, Federated could offer the scale that big marketers demanded, and the writers wouldn't have to worry about the revenue side of things. At its peak, Federated served hundreds of sites and was the financial engine behind some breakout hits, such as the "mommy blog" known as *Dooce*. *Boing Boing*, where Battelle had already been helping out in a role he called "band manager," was a mainstay, as was Scott Beale's *Laughing Squid*. Federated Media did well enough that Battelle was able to raise another $50 million and even get a payout for himself in early 2008.

There were some success stories in tech-focused digital media too. A blog called *TechCrunch*, launched by an overbearing Silicon Valley attorney with a knack for writing named Mike Arrington, had become a must-read for Web 2.0 entrepreneurs. Kara Swisher, a former *Washington Post* reporter who'd been early to the tech story, brought good sources and a lot of panache to *All Things D*, the *Wall Street Journal*–owned blog and tech conference that she built along with powerhouse product-reviewer Walt Mossberg. She and her then-wife, the onetime PlanetOut CEO Megan Smith, were a San Francisco power couple.

The stylistic innovations of blogging didn't always sit easily with old-line newspapers like the *Journal*, but they were trying to adapt. "We were fast, we were funny, we were putting out the same good reporting, but we got to make it fun," Swisher said later. She and Mossberg would later spin out the operation into their own company, called *Recode*.

For established media outlets, there was a growing realization that digital advertising was not going to make up for the collapse of print advertising; it just didn't work the same way. Most new media companies were also finding it tough to make ads pay. Salon.com, founded back in the 1990s by *Examiner* writer David Talbot and a handful of colleagues, had made an early splash as a pioneering "digital magazine" and won backing from Adobe founders John Warnock and Chuck Geschke, as well as tech investment banker Bill Hambrecht. It had even gone public during the dot-com bubble, but the shares barely maintained any value and it struggled constantly to pay the bills. Experiments like the *Bold Italic*, funded by the publisher of the *San Jose Mercury News*, or Patch, a startup trying to build a national network of local sites, weren't getting much traction either. My own new media venture, a local and regional online magazine called *New West* that I'd started after moving to Montana in 2002, fell victim to the same trends, as well as the Great Recession.

The dollars that were moving away from print weren't going to new digital publications, it turned out. They were going to Google, which had figured out highly effective advertising products based on charging not for viewing an ad, but on how often people clicked on it. And they were going to Facebook, which had learned how to precisely target its advertising by systematically monitoring people's behavior on the Internet. Both were frighteningly more effective than print—or banner ads.

In local news especially, it wasn't a pretty picture. The city had lost hundreds of journalists over the course of the decade, and there wasn't much capacity for the in-depth daily news coverage and deep investigations that would hold the city's often-fumbling government to account. In the surrounding counties, the problem was even worse: A few of the richest towns, like Palo Alto and Tiburon, had healthy community newspapers, but the *Mercury News* and the *Oakland Tribune* were in bad shape, as were suburban dailies like the *Contra Costa Times*.

In the view of civic-minded leaders who'd grown up with a newspaper habit, something needed to be done. It was a perfect project for Warren Hellman.

Hellman was something of a journalism fanboy, a voracious consumer of news who enjoyed talking to reporters and thought what they did was important. And in 2009 he was irritated with the declining *Chronicle*, which among other failings had snubbed his beloved San Francisco Ballet. The future of news, it was already clear, would not be in daily print newspapers; maybe Hellman could play a role in what came next. He'd already been thinking about it when Tom Hall, a newspaper union organizer, did what many others in San Francisco did when they had a problem: He went to see Warren Hellman, in the hope that he might have an idea. And he did.

What eventually emerged was a complicated plan for a nonprofit news organization that would operate in collaboration with the UC Berkeley School of Journalism; KQED, the local public radio station; and, most importantly, *The New York Times*. Not convinced that print was totally dead, *The Times* wanted to support the circulation of its national edition and was looking for local partners in key markets, with the Bay Area at the top of its list.

The project needed an editor who had local knowledge, startup and leadership experience, new media credentials, and most importantly in the circumstances, enough old-media chops to satisfy *The New York Times*. I fit the bill—a major stroke of good fortune for me personally, with my Montana business circling the drain—and I was hired at the end of 2009 to create what we'd call *The Bay Citizen*. In May 2010, we began producing two pages of Bay Area news, twice a week, for the print editions of *The Times* that served Northern California, along with a daily news website. In the teeth of a media industry crisis, I was able to hire two exceptional deputies, Pulitzer-winner Steve Fainaru and Salon veteran Jeanne Carstensen, and build an all-star staff. The local chattering classes—always a bit too obsessed with what the tastemakers in New York thought of their city—were immediately devouring every word.

Hellman was barred from any role in the editorial, at the insistence of *The Times*, which was allergic to the idea of a wealthy benefactor having a say in what went into their precious pages. Our board of directors didn't offer much reassurance on that front: After the deal with KQED collapsed over various issues and the dean of the Berkeley J-School had a personal crisis and resigned, it consisted of just a handful of Hellman's friends and associates. Hellman could be testy when crossed and at one point created an incident by berating a reporter for pursuing a story close to him, though he learned the error of his ways and never again tried to meddle. But the organization was a creature of the city's philanthropic elite: Socialite Dede Wilsey, the grand dame of local donors—and the mother of entrepreneur Trevor Traina—was among those who would kick in a seven-figure sum, as was Bob Fisher of The Gap. The high-dollar philanthropy had a favor-trading quality: You give a million to my pet project, and I'll give a million to yours.

Hellman sometimes enjoyed tweaking his fellow moguls. I can still see his bemused smile when *The Bay Citizen* dinged the Gettys by revealing the unlicensed preschool for the super-rich that operated out of their compound—a story that prompted quite a scene when the fire department showed up at the mansion the morning it was published and (temporarily) shut it down, followed by frantic parents scooping up their little ones while trying to hide their faces from our photographer. (We opted not to run the pictures.)

A few months after that, Hellman buzzed my office phone. "Dick Blum is on the line, and he's upset," he said, chuckling. A student reporter at UC Santa Barbara was looking into allegations that Blum, a wealthy private equity investor who sat on the UC Board of Regents and was married to Senator Dianne Feinstein, might have directed UC endowment funds to favored investments. He'd actually written a story about it for the school paper, but it wasn't fully baked, and it had quickly vanished from the Internet after complaints from Blum. Given Blum's stature and power, and the challenges of proving the allegations, it was a very tall order for a young reporter, but also a worthwhile story to pursue, and we'd agreed to help him out. I was quite certain this was why Blum was calling.

Warren transferred him over, and after brief pleasantries Blum told me breezily that none of what the student reporter was looking into was true, and that if we published a story with anything resembling those allegations, "certain things are going to happen." Like a lawsuit, was the clear implication.

"Are you threatening me?" I asked, taken aback by his approach. I'd been threatened over stories many times, in various ways, but his mob-boss-style message, delivered personally, was a first. Normally the rich and powerful had their lawyers handle the bullying.

"No, I'm not threatening you at all," he responded. "I'm just telling you, for your own information, that if certain things happen, then certain other things are definitely going to happen." His tone had turned smug and cold.

I told him I considered it a threat, and wildly inappropriate, and hung up. I was in disbelief that I'd been spoken to that way by the spouse of a US senator. In light of how that conversation transpired, I'll confess I was disappointed when we weren't able to stand up the story.

Like so much of what went on in City Hall or in the boardrooms of downtown, *The Bay Citizen* was oddly disconnected from the tech industry, even though by then it was clearly the future of the city's economy. We nominally covered the nine counties of the Bay Area, including San Jose and Santa Clara, and bringing Silicon Valley money and technology expertise to our board of directors seemed like an obvious step. But Hellman wanted to keep *The Bay Citizen* close and resisted expanding the Board, even though that was a routine step for a nonprofit looking to build a donor base. It wasn't at all clear where the money would come from beyond the $5 million Hellman had committed and the initial favors from his friends; *The Times* didn't pay us anywhere near enough to cover the work we put into their pages.

I thought we might be poised to break through Hellman's resistance in the spring of 2011 when he asked me to come along for a fundraising visit to Salesforce CEO Marc Benioff. Soliciting his fellow moguls was one of Hellman's well-practiced skills, and Benioff would be a great place to start in bringing the tech world closer to *The Bay Citizen*. Twitter might have stolen the spotlight, but Salesforce was still the biggest tech company in the city, and its success was pushing Benioff quickly up the ranks of the local mogul league tables. The company was adding workers at a rate of about seven a day, and had just bought fourteen acres along the waterfront in Mission Bay, with plans for an opulent campus housing more than four thousand employees.

Hellman had known Benioff since he was a kid. Both of their extended families were among the city's prosperous Jewish clans and members of Temple Emanu-El, the beautiful Moorish synagogue in Presidio Heights that's home to one of the oldest and largest reform congregations in the country. Benioff's grandfather, Marvin Lewis, was an influential city supervisor, credited with bringing the BART commuter rail system to life, and later in his career gained fresh fame as a trial lawyer. (Among his exploits was convincing a jury in 1971 to award a young woman $50,000 for a cable car accident that she said had resulted in nymphomania, a case the press joyfully dubbed "a cable car named desire.") Marvin Lewis was a friend of Warren Hellman's father, and both families were major philanthropists, giving heavily to Jewish community groups and many other causes. Benioff sometimes cited Hellman, along with his grandfather, as role models for businessmen who gave back.

Benioff had cut his entrepreneurial teeth while still in high school and become a top salesman at Oracle before launching Salesforce out of a Telegraph Hill apartment in 1999. He pioneered the obvious-in-hindsight idea that companies could avoid all the hassle and risk of buying expensive business software and instead pay a fee to tap into software running at a remote data center. There is not much that's sexy, or inspirational, about enterprise software, as it's called; it's hard to claim you're changing the world by making salespeople more efficient, the core promise of Salesforce's products. The company was barely on the radar of the SoMa Internet crowd. But cloud software was a major innovation, and Salesforce had hit the billion-dollar mark in sales after just five years.

Benioff was now eager to build his personal brand. He was an obvious target for our entreaties.

Warren was busy and full of life at seventy-six—he still started many mornings with a 16-mile run through the Presidio—but on this day he seemed tired and distracted, his usual nervous energy just a little bit off. In hindsight, I suspect the leukemia that would be diagnosed later that year, and kill him not long thereafter, was already taking its toll, but at the time I didn't think much of it, other than noting that Warren was getting old. I felt protective of him. He reminded me of my father, an elderly but handsome man, slightly hunched, with a prominent Jewish nose and craggy skin and strong opinions on most things—and now growing a little bit frail.

Benioff was running late, and Warren and I waited awkwardly for almost a half hour on a small, low-slung couch outside his office, my irritation on Warren's behalf slowly mounting. Finally Benioff, tall and commanding, stepped out of an elevator and strode across the waiting area. We stood to greet him, he said hello, and then made us wait again as he stepped into an adjacent office.

Finally Benioff emerged and ushered us into a meeting room. I was prepared to give a short presentation on *The Bay Citizen*, but we never got to that.

"What are you using for your publishing software?" he demanded at the outset. "You should use Salesforce!" This was unexpected. Salesforce did not make the kind of software we needed, something called a content management system, or CMS.

"I didn't think Salesforce offered a CMS," I said, as mildly as I could. Benioff was undeterred, standing and gesticulating with passion as he lectured us about how flexible Salesforce was and how some new features made it perfect for a CMS. "It can do all these things," he insisted.

I was literally speechless. The simple truth was that Salesforce software was not even remotely suitable for our purposes, and Benioff had to know that. What was his game?

We eventually got through that conversation, and Benioff then pivoted to telling us how the structure of *The Bay Citizen* was all wrong and *The New York Times* deal was a big mistake. "Why would you tie yourselves to *The New York Times*? They're just going to tie you down," he declared. He wasn't necessarily wrong about that—delivering *The Times* pages consumed a lot of our resources—but his demeanor was oddly combative in the context. Then he wanted to know how we were going to distinguish ourselves from the *Chronicle* and others, and how we would produce journalism that made a difference with only a handful of reporters. These, again, were fair questions, but he didn't seem very interested in the answers. I tried to explain what we had in mind, but he was impatient and peremptory. Warren was silent.

Finally Benioff said, "Yeah, I'm not going to support it," hardly a surprise at that point, and we feigned good cheer and were on our way.

Warren wasn't nearly as annoyed by all of this as I was, or at least that's what he told me. And while I thought Benioff's behavior was straight-up rude, I could also see the mogul logic at play. Warren was the old guard, and Benioff didn't see much reason to support the personal media project of an aging investment banker. The fact that Warren was practically family only seemed to accentuate the standoffishness. Benioff didn't need Warren Hellman or his circle to help him build influence, as he was clearly going to have enough money to buy it on his own (and in fact he would later buy *Time* magazine). He certainly didn't like sharing the spotlight. Hellman might have been the business community's most influential civic leader, but that was yesterday's business community, the one that ran things before the tech industry arrived. It was Benioff's time now.

As Web 2.0 generated a new crop of Internet companies, the smartphone was about to turn everything upside down yet again.

Apple introduced the iPhone in 2007 and the first cell phones using Google's Android software came in late 2008. While the early models lacked app stores, maps and many other features that would come later, the possibilities presented by a high-powered pocket computer, with a camera and an Internet connection and satellite-based "geo-location," were extraordinary.

Garrett Camp, the Canadian who'd already made millions on the sale of StumbleUpon to eBay, was among the first to see what could be done. As was often the case with breakthrough inventions, his motivation was personal. He'd settled happily in San Francisco and was a fan of the nightlife, but was endlessly aggravated by a problem that locals had complained about forever: You could never get a cab when you needed one.

Like many cities, San Francisco had a "medallion" system, which limited the number of cabs on the streets, lest offering rides become a dangerous free-for-all where nobody could make any money. The result was a handful of fleets controlling the medallions, with no incentive to respond to supply and demand, or even keep the cars clean. The fleet owners refused every effort to establish a central dispatch, leaving patrons to call the individual cab companies in sequence and hope, often in vain, that one of them might show up. Willie Brown's failed effort to reform the taxi system was one of his biggest defeats. Newsom's reform efforts also went nowhere.

Ed Lee, appointed mayor in early 2011, would be overtaken by events before he could even try.

Camp's idea was for a smartphone app where you could order up "black cars," which were operated by a different set of companies than the taxis, and had to be booked in advance. They were a lot pricier than a cab, but Camp saw that with the city's growing population of flush young people, that could be a feature and not a bug—the fashionable way to get around town. He soon teamed up with a maniacal young serial entrepreneur named Travis Kalanick, who'd been holding what he called jam sessions with like-minded enthusiasts in his Twin Peaks apartment, mind-melding about the endless array of opportunities at hand with the Internet and the smartphone.

Kalanick's relentless drive and win-at-all-costs approach would soon make him an exemplar of a certain kind of San Francisco entrepreneur. Determined to ignore political obstacles and regulatory realities, he couldn't hide his contempt for those who might stand in his way, declaring at a conference in 2014 that the enemy was "an asshole named taxi."

In Uber's home city, that attitude wasn't usually appreciated, but in this case most residents could only mutter in agreement.

UberCab, as it was then called, launched in San Francisco in 2010 after signing up a few dozen black-car drivers, who were already vetted and licensed for livery services. The city quickly moved to shut it down, sending inspectors to the tiny office Uber had rented in North Beach and threatening the company with fines that might smother it in its crib. Uber was immediately at loggerheads with the city's top taxi regulator, a veteran civil servant named Christiane Hayashi, who was already battered from conflicts with drivers and fleet owners and found Kalanick obnoxious and arrogant.

But after the company agreed to stop marketing itself as a cab service—it had already decided to shorten its name to Uber—the city and state backed down. Uber was able to continue the black-car service.

The viral growth was phenomenal.

Soon after, Sunil Paul's Sidecar and Lyft, originally founded by Logan Green and John Zimmer as a long-distance carpooling service, pushed the envelope much further by letting unlicensed private drivers pick up paying passengers, under the guise of "ride-sharing." Lyft devised an elaborate ritual featuring fist bumps and cars festooned with pink mustaches, both to make people more comfortable getting into a stranger's car and to make it look more like a carpool than a taxi. It was an odd thing at first, but San Franciscans were open to interesting or helpful disruption—especially on something like local transit that could use a lot of improvement—even if it seemed illegal on the face of it.

Kalanick expected Lyft and Sidecar to be quickly banned, and actively campaigned against them. But when that didn't happen, he just as quickly joined them—and ride-hailing exploded.

The city's taxi industry fought the newcomers at every turn as the city struggled with how to respond. Hayashi had been exploring some other ideas, including an app called Taxi Magic that predated Uber and was designed to enable existing taxis to be called via a smartphone. Jason Elliott, Newsom's policy advisor, recalls her showing it to him at a 2010 City Hall hearing on taxi issues. "I was standing outside room four hundred [in City Hall], and she showed me Taxi Magic on her phone. She's like, 'This is a centralized dispatch,'" Elliott recounted. "But the taxi companies in that meeting basically fought it and said, over our dead bodies, we will never participate in centralized dispatch." It would, in fact, be over their dead bodies.

Ed Reiskin, the head of the city's Metropolitan Transit Agency, believed the new services should be regulated like taxis, with background checks for drivers, specific insurance requirements and, most importantly, limits on the overall number of vehicles. Plus, they needed to fit into the city's broader transit-first policies. The companies were never going to agree to most of that. But the MTA was a regulator, not a market innovator, and the city already had a thing called taxis that had to follow certain rules.

Yet the industry's refusal to modernize, or even provide a single number for people to call, had undermined that simple logic. Even some in City Hall thought the MTA's resistance to app-based ride-hailing was wrongheaded. The agency's leadership "fundamentally thought that they should disappear," said one insider. "Meanwhile, I'm thinking to myself, 'Everybody I know, myself included, is definitely going to choose ride-share over this subpar taxi service.'"

Uber and Lyft had a couple of powerful cards to play with the MTA, and this was one of them: Given the pathetic state of local taxi service, blocking a promising alternative wasn't going to play well with voters. Its second advantage was that it fell between the regulatory cracks. The city had authority over taxis that picked people up on the streets, but the state had authority over roadways and licensed livery drivers. The offices of the state regulator, the Public Utilities Commission, were on Van Ness Avenue in San Francisco, and most of the commission members lived in town; they were all-too-familiar with the cab problem, and the city government's failure to do anything about it. So was Gavin Newsom, now biding his time as lieutenant governor while plotting to run for the state's top job on a tech-forward platform, often working out of an office in central SoMa.

Lyft hired lobbyist Susan Kennedy, a savvy former PUC member who'd been chief of staff for Republican Governor Arnold Schwarzenegger, despite being a Democrat. She all but authored a new set of state rules for what were dubbed "transportation network companies," preempting San Francisco's efforts to regulate the upstarts and giving Uber and Lyft almost everything they wanted.

The political and legal battles over this new industry, which Uber would fight with almost gleeful abandon around the world, had only just begun. But the early wins in San Francisco would prove to be pivotal precedents that shaped the debate far beyond the city's borders.

Airbnb, too, sprung in part from a local reality. Hotels in San Francisco tended to be extremely expensive, and there were always a lot of young people coming to the city who couldn't afford it. At the same time, with rents so high, there were plenty of residents looking for extra cash. Soon Airbnb, like Uber, was proving the power of a two-sided marketplace, where hosts would be drawn in if prices were high enough, and travelers would come if they were low enough. Unlike in a traditional marketplace, strong demand would *create* more supply, or so the theory went.

Airbnb was far from the only place you could go to rent a room for a few days or a few weeks; Craigslist had featured such listings for years. But Brian Chesky and his cofounders had a strong design sensibility and were extremely aggressive, building critical mass early on by shamelessly poaching listings from Craigslist. As the journalist Brad Stone describes in his book about startups of the period, they turned out to be masters of what would later be called growth hacking.

The company cultivated a casual vibe, casting itself as a way to make friends while traveling more cheaply, or earn a little extra cash on the side. But Airbnb also made it easy to convert apartments into short-term vacation rentals, which in major cities and tourist destinations—and San Francisco was both—could be extremely lucrative. No sooner had the site achieved breakout success in the early 2010s than a new industry of borderline operators began renting out residential properties with the express purpose of offering short-term lodgings via Airbnb.

Once they realized what was happening, San Francisco officials began demanding that Airbnb crack down: Renting out apartments for less than thirty days was plainly illegal in the city, for starters. And it didn't make a lot of sense to suddenly allow residential apartments to serve as unregulated hotels. At the very least they should be paying the 14 percent lodging tax.

San Francisco had to cede to the state when it came to regulating Uber and Lyft, but it clearly had the authority to set lodging rules. The city demanded information from the company and wanted to require hosts to register, at a minimum. But Airbnb dug in its heels, insisting it was only an intermediary serving independent entrepreneurs and travelers. Chesky argued passionately that Airbnb was a solution to high rents, not a cause, allowing travelers to share the burden, as it were.

Airbnb would prove to be especially polarizing as the housing crisis deepened, but the company wasn't going to give ground in San Francisco. As in the case of Uber, similar conflicts were playing out around the world, and local officials and housing activists everywhere were watching developments in the birthplace of it all, trying to figure out what they should do next.

Ron Conway was prescient in recognizing that startups getting into businesses like taxis and room rentals were going to need political cover, and fast. It wasn't only taxi drivers who were agitated. Airbnb didn't look like anything good to hotel owners, or their unionized workers. The city's labor leaders were waking up to the threat posed by the "gig economy," which was advertised as a liberating new lifestyle but could also be a way for companies, or even entire new industries, to dodge fair wages and benefits. All manner of things that would normally require some kind of license or permit—a pedicure, a fresh-cooked meal, someone to do a little drywalling—might now be summoned by an app, chipping away at the established practices of a variety of trades.

Conway's investing style involved finding ways to get a piece of the most promising companies, even when he missed his chance in the earliest rounds. He now held stakes in many of the city's hot Internet companies, including Airbnb, Sidecar, Dropbox, Square and Slack, and local policy decisions loomed as almost existential for some of them. If San Francisco were to curtail Uber or Airbnb, or later inventions like rental scooters or delivery robots, it would be much more than a humiliation for the companies and their investors: It could prompt other cities to take notice, and bolster opposition in those places too. The initial ground rules and legal precedents, were being established for new forms of commerce and human interaction. Conway had very clear ideas on what Ed Lee was supposed to do.

Conway put together a new lobbying group, called SF.Citi. "Technology companies needed to take a 'One City' approach and build a shared sense of community and civic responsibility in San Francisco," he declared later in describing the purpose of the new organization. The group would back efforts like a "clean streets" initiative where tech company employees would volunteer to pick up trash, and Conway by all appearances was sincere in exhorting his companies to "give back." SF.Citi's first task, though, was less high-minded: assuring the passage of a 2012 business tax overhaul, which benefited the tech industry at the expense of old-economy companies. The Chamber of Commerce, and other business lobbying groups like Advance SF, were dominated by banks and hotels and real estate interests, alongside perennial power player PG&E. They didn't like the tax changes, which replaced the payroll tax with a so-called gross receipts tax, nor were they interested in playing second fiddle to the likes of Ron Conway. Some of them could see pretty clearly that for their own businesses, the Internet economy was a mixed blessing at best.

But Conway had persuaded Lee to brand San Francisco the "tech capital of the world," and there was no doubt who had the upper hand now.

Twitter moved eight hundred employees into what was now called Market Square in the summer of 2012, and the company had reason to cheer: By all accounts, architect Olle Lundberg, and new owner Shorenstein, had done a spectacular job in remaking the old Furniture Mart. Lee was delighted too, touring the site along with Supervisor Jane Kim on move-in day in the spring of 2012 and donning a hard hat for an inspection of the not-quite-finished rooftop garden.

Soon the mayor was showing up at the Twitter building so often that it became a joke among the company's staff. He palled around constantly with Conway. The way Lee saw it, cozying up to tech was a no-brainer. The city was still recovering from the Great Recession, and there weren't a lot of sources for new, high-paying jobs. The big companies down south were hiring like mad, but that threatened to leave San Francisco a bedroom community for Silicon Valley commuters. "Job creation is everything," declared Conway. "Jobs equal prosperity." What city in its right mind would reject that premise?

Lee and other city officials continued to correct anyone who would listen that it was not the "Twitter tax break" but rather the "Central Market/ Tenderloin Payroll Tax Exclusion," and sure enough, there was a flurry of new companies moving to take advantage. One Kings Lane, a luxury goods retailer founded by Mark Pincus's then-wife, Allie, moved into the Twitter building. A social network for businesses called Yammer, led by PayPal mafia alumnus and future Donald Trump acolyte David Sacks, would take the whole third floor. Zendesk, a Danish customer service startup that had moved to the city in 2008, rented a historic building at 989 Market Street, mainly for the bargain-basement rent of just $5 a square foot—less than a tenth of the prices downtown. Happily for the city, Zendesk brought its Scandinavian social justice values to the challenge of revitalizing Mid-Market, and would be the first to implement a community benefits agreement.

"All of us at Zendesk are proud to be citizens of a city that has the vision and ability to execute on making San Francisco a better place to live and work," CEO Mikkel Svane enthused.

For most of the new arrivals in Mid-Market, it wasn't about the taxes in the end. Uber would move into the old Bank of America data center at 1455 Market Street, a massive, thick-walled building that lacked important office amenities—like windows—when it was acquired by Strada Investment Group and Hudson Pacific Properties shortly before the tax break was passed. Strada's principals, Jesse Blout and Michael Cohen, had both worked in the mayor's office under Newsom, and they quickly made the building into a go-to space for startups, even though it wasn't in the tax-break zone. Dolby Labs, a homegrown, family-owned company that was responsible for numerous breakthroughs in audio technology but had little relationship to most of the local tech industry, bought the building at 1275 Market for its new headquarters; that wasn't in the tax-exclusion zone either.

"The exemption excluded all of the large office properties in Mid-Market with the exception of the Furniture Mart," noted city economist Ted Egan. The draw for companies moving to Mid-Market wasn't the Twitter tax break, said Egan. It was Twitter.

Like an anchor tenant in a mall, Twitter's commitment guaranteed the critical mass needed to make it all work. A gourmet grocery store soon opened in the building. A clutch of new restaurants sprang up nearby. The old Grant Building, where Critical Mass got its start, was renovated into a Yotel, and the upscale Proper Hotel would soon open up on a once-desolate corner. The Burning Man organizers joined the influx, taking space in a sixteen-story building at Sixth and Market.

For the first time in many decades, Mid-Market was attracting investment and office workers, and you could feel the change on the streets. The American Conservatory Theater, which had deteriorated into a squathouse, would be magnificently renovated. The Warfield Theatre building was spiffed up too, and small arts nonprofits began to reclaim vacant or underused spaces in the neighborhood. The idea that Mid-Market might once again be an arts and entertainment hub didn't seem totally far-fetched.

Still, cleaning up Mid-Market was a daunting task, starting with the fact that there wasn't any agreement on what "cleaning up" meant. Literally cleaning the trash from the streets and sidewalks was a good start, but the consensus ended there. There were a lot of homeless people in the neighborhood, but advocates resisted proposals to force them off the streets. There were a lot of drug dealers and drug addicts, but little appetite for trying to put them all in jail, and few ideas on what else to do. Most importantly, perhaps, there were simply a lot of poor people, many with disabilities, living in the subsidized housing, shelters and single-room occupancy hotels that dotted Mid-Market and the Tenderloin. Lacking living rooms or other social spaces, people hung around on the streets. The city had committed to a new police substation and other security measures, but law enforcement was only part of the challenge.

Zendesk had tapped Tiffany Apczynski, once a metro reporter for the *Examiner*, to help develop its community benefits program, and she was excited about the challenge. But she knew her way around the city, and had a prescient warning for her CEO: "This is gonna get spicy, real fast."

In nearby SoMa, the decrepit warehouses and abandoned factories that had been hothouses of innovation in the 1990s were finally succumbing to redevelopment. Blocks of cavernous buildings had been torn down to make way for the baseball stadium and the condos that would go up along Townsend and King Streets, while others were demolished to allow the expansion of the Moscone Convention Center. South Park was being colonized by venture capital firms, with the small brick buildings and crumbling row houses converted, one by one, into sleek modernist offices with plenty of glass and exposed wooden beams. Kleiner Perkins would open an office there. So would a new venture firm that was shaking things up: Andreessen Horowitz, propelled by the brand name of Marc Andreessen, famous for his role in creating the Web browser.

When Andreessen arrived in the Bay Area from the University of Illinois in 1994, he was something of a type—a blond, wide-eyed boy genius, talking fast, determined to change the world through a seminal invention. He'd settled in as a man of Silicon Valley, rather than San Francisco, a hardworking big thinker and hypercompetitive entrepreneur who enjoyed his drinks and stayed up late, but wouldn't have known his way to a rave. He'd married Laura Arrillaga, the daughter of a developer who'd made a vast fortune converting the orchards of the Santa Clara Valley into office parks and strip malls, and the couple lived in Atherton, the bucolic hillside community where the very richest in Silicon Valley made their home. Now balding but still a fast talker, with a tight smile that often looked like a smirk, Andreessen had little patience for liberal pieties and a powerful contrarian streak, though it was only later that his politics would veer to the far right.

Netscape, which Andreessen had cofounded with veteran entrepreneur and investor Jim Clark, had been a big success out of the gate, but the company never figured out how to make money consistently from software that was, in the end, mimicking open-source code developed at the University of Illinois. The Netscape Moment, when the company went public back in 1995, may have been a turning point for the industry, but the company's stock price never exceeded its first-day pop, and in 1998 it had been sold to AOL, then gradually faded away. Its influence lived on, though, in ways large and small—not least in the second act of Marc Andreessen.

He'd tried a couple more startups after Netscape, including a social network, and managed to sell a company called Loudcloud to Hewlett-Packard and get a good payout. Still, he'd largely been eclipsed by the next generation of Internet entrepreneurs when he dove into what would prove to be his true calling.

Andreessen had broken through as a public intellectual in 2011 with a landmark essay called "Why Software Is Eating the World," which correctly foresaw the sweeping impact of new technologies on every aspect of life and business. That came a couple of years after he'd joined with Ben Horowitz, a sly and slightly unconventional engineer who'd been a product manager at Netscape and Andreessen's partner at Loudcloud, to launch a new venture capital firm. Horowitz was the son of a renowned Berkeley professor who'd famously flipped from leftist stalwart to Ronald Reagan enthusiast in the 1980s—presaging almost a little too neatly some of the political gyrations to come.

Their new venture, Andreessen explained to me in an interview aboard his shared private jet in 2012, would take a different approach than its peers, investing heavily in a wide range of support services and doing everything it could to be the most "founder-friendly" investor in the business. He offered a rapid-fire history of venture capital as he explained how all the money was made on just a handful of companies, waving his arms and drawing little charts and graphs to illustrate as he spoke. The key to success, the numbers showed quite clearly, was to make sure you were among the backers of the tiny minority of entrepreneurs who made it really, really, really big. One early investment in an Amazon or a Google could make a career; the "power law" of venture capital, as it came to be known, applies more than ever today.

Andreessen Horowitz arrived at an extraordinary moment. The second Internet boom, animated by the smartphone, was just beginning, and a confluence of technological developments and global economic forces would make it bigger, by far, than even the vertiginous dot-com era. It would violently shake every corner of the city in a way that would make the ructions of the late 1990s look like a mere tremor. It would go on to upend the world.

PART THREE

Prosperity and Its Discontents (2012–2019)

CHAPTER 7

The Big Boom

The turnaround in San Francisco's economic fortunes unfolded with breathtaking speed.

By 2012, the biggest city-based startups—Twitter, Salesforce, Zynga, Airbnb, Uber and Lyft—were hiring as fast as they could, and so were many lesser-known local companies that were on their way to becoming global brands, including Reddit, Instagram, Etsy, Square, Pinterest and Dropbox. The city would add about one hundred thousand jobs between 2010 and 2014, an increase of almost 20 percent, with unemployment falling by more than half, to just 4.5 percent. Retail sales citywide leaped by 16 percent in 2013 alone, while construction spending soared 27 percent to more than $5 billion in 2015—its highest level ever, and some four times what was spent in 2009.

The new tech wave was starting to create immense personal fortunes for entrepreneurs who had the right combination of talent, luck and timing. Mark Pincus's Zynga went public at the end of 2011, and he and some other investors (including Reid Hoffman) sold additional stock early the following year at a slightly higher price, earning Pincus nearly $200 million for 15 percent of his shares. (The company would soon run into trouble due to changes in Facebook's approach and the quick rise of mobile games, and Pincus and others would face a shareholder that was eventually settled for a little over $11 million.)

Twitter would go public in 2013 at a market value of $14 billion—modest by tech industry standards even then, but solidifying its place as San Francisco's signature Internet company and endowing Ev Williams, the largest shareholder, with a $2.4 billion fortune. Jack Dorsey's stake was worth about a billion.

The giants down south were producing huge piles of cash. Facebook went public at a valuation of $100 billion, making Mark Zuckerberg one of the world's richest people and vaulting a handful of early employees and investors—most prominently cofounders Dustin Moskovitz, Eduardo Saverin and Chris Hughes, as well as Sean Parker, the Napster entrepreneur who was Facebook's first president, and Jim Breyer, a veteran VC—into the billionaires club. Peter Thiel, still living in the city and already very wealthy from the sale of PayPal, would earn yet another nine-figure fortune and hold a seat on the Facebook board, though an early sale of a chunk of his holdings cost him an even larger fortune.

Google, meanwhile, added twenty-one thousand jobs globally in 2012 alone, and now had more than fifty thousand employees—the first of the Internet startups to achieve true "big company" status. Google didn't provide a breakdown, but people with knowledge estimate that about half were in the Bay Area, and a third of those—or around eight thousand people—lived in San Francisco. Fleets of buses now carried young engineers, product managers, designers and sales executives from their San Francisco homes to the sprawling campus in Mountain View—highly visible symbols of the new economic order.

It was a quirk of history that the arrival of the smartphone coincided with the era of near-zero interest rates that followed the Great Recession, but the circumstances had a powerful compounding effect. An enormous new investing opportunity had appeared just at the moment when rock-bottom interest rates changed the risk-reward calculations of the money managers who controlled the trillions of dollars sitting in government treasury accounts, pension funds, university endowments and family investment offices around the world. With so little return to be had in safe government bonds, it made sense to allocate more money to riskier but potentially more rewarding venture capital funds.

Andreessen Horowitz and a few other high-profile investors saw that capital was now so cheap and plentiful that startups whose fortunes once rode on building a better mousetrap could deploy money as a competitive weapon, raising once unheard of sums to grow at breakneck speed and crush any rivals.

The dynamics of startup investing were shifting completely. The risk-loving Japanese entrepreneur Masayoshi Son, who'd already made and lost a fortune in the dot-com bubble, had persuaded Saudi Arabia and the United Arab Emirates to back his $100 billion "Vision Fund," and he was spreading the money around as fast as he could. Mutual fund companies like Fidelity Investments and hedge funds such as Tiger Global were becoming prominent startup investors too.

The dollar figures and the business ideas often looked crazy, and sometimes they were: The Vision Fund famously invested $300 million in a dog-walking app (it didn't work out). A $700 machine for making juice, it was revealed by a Bloomberg investigation, didn't squeeze the special juice packets any better than you could with your hands. The "app economy" was all the rage, and for a brief period it subsidized a slightly comical range of personal service businesses—an app for your laundry, a fresh-cooked meal at lunch, a roving massage service—that looked like nothing so much as what a twentysomething tech bro would want a woman to do for him.

Andreessen Horowitz, as well as Masayoshi Son, drew some mockery for paying high prices to get a piece of companies that had already proven themselves—including Facebook, Uber and Airbnb—but the strategy paid off. Andreessen's first Airbnb investment, for example, came in 2011, when the company raised $115 million in a deal that valued it at $1.3 billion. It looked pretty rich at the time given the company's steep losses, but turned out to be a bargain. Airbnb would raise nearly $3.5 billion from private investors over the course of the decade before finally going public in 2020, and by 2025 had a market value of around $80 billion.

The cash-driven approach to market dominance repelled many early Internet entrepreneurs, both the technical purists who were more interested in inventions than consumer products, and the idealists who wanted to steer tech development toward the greater good. Even among the business-minded, the new necessity of "go big or go home" often seemed like a narrow choice. "Blitzscaling," coined by LinkedIn's Reid Hoffman, became the term of art for all-out growth, and even some of his friends were distressed.

"This whole 'blitzscaling' idea, much as I love Reid, I think it was a truly pernicious idea," said Tim O'Reilly, the Web 2.0 champion, in an interview years later. "You drive all the competition out with this massive infusion of capital. It became a game where it's not the market that chooses the winners, but the VCs who choose the winners."

The second tech wave would also change the fortunes of San Francisco's real estate industry, itself a highly speculative, boom-and-bust business dating to the gold rush days. Tensions over development and land use had roiled city politics on and off for decades, and the 2010s would bring a massive building boom—and a lot of strife.

Property prices had been climbing in the city, albeit unevenly, since the 1980s, but the Great Recession staggered the office sector especially. Paul Paradis, the Bay Area boss for Texas-based developer Hines, was in the midst of trying to get his biggest-ever project off the ground—a SoMa skyscraper that would be the tallest in the city—when the financial crisis hit and sent him scrambling to salvage the plan.

Hines had been building office towers in the city for decades, and Paradis was coming off the success of 560 Mission Street, one of the signature buildings in the expansion of the financial district across Market Street and into SoMa. His new project was part of the city's plan to replace the grim, decrepit Transbay Terminal bus depot with a sparkling new building that would accommodate not just buses but also trains—Caltrain, the commuter rail line serving Silicon Valley, and the future high-speed rail that was supposed to connect San Francisco and Los Angeles. The terminal was to be paid for in part by selling an adjacent lot to a developer, along with permission to build an extra-tall tower.

A design and development competition was launched in late 2006, and ten months later Hines was named the winner, with a design from the firm Pelli Clarke Pelli, led by the renowned Argentine-American architect César Pelli. Paradis, who grew up in a French-speaking family in Maine and studied classics at Harvard before falling into the real estate business, came up with the idea for a rooftop park, now the defining feature of the transit terminal.

"This part of downtown has always been open-space-starved," Paradis recounted later. "So I worked with our conceptual construction experts to figure out if a rooftop park was economically reasonable, and came to find out that it was."

His colleague, Merredith Treaster, thought it would be a competitive edge. "We wanted to have a building that's a nice complement to the city, not a screamer, like 'look at me,' " she said. "And we suspected the other designs didn't prioritize green space." The park, along with an offer of $350 million for the lot that far exceeded rival bidders, helped Hines's office-only design prevail over three other proposals for mixed-use buildings.

Winning the bid was one thing, though; getting it built and occupied was something else. Hines was in the course of negotiating the final, detailed agreements in late 2007 when the real estate market collapsed, marking the start of the Great Recession.

"Everyone's hair was on fire—the industry was in deep trouble," Paradis recalled. "I spent a couple of years dealing with lenders, and the Transbay tower was basically put on hold. The timing was too fast, the land value was too high—it was no longer a deal that worked in the new environment." He told the special agency set up for the project, the Transbay Joint Powers Authority, that he'd need to renegotiate. Now he had to find ways of cutting costs.

The obvious way to make it cheaper was to make it shorter—but that was a no-go. The tower was part of a larger development plan for the area that was many years in the making, noted Joshua Switzky, the city planner at the center of the discussions, and it was supposed to be a beacon, a symbolic marker of the city center and its transit crossroads and the tallest building in town. So Paradis got creative, focusing first on the elevators. Originally, he explained, the design included a complex elevator system with a transfer floor, which added a lot of expense.

"We wanted a simple elevator system where you got in here and got out there," Paradis said. "The way the building was set up, we only had a certain amount of space in the core, and our elevator consultant calculated how many elevators he could get into service, and how many floors we could have, and the answer was sixty-one." That would leave the building several hundred feet short of the thousand-foot height that the planners were looking for. (The then-tallest building in the city was the Transamerica Pyramid, at 850 feet.)

"So we got together with César Pelli and his gang. 'We've got a problem here, what are we going to do?' We said, okay, let's make the ground floor super-high so we have a nice, big, tall lobby. Let's expand the ceiling height on each floor so that it's very generous. People will love that. And let's include on every floor an under-floor air delivery system, which will add a foot to each floor, and that's a really cool way to deliver air into the building."

This would boost height while keeping the number of occupants at a level that the simple elevator system could handle. But the tower was still too short. He went back to the architects.

"César said, 'We'll just make a top.' So he designed the top, which is completely decorative and serves no function other than to meet the height requirement of the city." There'd be a six-story-tall atrium at the top of the building, and the outside would be enlivened with a light show, one that would help Paradis satisfy the city's public art requirement. He also got a big break on price of the land, ultimately paying $192 million.

With the redesign and renegotiation done, Paradis turned to financing the project, which was being built on spec at an estimated cost of around $1 billion. Just days before critical payments were due, Paradis signed Boston Properties, whose local holdings included the landmark Embarcadero Center, as a partner.

Now all he needed was a tenant. Paradis, along with his boss, company founder Gerald Hines, had visited Marc Benioff at Salesforce to pitch him on the tower, but it didn't go well. Benioff, in fact, had something else entirely in mind. He'd spent $278 million for the fourteen acres in Mission Bay, and his architects had conceived a lush complex of offices and amenities, including a swimming pool and a ferry dock, along with ample open space, some of it for public use. Benioff's CEO suite would include its own pool on an outdoor terrace. Total cost: about $2 billion.

Salesforce shareholders, though, were unimpressed with the plan. The stock price was down, and the new headquarters seemed needlessly extravagant; there'd been too many instances of a trophy headquarters building marking the start of a company's decline. No, it turned out, Salesforce investors did not want to pay for Benioff's private pool, no matter how much money he might have made for them in the past.

The company had spent a couple of years on the project and was just days away from final approvals when it abruptly pulled the plug, to the shock and chagrin of city officials. Salesforce had already begun leasing high-rise space downtown and now it was back in the mix for the new tower. (The Mission Bay property was sold and would become home to the Chase Center arena, where the Golden State Warriors play, as well as the headquarters of future tech titan OpenAI.)

Construction was underway at the Transbay site when negotiations with Salesforce began—it would take a year to build the foundation, and the work could then be paused if no tenant had been found. Salesforce wanted naming rights, and Paradis insisted that they take at least half the building in exchange for that. The deal was done in 2014, and the building would open three years later, winning a complimentary review from

John King, the *Chronicle*'s respected architecture critic—and a lot of complaints about it being a little too symbolic of the city's very male tech elite.

Salesforce Tower was only the most visible symbol of a real estate market that had turned overnight. The near-zero interest rates that were driving money into venture firms were also reigniting the property business, and big global investment funds were now on the hunt for San Francisco real estate. Shorenstein, after acquiring the Twitter building for $120 million in 2011 and spending about $300 million on renovations, sold 98 percent of its interest to JP Morgan for $900 million four years later—an exceptionally lucrative deal for the hometown company. Blackstone, the New York investment giant, was spraying money all over: $600 million for 49 percent of One Market Plaza, $510 million for 555 Market Street and $245 million for Embarcadero Square, a low-rise complex in Jackson Square.

A flurry of new high-rises got underway, mostly in SoMa, and would open in the latter part of the decade. There was the lavish, 70-story mixed-use tower at 181 Fremont Street, with the office portion leased entirely to Facebook's Instagram unit and a penthouse condo initially priced at more than $40 million; the 43-story Park Tower at 250 Howard Street, next to the new transit terminal, also leased entirely to Facebook; a 26-story modernist structure at 222 Second Street, all of it to be occupied by LinkedIn; and the 30-story building at 350 Mission Street, leased by Salesforce.

A similar dynamic was sweeping the residential real estate market. Across from the Twitter building, a black-glass tower called the Nema boasted of its "tech-forward design" as it solicited the well-paid young professionals now working in Mid-Market. Down by the transit center, in the corner of SoMa that real estate brokers were now calling the East Cut, there was the Millennium Tower, conceived as the ultimate in luxury apartment living and boasting celebrity tenants like Joe Montana and the financier Tom Perkins, one of the godfathers of venture capital (and briefly the husband of romance novelist Danielle Steel after her marriage to John Traina ended). The Millennium condos sold out quickly, though the building's luster was permanently dimmed when it began tilting ever so slightly in 2016, resulting in a festival of headlines and litigation, and eventually a $200 million fix.

The core of the city's housing market—small and midsize apartment buildings built before 1979, and thus subject to rent control—wasn't immune to the speculative frenzy. A San Mateo businessman named Yat-

Pang Au had begun buying older rental buildings in 2006, and in 2011 he acquired Citi Apartments, with some two thousand units, from the Lembi family, long a major residential landlord. He kept expanding, using entities financed by global investment funds to acquire properties that would then be managed by his company, Veritas. By 2017, Au would control more than five thousand apartments, making him the city's largest residential landlord. The Mosser family, another big apartment owner, was also in growth mode, trying to squeeze more money from old wooden buildings. The business model of Veritas and its ilk was built on the idea of finding ways to hike rents or otherwise make more money on rent-controlled apartments, a formula all but guaranteed to generate resentment.

For the richest of the new tech elite, the destination of choice was the Pacific Heights Gold Coast, as it was called, a handful of blocks along Broadway and Vallejo Streets that boasted the city's most extravagant homes. Trevor Traina, the rare local blue-blood in the entrepreneurial set, helped show the newcomers the ropes in the exotic neighborhood, built mostly by the barons of the industrial age. By the middle of the decade residents included Mark Pincus, David Sacks, the Birches, Michael Moritz, Oracle's Larry Ellison and Apple design chief Jony Ive, living alongside established local aristocrats like the Gettys. The city's two mega-rich politicians, Dianne Feinstein and Nancy Pelosi, were nearby too, the former occupying a particularly elaborate manse on the Lyon Street steps, adjacent to the lush greenery of the Presidio. *Vanity Fair*, then in its prime as the arbiter of such things, reported in 2013 that the nouveau tech moguls had "put some Old Guard noses out of joint," but the various subcultures of the city's ultra-wealthy had a lot more in common than not.

For all their flashy deals and ever-expanding war chests, venture capitalists were hardly alone in stoking the San Francisco startup economy of the 2010s. Entrepreneurs who'd made money during earlier waves of the Internet boom were major movers too, with the recirculation of windfall fortunes creating a flywheel effect that powered the industry, and the city's economy along with it.

Ev Williams was still Twitter's largest shareholder and remained on the board of directors when he launched a new publishing platform called

Medium in 2012, installing it in a classic flatiron building a few blocks down Market Street from Twitter, near Union Square. Jack Dorsey, fiddling with a few associates in an apartment at Mint Tower, had come up with a way for small merchants to take credit card payments by plugging a compact card-swiper into an iPhone. Square, as it was called, proved a hit, and the company was soon expanding into related services and would go public in 2015. The success proved Dorsey's bona fides, and he was invited to return as Twitter CEO that same year, even as he continued to run his new company, later called Block, from the old data center building at 1455 Market Street, a block up from Twitter.

Stewart Butterfield, the cofounder of Flickr, had returned to his passion of building video games, but after the first game's sales prospects didn't look promising, he repurposed the company to focus on a chat feature that seemed useful. He decided to call it Slack, and only a few years later his new company was occupying an entire twenty-story building in the financial district. Justin Kan's Justin.TV became Twitch, and was sold to Amazon for nearly a billion dollars. Twitch cofounder Kyle Vogt would go on to launch the self-driving car startup Cruise.

Even as tech money was recirculating into new startups, it was also finding its way into the entertainment and culture economy.

Michael Birch and his wife, Xochi, with three small kids and all the demands of startup life, had never built much of a social life in the city before selling Bebo, the social network they'd founded after moving from London to the Bay Area in the early 2000s. With their epic $590 million payday from AOL, they'd now be making up for lost time. First they bought a contemporary four-story mansion on the Gold Coast and hired the celebrity designer Ken Fulk to outfit it in eclectic and exotic style, a proper British pub among its features. Fulk was the go-to decorator in San Francisco for everyone from the Gettys on down, and the new tech moguls would have plenty of work for him.

The Birches were also looking for a different type of real estate, and in late 2008 bought a crumbling former marble-cutting factory on Battery Street between Broadway and Pacific, just across from the old *Industry Standard* building, with the idea of creating a London-style social club. Birch recounted later how a new wave of such clubs, which unlike their traditional brethren were run as businesses rather than member-owned cooperatives, had grown popular in the British capital in the 1990s, and he thought something like it was needed in San Francisco. He and his wife were eager to do something outside of tech.

"A lot of people told us it wouldn't work in San Francisco, it's very different from London," he recalled. "I'm like, I think people have the same human needs no matter where they are, right?"

San Francisco had its own long-established clubs for the upper crust, most famously the Bohemian Club, which had a building in lower Nob Hill but was best known for a secretive annual retreat at a compound north of the city, near the Russian River. Few would turn down an invite there, but it was very exclusive, and with a distinctly old-money, males-only, captains-of-industry vibe. The Olympic Club had a distinguished headquarters near Union Square, as well as a golf course out near the ocean, but it was also a bit stuffy and old-school. The Birches envisioned something more casual, more modern, geared to a newer generation.

Michael Birch avows that he didn't want his establishment to be dominated by tech people, and the couple spent a lot of time at the start hand-screening potential members, seeking whatever measure of diversity could be had at a members-only club priced at $2,400 a year. They again hired Fulk, and over the course of five years, the marble-factory ruin was transformed into a sleek warren of brick and glass, with striking artwork, classical furnishings and amenities including a gym, guest rooms and an airy restaurant and bar at the center. It opened in late 2013 with 1,400 members, and despite the Birches' counter-programming efforts it quickly assumed its place as a go-to spot for the tech crowd. A sniffy *New York Times* story was a sure mark of its arrival: "If there is a place to see-and-be-seen in the web startup capital at the moment, the Battery may be it, at least for those who update their LinkedIn profiles religiously and count 'ideating' among their daily rituals."

Entertainment and nightlife would prove a popular second career for a number of successful Internet entrepreneurs.

Jonathan Nelson, who'd sold Organic Online to global advertising giant Omnicom, was ensconced with his family in Dolores Heights and became a partner in the nightclub Slim's, an anchor of the boisterous nightlife strip on Eleventh Street, at the southern edge of SoMa. It had become a popular stop for touring troubadours, as well as locals like Carlos Santana, Jonathan Richman and Boz Scaggs, who'd cofounded the club in the late 1980s. Investors now included Warren Hellman, along with Nelson and several others. The group also owned another storied music venue, the Great American Music Hall in the Tenderloin, where Jonathan Steuer and Brian Behlendorf had met before launching HotWired.

Jamie Zawinski, a well-known programmer and open-source software pioneer known on the Internet simply as jwz, had been a key player in the early days of Netscape, and in the creation of the free Web browser known as Mozilla and the San Francisco nonprofit Mozilla.org. In 1999, he'd bought the DNA Lounge, a run-down club just up the street from Slim's, and then spent two years on renovations, battling the city's permitting bureaucracy and local residents in a process so arduous that he documented it in a blog called *DNA Sequencing*.

Nelson had also run into issues with the city, mostly relating to the very–San Francisco phenomenon of people buying condos on a block filled with live-music venues and then complaining about the noise. One unhappy resident escalated matters by making a "citizen's arrest" of Slim's manager, and, to Nelson's amazement and fury, District Attorney Kamala Harris had agreed to look at the case.

He called the DA and got one of her deputies. "I said, 'What's going on?' and she said, 'Well, you guys are disturbing the neighbor.' And I said, 'I put in $100,000 in soundproofing, what do you want me to do?' And she said, 'Well, the neighbor might have a point.' "

Nelson was pissed. The Committee on Jobs, a group dedicated to boosting the economy, which he chaired for a time (it later morphed into Advance SF), had done a lot to help elect Harris back in the early 2000s in the hope that she'd be good for the business community. "She was not the obvious choice to be elected DA," he recalled. Nelson felt she should have helped a local business on a frivolous lawsuit, rather than trying to score points with it. "That's the way it's supposed to work, right?" The complaint didn't ultimately go anywhere, but Nelson wasn't alone in the view that Harris was an ally of convenience.

By 2012 the noise complaints were becoming a constant hassle. A reprieve would finally come in classic City Family fashion.

Warren Hellman, a partner in Slim's, was dying of cancer. His doctors believed it was treatable when he was first diagnosed, but the chemotherapy had instead made him sicker, and now the end was near. He was taking calls from people who wanted to say goodbye, Nelson recounted later, among them Willie Brown. "Willie calls him and says, 'Thank you for everything you did for the city. Is there anything I can do for you?' And Warren says, 'Yeah, take care of these noise complaints.' It ended right there."

Hellman's death in December 2011 would leave a big void. He bequeathed enough money to keep the Hardly Strictly Bluegrass festival going, and Nelson and others would continue to run the music venues, but there was no substitute for his unique role as a moderator among the city's political factions. Nor could *The Bay Citizen* survive without him. The then-novel idea had been to generate money from memberships, grants, charitable contributions and syndication of our content, but he'd run it as a personal project, and its donor base never extended very far beyond his friends. The publication was merged into the Berkeley-based Center for Investigative Reporting, a nonprofit where Phil Bronstein, the former *Examiner* and later *Chronicle* editor, was chairman of the board. Hellman had always liked Bronstein. But CIR quickly ended the flagship *New York Times* partnership and *The Bay Citizen* was officially shuttered a year later, another tombstone in the crowded graveyard of local news.

Doug Dalton, another early Netscape alum, also took the plunge in the nightlife business, inspired by how things were done in New York. He'd had a classic dot-com journey: learning about the Internet at the start of the 1990s as a young engineer at a telecom company, then leveraging that into a job at Netscape, then following his boss to a dot-com startup. In this case, it was an educational-services company called Knowledge Universe founded by Michael Milken, the notorious junk-bond king who'd done time for his dubious dealings on Wall Street in the 1980s. Milken's dot-com strategy included buying up many small companies, and he was ruthless in his approach, Dalton says, so much so that Silicon Valley VCs advised him that he was in business with the wrong guy if he wanted to have a career in the Internet industry. He then landed at a startup called Gloss.com, which had offices in the old Hamm's brewery building and was selling cosmetics on the Internet. To Dalton's everlasting good fortune, Gloss.com was acquired by cosmetics giant Estée Lauder after the bubble burst, with the idea of having Dalton and his team build an in-house online operation. CEO Fred Langhammer and the Lauder family thought their new tech leader should understand their world a little better.

"They had put in my contract that they wanted me to go to a variety of events with them," he recounted years later, still incredulous that his job duties included attending spectacular parties. "I went to the Met Gala. I went to some of the coolest places I've ever been in my life. I got to see all the coolest bars, and that's where I would say my transition from tech to nightlife occurred." He'd been a fan of music clubs since his high school days in DC, especially one called the 9:30 Club, and during the dot-com boom he'd helped his friend Brian Sheehy throw house-music parties on boats in San Francisco Bay. They started kicking around the idea of opening a bar. Sheehy worked for the upscale hotel group Kimpton, and they wanted to focus on what Dalton called "premium experiences."

It was an instinct in line with San Francisco's historic aspiration to be a cultural rival to the big cities of the East.

"Coming from New York, I'd seen these wonderful experiences that I wanted to replicate here. San Francisco had Michelin-star restaurants and hotels with great bars, and then really shitty dive bars," he recalled. There seemed to be an opportunity in the middle somewhere, and the early 2000s weren't a bad time to start: Dalton recalls Langhammer advising him that bars (and cosmetics) did just fine in a soft economy. Dalton's vision included "an imposing wall of alcohol," with high-end brands arrayed on floor-to-ceiling shelves behind the bar. They'd open their first establishment, Anu, on Sixth and Market, followed by Swig, on Geary Street near Union Square.

Dalton was working at another tech job at the time, and was disappointed to discover that owning bars actually hurt your credibility in the tech world, where it was seen as a sign of unseriousness. But he was hanging around with all the Web 2.0 cool kids: He counted Mark Pincus and Ev Williams as friends, along with Kevin Systrom, founder of the fledgling Instagram, and the beloved Tony Hsieh, cofounder of shoe retailer Zappos, who was a hero to many for his radical ideas on remaking the workplace. Another buddy was Kevin Rose, who came briefly to prominence with a social news site called Digg, appearing on the cover of *BusinessWeek* under the misleading headline "How This Kid Made $60 Million in 18 Months." (The fortune was strictly on paper and never realized.) When Dalton opened a third bar, Bourbon & Branch, in a former speakeasy in the Tenderloin, he was thrilled when real pop-culture celebrities, like the rock star Dave Matthews, began showing up.

Dalton finally quit the tech business in favor of his new métier, and he and Sheehy would open a dozen more spots in the city as the boom heated up in the 2010s, most of them featuring that "wall of alcohol," but otherwise with their own themes.

For the big-time attractions at the very top of the media-entertainment complex, newly flush tech investors were becoming a major force. Joe Lacob, a low-profile venture capitalist at Kleiner Perkins, had done well on deals including Autotrader.com, and in 2010 he teamed with Hollywood producer Peter Guber and a few other partners to buy the Golden State Warriors—snatching the team from the grip of software magnate Larry Ellison, who wanted it badly but misjudged the sale dynamics and made his bid too late. Then a middling team playing in a spare but beloved arena in Oakland, the Warriors had just drafted a young sharpshooter named Stephen Curry and would soon be developing plans for a glitzy new home in Mission Bay. Together with the baseball Giants, who'd capture three World Series titles over the course of the decade, the Warriors' winning ways in the 2010s were a fitting complement to the triumphant tech economy. Stars like Draymond Green, with long-term contracts worth hundreds of millions of dollars, were now dabbling in startup investing too. Joe Montana, the retired 49er quarterback who was among the city's preeminent sports heroes, learned startup investing under the tutelage of Ron Conway. In 2015, he launched a venture firm of his own.

Larry Ellison's sporting interests weren't limited to the Warriors. He was an accomplished sailor and had spent many millions competing in the America's Cup, one of the sporting world's oldest and oddest events, finally winning the trophy in 2009. The cup-holder had the right to choose the venue for the next regatta, and since San Francisco Bay is a celebrated sailing venue, the city was the obvious choice for Ellison. Hoping to generate fresh interest in a sport that was rather boring to watch, his team came up with a radical new specification: seventy-two-foot catamarans, equipped with rigid "wing" sails, that would rise up out of the water on narrow metal hydrofoils, reaching speeds of more than fifty miles an hour. It was impressively high-tech and held the potential for much more exciting races.

Ellison, though, thought the city should subsidize the event, as previous venues like Newport, Rhode Island, and Valencia, Spain, had done. But unlike those pleasant but out-of-the-way places, San Francisco didn't need the money and the media attention that the races would bring, and nobody could understand why Ellison wouldn't pay for his own party.

The city eventually agreed to put up about $40 million for infrastructure and services, to be raised from private sources. The fundraising task fell to philanthropist and former city official Mark Buell. It wasn't much fun.

"Raising money for the third-richest man in America for his boat race was not an easy lift," Buell said later, with some understatement. At the Board of Supervisors, Jane Kim and Aaron Peskin demanded regular updates from Buell on the private fundraising, lest the city be on the hook if it fell short. Buell had been worried from the start that Ellison and his team were underestimating the financial challenge, and his worst fears were confirmed after a Chamber of Commerce presentation from Ellison's deputies on what it would all look like. "I said to the guy afterwards, who are your lead sponsors? Who are going to be the big names? He couldn't answer the questions on the economics. That's when I realized I had signed off on something that wasn't going to work the way they said it was. They had no clue."

Outside of sailing circles, public sentiment ranged from indifference to contempt. In the event, though, the competition itself was spectacular, and a very San Francisco exercise in technical ingenuity, risk and excess.

The boats would be more accurately described as sailing machines. Tilted against the wind, twelve thousand pounds of advanced composite materials riding a thin, twenty-three-foot-long hydrofoil blade with the crew of eleven mostly atop the opposite hull, they were a surreal sight hurtling across the Bay, hard to keep up with even in a speedboat that had to pound across the waves as the giant catamarans glided above. Controlling the vessels took enormous strength and skill, and the crews were Olympic-level athletes, grinding winches and making split-second decisions amid the wind and the spray while trying to avoid toppling overboard. It looked dangerous, and it was: A British sailor was killed in practice when his boat flipped end over end and trapped him under the wing sail.

The regatta was also marked by a cheating scandal that resulted in a penalty for Ellison's team, and then a remarkable comeback where it won the last six races to defeat archrival New Zealand, which had been just one win from victory. But the intrigue, the competitive fervor and the technological marvel of the sailing machines weren't enough to win the city's heart. Attendance was far below what organizers had hoped, and there was predictable acrimony when the private fundraising fell short. The feeling was mutual: Ellison had soured on the city long before the competition ended, and the America's Cup would not be back. The high-tech catamarans were ultimately judged too dangerous and would never race again.

The gaudy excess of the America's Cup was fitting for a moment in US history where the rich were getting richer very fast, and pushing income inequality to a level not seen in a century. There would be plenty more extravagance: Sean Parker spent $4.5 million in 2014 on a medieval-themed wedding, held in a redwood grove in Big Sur bedecked with three hundred thousand flower petals, faux stone walls, a specially built bridge and erstwhile castle ruins. He was pilloried on social media for allegedly lacking a permit and damaging the redwood forest and ended up paying a fine, though he was livid about being cast as a bad person and insisted he'd done everything by the book. Inviting magazine photographers to the over-the-top event was perhaps not his best move.

Travis Kalanick of Uber would take thousands of company employees to Las Vegas for a blowout featuring a private performance by Beyoncé, the booze-filled days enhanced for many participants by cocaine and other substances. He'd take heat, too, for the bacchanal, but it was the mood of the moment.

More quietly, but in the same spirit, adventurous libertines who were well-connected—or especially good-looking—were partaking in another sort of party circuit: invitation-only events at fancy private homes or discreet luxury venues featuring sexual adventures, sometimes along with drugs like ketamine, ecstasy, cocaine and LSD. The decadent happenings were semi-open secrets, connecting wealthy executives and entrepreneurs to the sex and drugs part of the counterculture, with politics left at the door. A party might feature sex "stations," where guests could partake in particular escapades, along with plenty of dark rooms and beds for "cuddle puddles," or more. The music would be carefully curated, as were the drugs, often in exotic combinations, with one popular "triple-play" mixing ecstasy, ketamine and LSD. The play parties, as they were sometimes known, often intersected with a growing interest in polyamory.

Inevitably, alternative lifestyles would be packaged into startup businesses too—sometimes with ugly results. A signature case of the hazards was One Taste, which aimed to create "a clean, well-lit place where sexuality, relationship, and intimacy could be discussed openly and honestly." It began as a communal living space in SoMa devoted to a female-focused practice called orgasmic meditation—one male resident described to me in detail the daily ritual involving extended stroking of the female residents' clitorises. It grew to have branches in multiple cities with all the language of a tech startup, but allegations of abuse would eventually lead to the founder and a top deputy being convicted of forced labor conspiracy, for coercing people into sex and unpaid work.

The sexist "bro culture" cultivated by Kalanick and many other young, male entrepreneurs could be especially toxic in the context of this libertine tolerance. Susan Fowler, a twenty-eight-year-old software engineer, triggered an uproar—one that eventually contributed to Kalanick being deposed as CEO of Uber—when she wrote about being greeted on her first day at the company by a manager who immediately propositioned her, explaining that he was in an open relationship but his wife was having more fun, and he needed to even the score. Andy Rubin, the General Magic veteran who'd created the Android smartphone software, was pushed out of Google in 2014 after revelations about allegedly abusive behavior toward a subordinate whom he'd drawn into his sex-obsessed lifestyle. Hundreds of employees participated in an unprecedented walkout several years later when *The New York Times* revealed that his exit had been accompanied by a $90 million severance package. Keith Rabois of the PayPal mafia departed as the number two at Square following sexual harassment allegations by a young man who worked for him, though he'd continue to be known for wild parties at his hilltop home in Glen Park.

Still, there was no general opprobrium for outré habits and tastes. No less a personage than Steve Jobs had endorsed LSD as a powerful, mind-liberating substance. Now "microdosing" of hallucinogens, which enthusiasts said could help clear the mind and increase focus, had become accepted in some circles for long hours on the job, alongside the stimulant Adderall, to say nothing of cocaine. The ravers were raving about ketamine, an anesthetic approved decades ago that turned out to have mind-altering properties at the right dosages, and was now being used both as a party drug and a mental-health treatment. New designer drugs were still emerging from local labs as the line between wellness regimes and recreational narcotics became blurrier. Apache founder Brian Behlendorf continued to host a website called Erowid, run by two individuals who called themselves Earth and Fire and devoted to good information on psychoactive substances, keeping it on his own server to better protect it from government agents with bad intentions. "Any proper assessment of the culture of the city would include a conversation about substances, positively and negatively," he said diplomatically some years later. Drug culture was very much embedded in the San Francisco startup scene.

Burning Man remained a totem of the countercultural melting pot, and like the city itself, it had grown dramatically with the infusion of the curious, adventurous and now-prosperous who'd come west for the tech boom. In 2011, the event sold out for the first time, and it would settle in at a population of about 70,000. Relations with the Nevada authorities had improved, mainly because of the financial windfall Burning Man represented for every public safety agency (and many businesses) in the vicinity, and permits and fees alone were running into the millions. Ticket prices marched ever-higher, but Burning Man now had serious cachet in the San Francisco tech world, from CEOs and venture capitalists on down, and tickets would be a scramble for the rest of the decade, even when they were going for $1,000 or more.

Black Rock City had an airport now, a temporary installation like everything else but with official Federal Aviation Administration recognition and the call letters BRC, so the rich could fly in on small private planes or buy a seat on a chartered commercial flight. Google's Sergey Brin continued to attend almost every year, as did by now old-timers like Mark Pincus, John Gilmore, Brian Behlendorf and Brewster Kahle, not to mention a newer generation, including Elon Musk, Brian Chesky, Justin Kan and many other stars of the Web 2.0 era.

Plenty of less-celebrated techies were drawn into the world of Burning Man too. Tim Chang, who'd go on to a successful career as a partner at the VC firm Mayfield, had finished his MBA at Stanford and was working as a junior associate at a venture firm when a friend from business school rallied a group to go to Burning Man in 2005. Chang had grown up in a conservative "Asian tiger parent" household in Michigan—"the only thing that mattered was grades," he said coldly years later—and that first trip to the Black Rock Desert was a revelation.

"It was so incredible," Chang recounted in an interview at The Portal, the Burning Man–inspired wellness club he founded many years later in Marin County, after retiring from venture capital. "I remember my very first moment there. There was a row of cupcakes on wheels riding by with people peeking out, and suddenly a life-size Viking ship with a fire breathing dragon came up, and it was just marvelous. It was like a return to our ten-year-old selves, before self-awareness and self-consciousness and all that programming kicked in. It was life-changing."

It was infectious too. Burning Man's remarkable growth and resilience could be traced to the inspiration it sparked among so many attendees to not only come back, but to do it better next year and bring more friends. "There's this pay-it-forward mentality" in introducing people to it, Chang said. "You get that sort of visceral, by-proxy feeling of experiencing the magic of the playa for the first time all over again." He'd attend for the next fourteen years and become one of the many Burners devoted to bringing the culture of Black Rock City from the desert to the default world, as regular society was known. He fashioned The Portal as a permanent Burning Man camp running year-round, in an effort to "spread more of the magic off-playa."

Tent structures of various sorts had once dominated at Burning Man, but as the event got bigger and more expensive, and camps got more elaborate, RVs were increasingly the preferred choice. In 2010, Chip Conley, who'd bought the Phoenix Hotel in the Tenderloin back in the 1990s and now had a collection of seventeen boutique lodgings in the city, took the amenities up a notch for a fortieth birthday celebration, with a plush camp called Maslowtopia featuring catered meals.

It was a precursor to the fancy "plug-and-play" camps to come—everyone still had to do some work and many stayed in tents. And Conley was a true Burner, having won some notoriety in 2009 when he posted pictures of himself in a tutu on the playa on his company's website—and after some deliberation, declined the request from his head of HR that he take them down. The story of the CEO who believed people should have the freedom to show themselves in full, whatever more prudish employees might think, made the national press. It also caught the attention of the Burning Man organizers, who invited Conley to join the board of directors.

Conley would play a key role in the organization's purchase of the Fly Ranch, where the 1997 Burn had been held; it would provide both a good operations and storage site and hedge against more trouble with the BLM. Elon Musk, Sergey Brin and Airbnb's Joe Gebbia were among those who kicked in to back the deal.

All the money, though, was complicating life on the playa.

A camp called Caravancycle had pushed the line on commercialization, marketing an all-inclusive experience for up to $20,000 where fancy meals, air-conditioned RVs, showers and private toilets were all provided, and wristbands were required to get in. David Rodriguez, a veteran Burner from San Francisco, thought it was all a little much. "This whole thing that you can pay a bunch of money and they're going to take care of you, that was against all of our ideas of Burning Man," he recounted. He and friends began brainstorming about maybe showing up outside as peasants with pitchforks. Eventually they came up with a plan: A group of six would dress up as maids and butlers and pretend they were employees.

"We lined up in formation and walked right in," recalled Suzannah Cowell, another participant. She had a brush, and set about cleaning people's boots, while another "maid" offered a refreshing spritzer. One "butler" had a beard brush, and others had dusters, and they made a show of giving people "service." "Oh my god, Jonny needs some service over here," Cowell playacted, and a couple of people would rush over. It went on for a while, and the attitudes of the recipients varied. One person muttered, "About time we got some service here," but mostly people took it with pleasure, with only a few suspecting a joke. Finally as they were leaving, someone approached and said, "Do you work here?" "We said no, and we told him, and he thought it was hilarious."

The Burning Man Project would later ban the practice of selling package tours or paying anyone to work at the event. It would eventually insist that RVs had to be driven into Black Rock City by whoever would be using them. It was all hard to enforce, though, and some of the policies were later eased. Conley didn't think fancy camps should be banned, but he acknowledged the tensions and how the event was evolving.

"In the early days, it was a little bit more, 'Hey, we're here for the art, we're here for the spirituality,'" he says. "And over time it became a little more, 'We're here for the party.'"

For the tech cohort, it was becoming a networking event. Back in 2001, Google founders Larry Page and Sergey Brin had famously agreed to hire Eric Schmidt as CEO in part because he was a Burner, and now it had become a place to prove your bona fides as someone who could appreciate teamwork and bonding and overcoming the elements together—the "group flow" of the playa could also apply to the office. Competition arose—how could it not?—over who could build the coolest camp or art project, or have the most beautiful people with the best costumes. It was the perfect recreational and artistic twin of the booming Internet economy.

Even as it matured, Burning Man remained a fundamentally unlikely operation, overseen by an organization riven with the kinds of issues you might expect when a collective effort of loosely affiliated artists and anarchists evolves into a world-famous event with a $20 million budget. The Burning Man Project now had around sixty full-time employees and a year-round office in Gerlach, in addition to acquiring the Fly Ranch. The six partners in Black Rock City LLC agreed in 2011 to convert it to a nonprofit, which would create more transparency and a more stable long-term structure. The transaction was completed in 2014 and they each got $75,000 for their shares.

When the Smithsonian Institution put on an exhibit of Burning Man art in 2018, it brought a measure of validation, even as it marked the drift toward the mainstream.

"I used to scoff at this idea that we're a movement, like that's such bullshit," said Candace Locklear, the PR pro who'd volunteered to wrangle the media in the early days. "Turns out, it's a global movement. It really is. The principles have inspired a ton of other events and festivals, so I think the vision was awesome." As for the corrupting influence of the money, she was on Larry Harvey's side. "When the billionaires started coming, the old school was like, 'Why are you letting them in, why do we have to have the one percent at our event?' And he would say, 'If five percent of our playa dust rubs off on them, then you have done something good to change their mind or open their mind to another way of life.' "

The spirit of Burning Man was infiltrating the city in a number of ways. Mike Farrah was working as an aide to Gavin Newsom in the mid-2000s when he went to the playa to see if there might be lessons to be learned on building infrastructure in difficult conditions. "That lasted about ten minutes," he said later. "I was completely blown away by the art, and it was unbelievable how they fostered community." When he returned to the city, he set about "greasing the wheels" for playa art to make its way to San Francisco. A lot of it originated in the city anyway, with a celebrated workshop on Treasure Island called Building 180, and another near Hunters Point called the Box Shop. "We had about twenty-five Burning Man artists do installations in San Francisco," recounts Farrah, who was also recruited to serve on the board of the Burning Man Project. The org launched a Civic Arts Program in 2005, and artist David Best created a "temple" in Hayes Valley that became a community favorite.

Ben Davis, a tall, cheerful Boston native who'd moved to San Francisco in the mid-1990s, had started going to Burning Man in the 2000s, and in 2010 he'd been at Maslowtopia with Chip Conley. Shortly after he returned to the city, he found himself sitting on a pier at the Embarcadero, contemplating the sweeping vista of the Bay Bridge. "The sun was coming up behind the cables, dancing, and it sort of showed itself to me as this canvas of light," he recalled later. "It was a visitation."

The Bay Bridge—really two bridges with an island in the middle, connecting San Francisco to Oakland—is San Francisco's overlooked masterpiece of civil engineering, with anchorages of unprecedented size and the spectacular "skyway" on the San Francisco side. Davis thought his "visitation"—thousands and thousands of lights amid the steel cables—might represent the bridge's best chance to shine on its upcoming seventy-fifth birthday, though he didn't yet have a clue how it might work.

After exploring the feasibility with some people he knew at Caltrans, the state highway agency, he eventually landed on the website of Leo Villareal, a well-known light artist who had a popular camp at Burning Man. Davis called Villareal with a question: "If the Bay Bridge was your canvas, what would you do?" Davis recalled. "We probably had four or five conversations, and then about three months into it, Leo turned in a forty-eight-second video that absolutely blew me away, and blew everyone else away." The concept featured twenty-five thousand programmable LED lights on the western span of the bridge that would form a shape-shifting display.

Davis had to raise the $8 million it would take from donors, and the first big gift of $100,000 came from Matt Mullenweg, the WordPress founder. Davis is still amazed that the project known as Bay Lights actually happened.

"There was no international selection committee, no judging panel, no more than a conversation with one artist that was very pure," he recounted. In March 2013, the playful, shimmering and seductive display that Villareal created came to life for the first time to raucous cheers, and a pledge from Mayor Ed Lee to help it outlive its two-year engagement. Bay

Lights would be the foundation of a nonprofit called Illuminate, which would bring all manner of light-art to the city in the years to come.

The digital economy was extending its reach across the city in every imaginable way. At the quiet Glen Park branch of the San Francisco Public Library, a twenty-six-year-old city resident named Ross Ulbricht was working at his laptop one fall day in 2013 when an apparent altercation erupted. He briefly looked up, and before he knew it an undercover FBI agent had abruptly shoved his open laptop across the table to another agent. More plainclothes officers quickly appeared, grabbing Ulbricht and handcuffing him while their colleagues carefully spirited away his computer.

The unassuming library patron, it turned out, was also Dread Pirate Roberts, the operator of a secretive online marketplace called Silk Road, where those in the know could buy illegal goods, especially drugs. Payments were made with a newfangled digital currency called Bitcoin, and Ulbricht's arrest would be the first time a lot of people had ever heard of it.

Launched in 2009 by a still-unknown personage who called themselves Satoshi Nakamoto, Bitcoin was an ingenious invention: an electronic form of money that was a self-governing system, with a limited number of new coins created by computers solving complex math problems and all transactions recorded anonymously on a public ledger known as the blockchain. The idea of a currency controlled not by government mandarins but by algorithms was extremely appealing to libertarian-leaning techies, even if the practicalities were daunting. Those government mandarins, for one, wouldn't be eager to see a currency that operated outside the laws of the global financial system. It would be much too easy to set up something like, say, Silk Road.

As the price of Bitcoin began to climb in the early 2010s, it looked like nothing so much as a glorified Ponzi scheme; there was nothing underlying the value of Bitcoin and other digital currencies besides people's belief in their value. Their main practical utility, beyond speculating on the price, was in the black markets of the Internet, and Ross Ulbricht could fairly claim to be the city's first multimillionaire crypto entrepreneur.

Ulbricht was convicted of drug trafficking and sentenced to thirty years. But the line between crime and commerce could be a blurry one when it came to crypto. (That would become especially obvious during the second Trump administration, and Ulbricht would see his sentence commuted in 2025.)

Yet crypto-boosters also believed there was a more traditional kind of opportunity at play. Blockchain technology held the promise of a new approach for moving money around, especially across borders, and there was in fact a serious need for that. The global banking system relied on an ancient messaging system called SWIFT, which was slow and had proven vulnerable to sophisticated hackers. Immigrants working in wealthy countries who sent "remittances" home to their families paid commissions as high as 15 percent. Surely there had to be a better way.

Chris Larsen's Prosper, the lending platform he launched in 2008 after selling E-Loan, had been a failure: The timing was bad, coming as it did in the teeth of the Great Recession, and the social dynamics of lending and repayment didn't work as anticipated. Now crypto looked like the next turn of the wheel, and in 2012 Larsen joined with two Canadian cryptographers who'd launched a company called OpenCoin, later changing the name to Ripple. The goal was to build a better system for banks to exchange funds based on the Ripple crypto token, known as XRP. By the middle of the decade the company had signed up some big international banks to try it out.

A lot of other entrepreneurs were also eyeing crypto. At first, Larsen said later, the main concern was a simple one: "Everyone is like, this is legal, right?" A turning point came in 2013, when the Federal Reserve signaled that it wouldn't simply shut the new industry down. Ross Ulbricht's arrest that year was in some ways part and parcel of crypto winning respectability: If Bitcoin was to become part of the global financial system, it wouldn't be a good look if it were mainly used by drug dealers.

Larsen called 2013 the "magic year" when crypto came into its own, with San Francisco as its capital. An online crypto brokerage called Coinbase, the brainchild of former Airbnb engineer Brian Armstrong, had launched in the city in 2012, fresh out of Y Combinator. The price of Bitcoin, still in the double-digits at the end of 2012, would spike to over $1,000 by the end of the year before falling back.

Over the next decade, crypto would remain highly controversial and appeared ready to disappear more than once. US regulators were less friendly than others, and a lot of crypto activity was driven overseas for a time. But crypto proved resilient. Speculators who were once sneered at as unscrupulous hustlers would eventually amass huge fortunes and, against all odds, win a privileged place in the Washington power structure. Skeptics still considered the whole thing a scam, but the emerging crypto fortunes were real enough for the moment, and would provide more kindling for the city's big boom.

The rush of money had a way of changing people's priorities. Still, the idea that the Internet was good for humanity, and maybe even for the city, held strong in many quarters. Rents were soaring, but so were incomes, and there seemed to be opportunities galore. Ed Lee, no longer a reluctant mayor, faced no serious opposition in his 2015 reelection campaign, with his economic record helping him overcome suspicions about his close ties to Ron Conway and the tech companies.

In the tech-forward view of the world, social media offered a way for otherwise isolated people to find their tribe and was playing a huge role in advancing gay rights and other social justice causes on what was fast becoming a truly global Internet. The "gig economy" of Uber and Lyft drivers, food delivery workers and other assorted "task rabbits" would liberate people from the tyranny of bosses and make everyone an entrepreneur. The "convenience economy" of personal services ordered up via an app would free us from drudgery. Learning a language, eating healthy food, finding entertainment, planning a trip, changing your career—you name it and there'd be an app for that, making our lives better. That was the story at least.

San Franciscans enjoyed a brief moment where all these new wonders could be sampled for a fraction of their real costs, thanks to the dynamics of venture capital "blitzscaling." Meal deliveries and Uber rides, among other app-based services, were essentially half off in the city as investors funded a frantic grab for new markets. It was good for the party spirit while it lasted, though it quickly became clear that gig jobs weren't any real version of running your own business.

More promising was the idea that tech could enable big, systemic changes of the sort that might restore citizens' shattered confidence in the public sector. That's what spurred Jen Pahlka, who'd been the producer for Tim O'Reilly and John Battelle's Web 2.0 conference, to launch a San Francisco nonprofit aimed at bringing tech talent into local government.

O'Reilly had begun eyeing ideas about "open government" in the mid-2000s, and after the election of Barack Obama in 2008, he, Pahlka and the rest of the Web 2.0 team began brainstorming an event they'd call Government 2.0. Obama had inspired a number of people in the San Francisco Internet community to see how they might bring their skills to the public sector. Now there seemed to be something of a movement, with early Web pioneers at the forefront.

Brian Behlendorf had been keeping busy on the speaking circuit after his own startup—a clearinghouse for open-source software—didn't get traction. One of his stops was the World Economic Forum in Switzerland, where he found himself in a lot of conversations about the social impact of technology and how open-source software might be a force for good. "Around this time, I also heard this politician saying some smart things about using tech to make government more transparent, more effective," recalls Behlendorf. "That was Obama."

Behlendorf was soon drawn into Obama's primary campaign and was quite impressed with how it was using the Internet for organizing, with decentralized groups built around issues and interests. (The Democratic Party's failure to maintain and build on that approach is considered one of its signature errors of recent history.) After Obama won, Behlendorf was invited to join the staff of the first White House chief technology officer, Aneesh Chopra.

Pahlka was in touch with Chopra, too, as she developed the new Government 2.0 event. "The question that was being asked was, 'If this version of the Internet can get a guy elected, can it also help him govern better?'" Pahlka recounted later. A conference only went so far though: Pahlka recounts a call in which Chopra conveyed Obama's thinking. "He said, 'The President is very supportive of this happening. And he has also asked that something come out of it that's not just a conference.' Nobody knew what to actually do."

Her revelation on that point had come in 2009, when she was talking with a group of old friends at a family reunion. The husband of a childhood pal worked for the mayor of Phoenix, and he was grousing about the futility of trying to build an app he thought would be useful without running afoul of the city's onerous procurement rules. He and his wife had both done stints with a much-touted program called Teach for America, where young professionals did one-year public service tours. "Then someone said, well, there needs to be like a Teach for America but for coders, and they can work on government apps," Pahlka recalled. "That day I wrote an email to people working in this context. Tim [O'Reilly] wrote back and said, 'That's a great idea, you should do that.' By the evening I was like, I'm gonna quit my job and make Code for America."

She did just that, and by the next year she'd cadged some donated temporary office space on Second Street and was working to recruit both the cities that were interested in help and the first group of coders. The pay was just $35,000 a year, and there were plenty of skeptics.

"People told us that nobody with any skills is going to want to work for government and they're certainly not going to work for $35,000," Pahlka recalled. O'Reilly tapped his network on her behalf and persuaded Mark Zuckerberg and Twitter's Biz Stone, among others, to take part in a recruitment video, extolling the virtues of serving the country. "We had 575 applications for 18 slots. It looked like a cool thing to do if you could afford to," she recounted. The next year there would be thirty fellows and eight cities, and the fledgling nonprofit moved into an old leather factory building on Ninth Street in SoMa.

After the November 2012 election, Pahlka got a call from Todd Park, who'd succeeded Chopra as White House chief technology officer, asking her to join his team. She demurred at first. But at the time she happened to be in the UK visiting an operation called the Government Digital Service, which she found deeply impressive, and she told Park that the US needed something like that. A few months later he called back and said he'd found a home for her idea, and now she'd need to come to DC and make it happen.

Pahlka took a one-year leave from Code for America, and in June 2013 joined the Obama administration as deputy CTO. Her boss and others were about to find themselves scrambling to recover from the disastrous launch of Healthcare.gov, the health insurance marketplace that was central to Obama's Affordable Care Act, and they started bringing in a lot of tech talent, especially from Google. Many would end up finding a home with Pahlka's creation, the US Digital Service, a White House office charged with bringing tech talent, user-friendly services and technological best practices to every corner of the federal government. It was destined to have influence in some ways that were intended, and others that definitely were not.

After Pahlka returned to Code for America, the nonprofit continued to grow, reaching more than a hundred full-time employees by the end of the decade, with scores of projects around the country. She was especially proud of a San Francisco effort where three Code for America fellows, who'd signed on for one year, stayed on voluntarily to continue their efforts at improving social services delivery. "They had started their own, rogue way that people could apply for food stamps, and they were advertising it around 9th Street with those flyers where you can tear it off at the bottom," she recalled. The notices directed people to a very simple website that collected basic information and then relayed it to the real application site for submission; it was a vast improvement over the existing process, which relied heavily on faxing documents.

The three fellows demonstrated the system for the head of the city's human services agency, who was both amazed and nervous, Pahlka recalled, since it amounted to a hack of the process. "He said, 'Totally not cool, you should not be doing that, but I love it anyway,'" Pahlka said. His advocacy made a huge difference. The system would ultimately be adopted statewide, and would also lead Code for America to shift its approach toward longer engagements so that fellows could see their initiatives through to completion.

Back in Washington, Todd Park had created a program explicitly modeled on Code for America, but for the federal government. He'd originally wanted Pahlka to lead the initiative, called the Presidential Innovation Fellows, and she'd declined initially, though she ended up overseeing it while building USDS. After Park left the White House, he was succeeded in 2014 by Megan Smith, the former PlanetOut CEO who'd later spent a decade at Google. Smith would champion the Fellows program and evangelize bringing tech to government with a barnstorming tour around the idea of "Innovation Nation."

From a certain techno-optimist point of view, the mid-2010s marked the apex of San Francisco's hopes and dreams for the Internet economy. It was producing innovative products and services and a huge amount of wealth, while underwriting a burgeoning culture industry and finding creative expression in the arts. It was holding out the hope of a better-functioning government. With an exceptionally smart and likable Black man in the White House, it seemed possible that capitalist exuberance and social justice could march hand in hand.

Behind this happy tableau, though, a lot of discontent was brewing. It wasn't long before it would begin disrupting the fun.

CHAPTER 8

The Buses and the Backlash

David Campos, who represented the Mission district on the Board of Supervisors, had come to the city in 1997 because it was a hub of innovation and opportunity. But his vision had nothing to do with technology.

An immigrant from Guatemala, Campos was fourteen when he crossed the border illegally with his mother and two sisters, destined for Los Angeles. He excelled as a student at Jefferson High, in the city's rough South Central district, and won admission to Stanford, though he was still undocumented. He worked his way through college and then Harvard Law, getting legalized along the way when his father finally got a green card, and landing a prestigious job with the Washington, DC, law firm Arnold & Porter. But his heart was in California, and with a boyfriend in San Francisco he decided to make the move. Even in the mid-1990s, his East Coast peers didn't seem to get it when it came to the West Coast, he recalled years later. "They think of the West Coast as, my God, that's like the Wild West, who'd want to live there?"

Campos saw something very different in San Francisco: a place where public policy and politics were churning with ideas and heading in the right direction. "The issues around LGBTQ, around human rights, workers' rights, it felt like a lot of things were happening in California and in San Francisco at that time," he recalled. "If you really cared about certain communities, if you were gay, if you were immigrant, if you were a person of color, San Francisco was a place to be."

Diminutive and soft-spoken, Campos wasn't an obvious future politician, but he was smart, personable and getting more interested in public affairs. He signed on with the prominent San Francisco law firm Howard Rice, known for its intellectual culture and engagement with politics, and after a couple of years he decided to make the switch to the public sector. One job offer came from the local US Attorney's office, which prosecuted federal crimes and was then led by Robert Mueller, later known as the special prosecutor in the Trump-Russia investigation. Another offer came from the City Attorney's office, which is most easily understood as a law firm representing the city of San Francisco, and at the time was led by a respected lawyer and politician named Louise Renne. As an assistant US attorney working for Mueller, Campos would have been barred from politics, a path he wasn't ready to foreclose. So he went to work for Renne and began climbing the ladder.

Along with Scott Wiener, another young, ambitious, gay lawyer who'd come to town around the same time, Campos in 2000 won a seat on the influential local Democratic Party committee known as the DCCC. He then became the general counsel for the San Francisco Unified School District, a power center in its own right. In 2005, the progressive majority on the Board of Supervisors appointed him to the Police Commission, which had considerable authority over department policies. It was a high moment for the local left: "San Francisco was a very politically vibrant, dynamic place, with a strong progressive presence on the Board of Supervisors," Campos says.

He'd win election to the Board himself in 2008. His earliest big battle was over then-Mayor Newsom's move to loosen San Francisco's "sanctuary city" policy after a murder involving an undocumented immigrant. Campos would continue to fight doggedly for immigrants and also get deep in the wrangling over Healthy SF, which would make San Francisco the only city in the country with its own universal health-care system. He advocated for affordable housing and for tenants facing eviction. He was no fan of the Twitter tax break.

"The fear we had at that time was that they were going to create all these jobs, but the jobs would be given to people from outside the city, who would then come in and take over," he said later. "And so they're coming in without building new housing, without creating any kind of local employment targets, and if you did that, without expecting anything in return from Twitter, you would have government subsidizing something that, in the end, would push out a lot of people."

The pressures over gentrification would only intensify as Campos's term wore on, with his district in the crosshairs. Echoing the dynamics of the dot-com bubble in the late 1990s, the Mission had become a preferred locale for the city's growing army of tech company employees and entrepreneurs, and this time around the changes were far more dramatic. The Latino population was in sharp decline, from about 30,000 in 2000—accounting for half the population of the district—to 22,000 in 2014. Evictions were on the rise, peaking at 237 in 2013, while affordable housing production crawled along, yielding just 122 units between 2011 and 2017. Many once-eclectic streets full of neighborhood businesses were becoming all but unrecognizable, with top-of-the-line restaurants and fancy boutiques of the sort that had never existed in the Mission.

Even Mark Zuckerberg now wanted to live there, or at least have a pied-à-terre. In the fall of 2012 he plunked down about $10 million for a house in the hilly, upscale western edge of the neighborhood, an area called Dolores Heights. He apparently wanted that place in particular—the real estate site Zillow estimated its value at just $3.2 million—and the price set a record for the Mission. He quickly launched a massive renovation that would take more than two years, ultimately yielding a nearly eight thousand-square-foot dwelling whose amenities included a giant turntable in the newly dug garage, so cars wouldn't have to turn around. Critics dubbed it Fort Zuckerberg for all the security measures, which to the great annoyance of neighbors often included two big black SUVs idling at the curb.

The Facebook CEO and his wife, Priscilla Chan, moved in in 2015, though it was never entirely clear how much time they spent there. They also had a compound in Palo Alto, where they had bought out all the surrounding homes, and were starting to accumulate real estate in Hawaii. The Mission, though, had become the heart of the burgeoning tech scene—and a dramatic representation of how the city was changing.

"[Zuckerberg's] arrival here, and his cohort's, isn't just a real estate choice: it marks the merger of tech culture and culture at large, a new hybrid in a neighborhood and a city whose key industry used to stand a bit apart," wrote Reyhan Harmanci, a shrewd critic who'd been the arts and culture editor at *The Bay Citizen* and was now writing for the preferred hipster outlet of the moment, *BuzzFeed News*. "Now, with most of the rough edges sanded off of the previously poor district, the tech guys *are* the Mission's mascots, the cool kids. Love it or hate it, the musicians and weirdos who won the neighborhood its hip reputation in the first place have mostly grown up, fled to Oakland, or both."

Mission activists weren't giving up so easily though. David Campos had taken the baton from Chris Daly as the supervisor leading the anti-gentrification activists, who were anchored in a handful of nonprofit community groups. During the springtime festivities for Cinco de Mayo in 2015, Campos called for a moratorium on all new housing construction in the Mission, saying it was the only way to give the district "a fighting chance."

The idea that new apartment buildings would push rents higher was—and is—a source of endless exasperation for housing advocates. Scott Wiener, who'd taken a more centrist path than Campos, was now on the Board of Supervisors and led the charge against the Mission moratorium, which was voted down twice. It was too drastic a step even for the progressive-leaning Board. But development in the district slowed dramatically in the face of all the political resistance: A proposed ten-story apartment building dubbed "the Monster in the Mission" by activists had become a symbol of the fight, and was ultimately abandoned. (As of this writing it was being revived as an affordable housing project, though opposition remains and no shovels have been turned.)

Yet the gentrification arguments weren't only, or even mainly, about the rent. Nothing would show that better than the theatrical protests targeting what were universally known as the Google buses—or, more commonly in many circles, the "fucking Google buses."

Cari Spivack, the mid-level Google employee who first created the company's commuter shuttle program, never imagined she'd be sparking a years-long political row over whether tech was destroying San Francisco's soul. Her motivation was simple and personal: She was sick of sitting in traffic.

A designer by trade, Spivack had been working at the networking company 3Com in the early 2000s when she saw the simple elegance of Google's website, then just a white screen with the Google logo, a box to type your query and a button that said, "I'm feeling lucky." Spivack thought its pure functionality was inspiring, and a friend of a friend connected her to a hiring manager at the company. She was brought on as a product manager, joining Google at a magical time when there were just a few hundred employees. It was a dream job—except for the forty-five-minute white-knuckle commute from her home in Bernal Heights to the Google building in Mountain View.

She tried taking Caltrain, the creaking, diesel-powered commuter railroad that connected Silicon Valley and the city, but with inconvenient stations and glacially slow and infrequent trains, it took forever. She tried carpooling, and that worked better, but the coordination was a constant hassle. "We're all leaving at the same time going to the same place on the same road—I thought there has to be a better way," she recounted later. A friend who worked at Genentech, the biotech pioneer based in the industrial city of South San Francisco, mentioned that the company had a bus that picked people up at the Glen Park BART station and dropped them at the office. Maybe Google could do that?

"Google was the type of place where you saw the patterns of problems, and just came up with solutions," she says. The company had hired her, in fact, for that very mindset. She was a product manager on the engineering team with no background in engineering. But nobody quite knew what product management was anyway, and she could teach herself programming. She had the quality that was judged "Googley," as the company would come to call it, and though a computer science degree from a prestigious university would later be all but required for many jobs, it wasn't like that at the time. Employees were encouraged to think creatively and use 20 percent of their time for their own projects, which could include almost anything—even commuter buses.

"I was yapping about it at lunch with people and they were like, 'Larry would love that idea,'" she recalled, referring to cofounder Larry Page. A few days later she mentioned it to him in the cafeteria line—the company still worked that way in 2004—and he said sure, figure it out. And so she did, researching the cost of a bus, where it would stop and trying to answer the critical question of whether anyone would actually ride it. Page liked the idea of reducing the company's carbon footprint, Spivack says, though Sergey Brin was doubtful that people would be willing to leave their cars behind in the city.

Spivack sent a company-wide email to gauge interest, and when almost every employee who lived in San Francisco responded, she figured she'd at least have enough people for one bus. It would pick up at Glen Park BART and at a parking lot near Candlestick Park, and it would make two trips a day in each direction. The company gave her a month to prove the concept.

"I think we had twenty-four people at the start and it just kept increasing and increasing and increasing, and we just kept going," she recounts. Google liked to call everything a "beta test," which Spivack saw as a wise way to avoid permanent commitments, but her initiative was clearly a winner.

"People started complaining because they lived in a different part of the city and they're like, 'We'd love a stop here, we'd love a stop there.'" She had never envisioned the buses serving the heart of the city, but she dug into her new role as a traffic planner with gusto, researching neighborhoods and streets and possible bus routes and stops, and developing rules on the fly. "I had some things that were really important to me, like number one, the bus should never take a left turn at a traffic light," she says, lest they provoke delays and angry drivers.

It wasn't long before some Google engineers volunteered to put wireless on the buses—a novel idea at the time, when Wi-Fi was still emerging and didn't exist on moving vehicles. Management liked the idea that everyone could keep working. The buses also chipped at the problem of not enough parking in Mountain View, which would become a major motivator later on as Google's head count soared.

Spivack, who was still doing the shuttle work on her 20-percent time, in 2006 would "roll off" the project, as they say at tech companies, and a full-time manager would take over. Meanwhile the other big Silicon Valley companies, that had a lot of employees living in the city, including Facebook, Apple and Yahoo, began to spin up their own shuttle services. By the middle of the next decade there'd be many hundreds of tech company buses serving the city, carrying ten thousand people a day.

The commuter shuttles had proliferated in the late 2000s without much fuss. The city was mired in the Great Recession, and with unemployment topping 10 percent and so many mortgages upside down, people had bigger concerns. By the early 2010s, though, it looked like 1999 tech mania all over again, and the buses seemed to be everywhere, with Google deploying the largest fleet. Operated by private firms like Bauer transportation, many of them were luxury coaches, sometimes double-decker, much larger than city buses, and they seemed out of place, appearing as outsized apparitions as they crawled across narrow, sloping neighborhood streets, physical symbols of the power of the tech industry and the priority it seemed to enjoy. Since one bus equals dozens of cars, residents in a very environmentally conscious city might have been expected to embrace them, or at least tolerate an efficient form of transit.

That logic did not carry the day.

The complaints began in Noe Valley, an unassuming but increasingly upscale enclave of old Victorians and low-rise apartments just west of the Mission. The tech shuttles, usually black, or white and gray, sometimes used the public bus stops, but they were not for the public. Their patrons stood conspicuously in lines on quiet sidewalks in their Allbirds and jeans, staring at their phones—walking symbols of a new reality in which rich young professionals, most of them white, were pushing families and the working poor from their homes. Even otherwise prosperous people who couldn't afford million-dollar homes or $5,000 rents were feeling the squeeze from what had come to feel like an invasion.

Noe Valley wasn't hip, by any stretch, not like the Mission or certain corners of SoMa, but it had become popular with young families and was also convenient for commuting south to Silicon Valley. As the shuttles proliferated, so did the complaints. Supervisor Bevan Dufty, who represented the neighborhood, led an effort to encourage the buses to use the bigger commercial arteries.

That would quickly backfire: Now the buses were even more visible, and they began using busier city bus stops along already congested commercial corridors, sometimes interfering with Muni service and further irritating neighbors. The city's Metropolitan Transit Agency, which was responsible for traffic as well as trains and buses, tried to engage the companies on possible solutions, such as adjusting routes and stops, but wasn't making much headway.

Gillian Gillett, who was in charge of transportation policy for Mayor Ed Lee, got a call one morning in the fall of 2013 from a colleague at the MTA, asking her to join a meeting with the tech companies. Gillett, tall and blond, was a data analyst and a commanding presence with intricate knowledge of the issues, and little time for posturing. She was often impatient with the city's lack of tech savvy, and specifically its lack of interest in new tools and new sources of data that would very obviously be helpful in solving transportation problems. She could be withering in her criticism of unreliable politicians and incompetent city bureaucrats, but at the same time was personally offended that attacking tech workers was now part of the playbook for local progressives. She was a techie herself after all, by profession and disposition, even if she'd gone to college at St. John's in Maryland, a bastion of the liberal arts.

On this day, though, it was the tech companies that were drawing Gillett's ire.

Google and Facebook and the rest did not need the city's permission to run the shuttles, since private buses are regulated by the state. And they hadn't asked. But from a transportation planning point of view there were obvious benefits to working closely with the city. If the MTA knew the shuttle schedules and routes, it could coordinate with the public buses, adjust traffic-signal cadences and move stops around to better keep the traffic flowing. The bus fleets were also collecting a lot of data on congestion and speeds on city streets—information gold for a data analyst like Gillett.

Carli Paine, a friendly MTA staffer with a degree in urban planning, had been charged with trying to coordinate with the tech companies and stem the growing number of complaints and conflicts with Muni. About fifty people from the various companies were now gathered in the big conference room on the third floor of the MTA building. Paine was politely trying to explain, again, why the city needed them to share their bus routes and schedules, but she wasn't getting anywhere. The firms had sent government relations representatives, whose jobs were essentially to refuse anything that wasn't mandatory, and they were stonewalling; the law said they could mostly do what they wanted, and that's what they were going to do.

When the Facebook representative began lecturing about how sharing bus schedules would be an invasion of privacy for the people on board—"What if one of the buses was attacked by a terrorist?" she asked—Gillett had heard enough.

"I stood up and said, 'Hey, I'm from the mayor's office. Are you kidding me?'"

She held up the latest issue of *Wired* magazine, which featured a map of the bus routes, cleverly derived from the social media posts of the Google and Facebook employees riding them. "Are you kidding me?" she demanded again. "Everyone knows where your stops are. Politicians campaign at them, right? Recruiters try to poach employees from other companies at the same stop, right? You've got these really big buses, right? It's not intellectual property we're talking about here. And you've got this very kind civil servant here who is trying to help work this out for everyone, including you."

Then she shut the meeting down, just fifteen minutes in.

"You're all going to leave because this is a bullshit meeting," she recalls saying. "You will not meet with the city until you come back with your engineers and your transportation planners." The meeting dispersed, and a few minutes later she got a text summoning her to the mayor's office. "What did you do?" Ed Lee demanded. "You just pissed off every tech company in San Francisco." She explained, and he told her to go home for the day. She retreated to her apartment in the Mission, not sure if she still had a job.

The following Monday, as she was walking to Ritual Coffee, there was a commotion at the intersection of Valencia and Twenty-Fourth Street. Protesters, some of them in clown suits, were blocking a bus, and pranking those on board by pretending they were city inspectors and handing out citations. Led by an art collective called Heart of the City, the protesters delayed the bus for about twenty minutes, while celebrating it all on social media.

Gillett's phone buzzed. It was the mayor: "Come see me now." It seemed the companies were suddenly anxious to meet. Lee told Gillett to get it done. "Within a week we had a meeting that was completely different," she said later with a wry smile.

The protests, though, had only just begun.

Sara Shortt, a social worker and activist, lived in a street-facing apartment on Valencia Street, where the buses were hard to miss: All of a sudden the towering luxury coaches were passing her window seemingly constantly, and as a bike commuter she was all too aware of them from the ground level too. Shortt had been active in housing-rights causes, and in the summer of 2013 she was part of a protest movement called eviction-free summer that targeted specific landlords over Ellis Act evictions, which allowed owners to turn rental buildings into the equivalent of condos, or single-family homes. The total number of evictions citywide had dropped to 1,269 in 2009–2010, but now was marching upward again, reaching almost 2,000 by 2013–2014.

As summer turned to fall and the anti-eviction protests continued, "it wasn't even so much organized tenant groups that were involved at that point," Shortt recalls. "It really was just all kinds of artists, queer students, lefties, who were feeling kind of run over by the gentrification and displacement and watching so many of their friends and neighbors getting evicted."

She doesn't recall exactly how discussions pivoted to targeting the Google buses, but it was an easy transition. "The Mission was ground zero for a major proliferation of Google buses, and it was ground zero for Ellis Act evictions," recalls Shortt. "People saw that." The word on the street was evictions were especially concentrated near the shuttle stops.

"The buses represented a huge class divide—the city privileging wealth, tech wealth," said Leslie Dreyer, an artist and activist who was also among the protest leaders, in a 2017 interview with the author Cary McClelland. As the buses proliferated, it was "on the tip of people's tongues. At every party: 'fucking Google buses.' "

The idea for the initial protests was to be lighthearted and funny, in contrast to the angry tone of many eviction protests. "These activists were all dead serious," says Shortt. "But we were in such an absurd historical moment."

Dreyer had called the MTA to find out the rules on using bus stops and learned that there was a $271 fine for unauthorized use. That was a good angle, and led to the idea of the fake citations and fines for "displacement and neighborhood impact." The first protest, in December 2013, was followed by a half dozen more over the next few months.

The protests drew global media attention and then imitators: Other cities were now seeing commuter shuttles too, and there would eventually be protests in Los Angeles and New York. Luxury vehicles with tinted windows and Wi-Fi whisking their wealthy occupants along what had been working-class neighborhood streets—they were just too perfect a symbol for all the disquiet and resentment the new tech boom had brought for those who weren't sharing in the prosperity. It was fair to argue that the companies should coordinate with the city, but the goals of the protesters went well beyond that.

Leslie Dreyer, for one, had no patience for bus passengers who pleaded that they were just going to work, and weren't to blame for the problems of capitalism. "People are used to making too much money and they don't feel personal responsibility for what's happening," she said in the interview. "They could start an initiative with their coworkers for housing justice, maybe consider giving away a huge portion of their income to mirror the median income of their neighbors." She also wasn't buying the argument that the buses benefit everyone by taking cars off the roads, since people who'd been evicted had to drive back to the city to get to work.

In any case, the city didn't have the power to ban the buses even if it had wanted to—that was a state matter—but there were certainly some practical steps that could be taken to ease the tensions. Gillett began working with the tech companies on how they could adjust their services, while figuring out a way to address one of the biggest complaints: Private buses were using public bus stops—which technically was not allowed, but was hard to enforce—and they should at least pay for the privilege. This was trickier than it seemed, because the city didn't have the legal authority to require a permit. The workaround was to make it a "voluntary" payment and hope nobody challenged it in court.

The companies also agreed to work with the MTA on routes and schedules, and many buses were shifted to bigger streets with stops further apart—changes Gillett attributes in part to the city's transit union, which didn't like the impingement on its turf. The unions in fact had successfully used the ruckus to organize the shuttle drivers and persuade the tech companies to go along with unionization, rather than find cheaper non-union operators—a key dynamic in forging the agreement with the city. Paine, the MTA staffer who Gillett had worked with, then led a big study on how it was all working, with MTA inspectors counting buses and passengers around the city, tracking Muni delays and categorizing complaints.

The shuttle buses were now less convenient, with fewer stops, and ridership declined accordingly—the protests had made their mark. Gillett had succeeded in forcing a deal but was frustrated by the outcome. "White people on buses," she quipped, has been a goal of transit planners everywhere and ought to be encouraged wherever possible.

The cultural conflict ultimately defied negotiation.

For many of the city's artists, musicians, writers and other creative professionals, the tech explosion of the 2010s and the money it brought had become a stultifying, homogenizing force, with a wealthy elite and the empty pleasures of consumption and luxury replacing just about everything else. The dot-com boom had brought money and tech to the city, but nothing like this: It was in the first half of the 2010s, according to dozens of longtime residents in a variety of trades, that San Francisco really began to seem like a different place.

■ ■ ■

Summer Lee was among the many local artists who loved the city for its egalitarian creative spirit, which set it apart from the more commercially minded arts industries in New York and Los Angeles. Growing up in San Mateo, she'd started coming to San Francisco in the mid-1990s, while still in high school, to partake in the city's unique lesbian bar scene. After college at Stanford, she'd gotten a job as a social worker with a group called Asian Women's Shelter, helping the homeless in the Mid-Market neighborhood. The first dot-com bubble was underway, and she remembers the trickle of companies making their way to the neighborhood. "Did they realize what neighborhood they were moving into? Are they going to step over all this suffering to get to their tech jobs?" she recalled thinking. In serving the city's most troubled residents, Lee was following the footsteps of her mother, who'd become deeply engaged with Buddhism and, after divorcing her husband while Lee was at Stanford, established an ashram in the Mission that cared for AIDS patients.

Five years of work in the streets "burned the shit out of me," Lee says, and she decided to switch tracks, enrolling at the venerable San Francisco Art Institute. With a gorgeous campus on Russian Hill that housed one of the greatest works of muralist Diego Rivera, the Art Institute, founded in 1871, had nourished generations of painters, sculptors, photographers, musicians and performance artists, many of whom went on to extraordinary success. Ansel Adams had founded the photo department at SFAI in 1945, and a "New Genres" department was created in the 1970s with the proceeds from the sale of long-lost photographs by Eadweard Muybridge, famed for his studies of motion. It was a beacon that had summoned Stewart Brand, the founder of the *Whole Earth Catalog* and The WELL, to come to the city and pursue a fascination with photography. Influential onetime students ranged from the Grateful Dead's Jerry Garcia to the performance artist Karen Finley to the photographer Annie Leibovitz to the robotics pioneer Mark Pauline, founder of Survival Research Labs. By the late 2000s, SFAI was wrestling with the convergence of art and technology, though not necessarily happily.

"This was the moment where the art world was like, 'We better start incorporating tech,'" recalls Lee, pointing to the sudden enthusiasm for iPads in museums. "There was a lot of cynicism. There was definitely a camp that said, 'I'm going to start using tech.' People were very thoughtful, but

I don't think we understood how much it was going to affect our lives." By the time Lee graduated in 2011, the second tech boom had begun, and Lee sensed that this one would be different. "The second one was where people started naming it, like, 'Oh no, another tech boom, they're back.' "

Ana Teresa Fernández, also an SFAI graduate, was more inspired by the stirrings of art and digital technology. She recalls being taken with the work of French artist Laetitia Sonami, who was teaching at the Art Institute and had created pieces where a long glove attached to sensors would evoke sounds as she moved. "Sound connected to body movement—it was an unexplored area," Fernández says, and she found it fascinating. A lot of people were playing with new forms of digital imagery, and she was even interviewed for a job at Adobe in 2008. "There was a bridge between arts and technology."

The economic and social impacts of tech in the city, though, were another story.

"Around 2010, 2011, when you started seeing the buses coming into the Mission, bringing all these people into the Mission . . . it really started dislocating people, and all the air went out of that bloom, the air of diversity. I go to the Mission now and I don't recognize it," says Fernández. "It used to be Mexican, Salvadorian, Peruvian. Now it's high-end sushi. Everything is upscale. It's diverse, but it's upscale."

The dramatic rent hikes of the early 2010s, she recounted, were draining the creative spirit of the Mission especially and stifling the social scene. "In the early 2000s we would just be moving through the city, you got to engage with all areas of the city because your friends lived there. Then all of a sudden people were pushed out of these areas. I no longer have friends in Baker Beach. I no longer have friends in Nob Hill. Everyone's restricted to the affordable area, and no one moves. Rent control anchored you like the golden handcuffs."

For those determined to make it as commercial artists, Los Angeles beckoned: It was much cheaper and a much bigger market. New York wasn't any less expensive than San Francisco, but it was no longer *more* expensive either and provided a lot more opportunity for someone looking to make a career. Portland was also a popular destination for those priced out. So was the East Bay. There were still refuges of cheap studio space in the city, such as the former warehouse buildings at the Hunters Point Shipyard, and city institutions like the Yerba Buena Center for the Arts provided support. Working artists could still be found in quirky places like the Odd Fellows Building on Mid-Market, owned by an obscure fraternal society. But for a lot of young creatives, the walls were closing in.

Even for established artists with ample resources, the market was changing.

The new rich weren't supporting the arts in quite the same way as their old-economy predecessors had, for reasons that were hotly disputed. A lot of wealthy tech entrepreneurs were quite young and weren't yet putting time into philanthropic endeavors—or if they were, they were opting for causes other than the opera and the art museums. Ev Williams, the Twitter cofounder, cited different cultural norms from, say, the finance industry, where expensive art was a status symbol. "Tech people aren't impressed by fancy art, so they don't buy it to show off to their friends," he says. "In fact, it can be the opposite, you can look dumb because people think that that art is overinflated and doesn't have true value."

Gordon Knox, who'd take over as president of SFAI in 2017, said many wealthy tech people brought a venture-capital perspective to philanthropy that wasn't helpful at all. "They're totally happy with the idea of losing scores of millions of dollars in scores of different enterprises, because if one in twenty works out, that's great. But as soon as you tell them it's actually for philanthropy—i.e., there is no return, this is money to help the community—then it's like, 'We don't do that.' The tech money is just so wildly un-philanthropic."

That sentiment was a common refrain in the city's philanthropic circles; it was easy to see from the donor lists that the bulk of support for art came from the city's old-money families. There were of course exceptions: The Minnesota Street Project, an ambitious contemporary art center in Dogpatch, was underwritten by tech investor Andy Rappaport. Matt Mullenweg, the WordPress founder, would become a patron of Ana Teresa Fernández and several other artists, and open a jazz club in North Beach (in addition to having backed the Bay Lights project). Williams was spending money too. And support for the arts wasn't always a fair measure of generosity: Marc Benioff, for one, had given heavily to schools and hospitals. Michael Moritz's Crankstart Foundation would become one of the city's largest philanthropies and a major supporter of social services programs.

For Knox and many others in the arts community, though, the question was whether the tech world really understood or appreciated what the arts brought, and what it all meant to San Francisco. SFAI's original benefactor was Mark Hopkins, of the Southern Pacific Railroad, and Knox says his vision went well beyond an arts school. "He put it together as a way of encouraging artists to exchange ideas, to create a culture of expression, innovation and communication, and that's really key to what the city became, and what the region started to reflect. It could have been a mining town, it could have been a port town, but it ended up being, instead, a town that has always supported and pushed the edges of how people think, and how they communicate and explore ideas, and that's very much what the arts do."

Summer Lee was a political activist as well as an artist, and when it came to housing, she saw a big problem in the explosive growth of one company in particular: Airbnb. She'd graduated from SFAI in 2011, just as the short-term lodging company was gaining traction and rents were starting to soar.

"San Francisco is a small city with a limited amount of housing stock," she said in an interview at her modest, ground-floor studio in the Bayview, where a swirl of paint and materials for a new exhibition covered the easels and workbenches. "If a significant percentage of it were occupied by tourists, or short-term stays, what kind of city would you have? Where would the funding come from for schools and libraries and basic services?"

Competition for apartments was fierce, not just in the Mission and SoMa but all over the city, especially in dense, older, upscale neighborhoods near the center city, like North Beach, Russian Hill and the Haight. A flush, powerful new rival for rental units was the last thing anyone wanted to see.

Airbnb was by then facing blowback in many cities, and ramping up a high-powered PR and lobbying operation. Ron Conway, the unlikely kingmaker and now self-appointed champion for all San Francisco tech startups, was getting seriously dialed into the local network of political consultants, many of them former City Hall officials, who knew people and could make things happen. This was the deep City Family, one might say, talented operatives working behind the scenes at firms like Barbary Coast Consulting, Ground Floor Public Affairs and Platinum Advisors.

They didn't come particularly cheap, but then again, their fees were puny next to what was at stake for the companies, which in some cases had little future if regulations went the wrong way. Y Combinator didn't teach you what to do when your startup got crosswise with local laws and city officials, and San Francisco was a crucial test case.

Airbnb was determined to maintain the principle that it was a marketplace, not a lodging company, and simply ignored the fact that it was illegal to rent out an apartment for less than thirty days in San Francisco. The Board of Supervisors, though, was confident they had public backing to regulate the company and, in 2014, passed legislation that legalized Airbnb but imposed some rules. Hosts would have to register with the city and pay taxes, and there'd be a ninety-day annual limit on the rental of entire units.

It could have been far worse for the company, with stiffer provisions dropped through months of negotiations before the bill went to a vote. Airbnb CEO Brian Chesky said he wanted to be a good actor: "We're not against regulation, we want fair regulation," he said at the time. "We're trying to help take this from an activity that existed under the table with Craigslist and bring it out of the shadows."

The sincerity of that statement would be tested repeatedly in the years to come, in San Francisco and around the world.

The Supervisors were proud of their compromise, but the new law was difficult to enforce without active cooperation from the company, and progressive activists were unimpressed. Led by the tireless Calvin Welch, they collected enough signatures for a ballot measure that would impose much stricter conditions on Airbnb, including a complete ban on renting full units for less than thirty days. For Chesky, the prospect of such rules in his home city felt like a mortal threat, and the day after the initiative qualified for the ballot, the company was mobilizing for war.

Chris Lehane, a veteran of the Clinton White House who'd built a nice business introducing Washington politicians to Silicon Valley innovators, was on retainer at Airbnb, and he was on his way to coach his son's baseball game on Treasure Island in the spring of 2015 when he got an urgent call, asking him to come by the company's 888 Brannan Street headquarters right away. Sporting his baseball uniform, Lehane arrived and was directed to the "Dr. Strangelove" room—all the conference rooms had themes, this one being a replica of the command center from the legendary dark comedy about nuclear destruction. He was further disoriented to find not the lawyer he'd been expecting to meet, but CEO Brian Chesky and the entire executive team, along with a couple of board members.

"They were freaking out," Lehane recounted later. "I did a lot of ballot work in California, statewide ballots primarily, and I said, 'Look, you're on the "no" side of the ballot. The typical way you win these ballots is, you spend a bunch of money, and the no side wins eighty percent of these. There's a pretty tried-and-true way to do these things.'" But Chesky wanted to see if there was a different approach. "Is there a way we can think about using this as an opportunity to rally and organize our host community and our users, our guests," Chesky asked, according to Lehane. That conversation eventually led to the creation of the first home-sharing club. "We organized the hosts. We communicated with the guests. We had hosts who were precinct captains and going to their neighbors. When we started off, we were close to twenty points behind. We ran a pretty strong narrative that talked about the economic benefits that the city and in particular these hosts were getting. That became the template for the campaigns we started to run around the world."

Such organizing would be an important dynamic in regulatory battles not just for Airbnb but for Uber, Lyft and other companies in what was sometimes known as the sharing economy. It's not clear how big a difference it ultimately made in San Francisco, though. The company had also employed Lehane's tried-and-true approach, spending more than $8 million to beat back the measure—an astonishing sum in a small city, amounting to $75 for every "no" vote. It lost by 55–45 percent. Whatever one might think about the merits of the issue, it was certainly a demonstration of the new Internet industry's political and financial muscle.

Some months later, the Supervisors passed another ordinance that would hold the company liable for unregistered listings, and Airbnb responded by suing the city. It was a cruel twist for the Electronic Frontier Foundation and others who'd been fighting for an open Internet that Airbnb was making a free-speech argument. The company contended that Section 230, crafted to shield online publishing platforms from liability for defamatory or otherwise illegal speech by their users, protected it from responsibility for unregistered listings.

"We got a preliminary ruling in our favor," recalled Aaron Peskin, who'd been reelected to the Board after a seven-year hiatus. "And then the judge put us all in a room together for a bunch of time, and we hammered out a deal." The lawsuit was settled in 2017, as Airbnb began to prepare for a long-awaited IPO, with the company agreeing to force hosts to comply with city law.

Peskin saw the company's actions as part of a pattern among tech start-ups. "They started out as a bunch of immature punks, and eventually started acting like adults." Whatever its disposition, though, Airbnb had succeeded in avoiding rules that would have crushed its business in the city, and later boasted that the number of San Francisco listings had only increased after the settlement, to 7,800.

The Airbnb battle was over in San Francisco, but it would continue to rage elsewhere. Summer Lee pointed to studies showing how damaging Airbnb could be in tourist communities and noted that the California Coastal Commission in 2023 had begun to look into the impacts on beach areas like Pacifica, a coastal town just south of the city. Airbnb, she said, represented "one of the most telling ways that tech 'solutions' gummed up life in San Francisco, bringing a lot of money to a certain sector of the economy while draining the milieu."

A broader battle over housing was heating up too. It was gradually dawning on residents and politicians alike that the city and the state weren't building nearly enough housing, in large part because environmental and neighborhood preservation policies had rendered new construction impossibly expensive and slow. It took twice as long to get a building permit in San Francisco as anywhere else in California, and comparisons with some laissez-faire red states barely fit on the graph. A new, tech-aligned YIMBY movement was beginning to gain strength, advocating for new housing whenever and wherever possible. It would take a few more years before state legislators—led by Scott Wiener, who was elected to the state senate in 2016—would pass a series of laws aimed at forcing localities like San Francisco to green-light more housing.

San Francisco was getting richer and richer as the 2010s marched on, but it was harder and harder to ignore the persistent problems in the streets. Homelessness was creeping up again, with the official count crossing 7,500 by 2017, almost where it was when Gavin Newsom took office in 2003. Even beyond that, there were too many people living in substandard conditions, especially in the cheap SROs and subsidized supportive housing units in the Tenderloin. For fifty years, the center-city neighborhood adjacent to Union Square had been treated as a containment zone for vice and crime, and now it stuck out more than ever as an island of poverty and need, all but abandoned by the city.

There was a busy street community and black-market economy operating across the Tenderloin and Mid-Market, as well as parts of the Mission and the Bayview. It was heavily centered on drugs and fencing stolen goods, with criminal gangs interlaced among the homeless, the addicts and the mentally ill. For years now, residents had grown inured to the problems in the streets, in part because the worst conditions were mostly in and around the Tenderloin, where they could safely be ignored.

Now, though, the Twitter tax break was bringing tech companies and their well-heeled employees into this mix. As artist Summer Lee had noted from her experience as a social worker in Mid-Market during the first dot-com bubble, it wasn't at all clear how they were going to fit in.

The so-called community benefit agreements that companies had to implement as a stipulation of the tax break were supposed to help with that. Tiffany Apczynski at the Danish-founded customer service startup Zendesk had been among the first to break the ice in 2011, after city officials suggested she start out by meeting with stakeholders in the neighborhood.

"We had a meeting in our offices, we just said, 'invite everybody,' and we had about twenty-five community leaders in that room," she recalled. "They were really suspicious of us. But we opened the door and let them tell us what they wanted. Those conversations were, 'Can we get better street lighting?' 'Can we have the garbage cleaned up?'" More intractable issues would come up later.

Zendesk employees were soon volunteering with community organizations around the neighborhood, on company time, and were especially helpful with a tech lab at St. Anthony's, a linchpin of the social safety net in the Tenderloin. Small grants were offered for community gardens and food programs, while Zendesk offices were opened up for art exhibits and neighborhood meetings. "We refused to make our campus an ivory tower, and we made sure everybody had to leave the building to go to lunch," Apczynski added, with pride. "Any catering we did do was local. We made sure it was from the Tenderloin."

At Twitter, it fell to Caroline Barlerin, who'd joined the company from

Hewlett-Packard in 2014, to carry the flag on community development. Twitter had been in 1355 Market Street for a couple of years already, but it had been slow to get its program going, what with all the boardroom battles and the IPO and the mounting political strife on the platform. There had already been protests outside the company offices, led by local groups who thought Twitter was shirking its neighborhood duties, even as Zendesk established itself as something of a darling. "When I got there, Zendesk could do no wrong and Twitter could do no right," Barlerin said later.

She set about talking to people, trying to build bridges and even inviting an anti-corporate activist group called "Market Street for the Masses" to hold a meeting in the Twitter building after their regular space had flooded. "The night before I was like, I'm going to get fired tomorrow because everyone's going to come in and they're going to unfurl the banners and handcuff themselves to the railings. This is the stupidest thing I've ever done." It didn't quite play out like that, but the sixty-odd people who showed up gave her an earful.

"There was a lot of yelling directed at me," Barlerin says. "People were saying, 'You put a gun to the head of the city and you're causing children to be unhoused and why don't you fix the damn elevators in the SROs? Don't you know they're broken?' "

Elevators aside, one tech-related need Barlerin and her colleagues identified was someplace where homeless people with children could go to use a computer, maybe get some training and relax a bit. Twitter wasn't willing to open up its own building, so across the street at the newly renovated Fox Plaza it debuted the Neighbor Nest, a combination day care center and computer lab that was operated by a local social services agency, along with volunteers from Twitter. Employees fanned out on numerous other volunteering tasks, led by CEO Jack Dorsey, who seemed genuinely committed to the service. Dorsey enjoyed being out and about as a regular guy in the city, commuting to Mid-Market by bus from his home in the ritzy Sea Cliff neighborhood, out by the ocean, long after he had the means to travel in higher style.

"Of all the volunteering projects that would be going on, he wanted to do the one that no one would sign up for, like cleaning the beds at the men's shelter," said Barlerin. "When Jack Dorsey signed up for it, we'd get ten more people to sign up for doing the physical work." Dorsey would later give $1 million personally to Code Tenderloin, led by Del Seymour, whose decades of social work in the neighborhood had earned him the nickname "Mayor of the Tenderloin."

Among the small group of women spearheading the efforts for the companies and the city, it was an optimistic moment. There was Tiffany Apczynski at Zendesk, Caroline Barlerin at Twitter, Joan Scott at Dolby and their counterparts at City Hall—especially Amy Cohen, of the Mayor's Office of Economic and Workforce Development, and Supervisor Jane Kim.

"People really had their hearts in it," Cohen said later. "Maybe it was seen as just a gesture by some people, but on the ground they were really doing a ton. I don't think there's a story like that anywhere in the country where companies put that much effort into engaging in the community. People were really working together."

Seymour, for one, while admitting his own bias, said the company efforts really helped, though he wishes there'd been more focus on jobs for those trying to escape the streets. "Employment should have been in the deal," Seymour said later. "It should have been mandatory hire [for people in the neighborhood]." Still, he says he had some success in helping around two dozen people get jobs with tech companies.

Jane Kim said later that her one regret was not insisting that companies forgo their own cafeterias and oblige employees to get lunch in the neighborhood, the way Zendesk had done. Twitter had made its building so plush and self-sufficient, with in-house catering and endless snacks, that there wasn't much reason to go outside. The tax break remained very unpopular in many quarters, despite the modest price tag—ultimately about $70 million in forgone revenue over eight years—and the neighborhood revival, however uneven, it had brought.

Yet even if the deal had been different, after a couple of years it had become pretty clear that all the new economic activity, all the efforts of the city and all the corporate volunteers couldn't solve a fundamental problem: How does a resident population that includes many homeless and indigent people, and a lot of drug use and mental illness, live tranquilly alongside upscale office workers and wealthy executives and investors?

"I'm preaching it, that everyone can coexist, but you can't have illegal activity or scary or violent activity coexist with everyday business," says Cohen. That was especially true later in the decade as fentanyl began to replace heroin and crack as the drug of choice on the streets.

The companies couldn't be expected to solve deeply entrenched social problems like homelessness and drug addiction, but they were clearly contributing to the growing chasm of income inequality that sat at the root of so many problems. San Francisco was committed to humane approaches to people who were suffering, but without backup from the state and federal governments and much more accountability from city agencies and their contractors, it was hard to make it work.

The companies in Mid-Market could—and sometimes did—simply leave the neighborhood when they got impatient with the progress. For San Francisco's City Family, though, and for the national Democratic Party that it did so much to shape, there was no such easy out. The city's inability to address highly visible disorder in the streets would have devastating political consequences in the years to come—for San Francisco, and for the country too.

Even as it worked to get onside with the city's Mid-Market efforts, Twitter was fending off a barrage of problems online. Like Facebook, the company had been forced to develop extensive "moderation" systems to try to block abusive behavior and content theft that would otherwise mar the service. But those efforts often seemed to create as many problems as they solved, and because it was all so messy and didn't contribute to the bottom line, it often didn't sit very high on the priority list. Trust and Safety boss Del Harvey was inventing policies and procedures on the fly and constantly begging for better tools.

Harvey had been hired back in 2008 to help with problems that were perceived at the time as annoying-but-harmless spam. It quickly became obvious, though, that hate speech, stalking and other forms of online abuse were a major issue. It had all become even more fraught as Twitter found itself the medium of choice for politicians and political movements—first the Arab Spring, and then the much closer-to-home Black Lives Matter. In 2012, picking up on a phrase coined by a manager in the UK, then-CEO Dick Costolo declared that Twitter was "from the free speech wing of the free speech party." The limitations of that simple-sounding philosophy would quickly become obvious.

Twitter was the hub for news and outrage over the 2014 shooting of Michael Brown in a St. Louis suburb, which gave rise to the Black Lives Matter movement. It became the place where the national social justice conversation got its oxygen. Jack Dorsey, himself a St. Louis native, had shown up in Ferguson, Missouri, where Brown was killed, and endorsed the idea of being "woke," long before the term became a weapon in the culture wars. Many Twitter employees, firmly in the tradition of San Francisco's Internet industry, were deeply inspired by the idea that their product might change the world for the better. Good policing of the platform would be part of that—advertisers also wanted to stay far away from any nastiness—but there was no playbook.

Harvey could be intimidating, with a severe visage and powerful physical presence that belied her five-foot-six stature and slim, muscled frame, but she also had a wicked sense of humor, and a fierce commitment to the mission.

"Twitter was in a lot of instances the first mover in terms of how certain types of trust and safety content was handled," says Harvey. "People worked at Twitter because they were like, 'Hey, you guys actually care about this stuff.'"

As Harvey would soon discover, though, caring didn't provide much defense against high-powered political attacks.

Craigslist had provided an unhappy preview of what was to come in the content moderation wars back in 2009, when it found itself in a pitched battle with the authorities over its "erotic services" section. It was pretty much what it sounded like but attracted little attention until Craigslist had spread far beyond its San Francisco roots and begun to reach into every corner of America. When a man killed a woman in a Boston hotel room in 2009 after responding to an online "escort" ad and attacked two others who managed to escape, he became "the Craigslist Killer." Breathless media coverage then brought political outrage and law enforcement scrutiny to what had been a feel-good community bulletin board—and an end-of-innocence moment for Craig Newmark. The activities being offered certainly appeared to be illegal in many cases, though under Section 230, Craigslist wasn't responsible for the ads posted by its users. Grandstanding politicians insisted the site was serving sex traffickers as well as willing adults, and a daunting array of eighteen attorneys general and a congressional committee hounded the company with threats and lawsuits. New incidents were emerging of crimes stemming even from routine Craigslist transactions like selling a car; the trust that enabled early online communities was breaking down as everything scaled up.

To Craig Newmark, the political attacks—not to mention the sensational news stories coming from outlets that blamed him for the woes of their businesses—were highly disingenuous. The company was a willing and helpful partner in law enforcement investigations, he noted, and was in continuous communication with various authorities on how to root out criminal activity. Furthermore, barring people from posting on his site would surely drive the activity to more dangerous corners of the Internet. For allies, including the EFF, who filed friend-of-the-court briefs in defense of Craigslist, it was an important matter of law and principle for people to be able to post freely, and have them—and not the website hosting the messages—bear responsibility for the conduct. A lot of libertarians didn't think sex work should be illegal in the first place.

Craigslist eventually backed down, first eliminating "erotic services" in favor of "adult services," and then after a new federal law was passed took that down too—with a black "censored" banner where it had once been.

For Newmark, looking back on it years later, it was all a depressing lesson—and a warning of what was to come.

"Trust and safety on the net isn't what it should be, which is a double-edged sword in all sorts of ways, because bad actors who are looking for PR can always complain about it," he said. There'd always be not enough policing, or too much, or something that people aiming to score points could make a fuss about. "I learned the hard way between 2010 and 2012 that no good deed goes unpunished, and that anyone who's fundamentally honest will never be able to successfully confront someone who lies for a living. In my naivete, I thought good work stands for itself. I was very wrong."

As the Internet became an increasingly central part of people's lives around the world, the arc of history was not bending toward the San Francisco architects of the independent Web.

San Francisco progressives had rallied around Bernie Sanders in 2016, though most local Democrats—and especially those in the tech industry—were all-in for Hillary Clinton, who had clinched the nomination before the California primary took place. She was the anointed heir to the beloved Obama and had inherited the support of Ron Conway and sf.citi and most of the local political donor class, including the Gettys. With

Donald Trump as the Republican nominee, and Sanders running further to the left than any national politician had dared in half a century, backing the prospective first woman president was a no-brainer for many Democrats and moderate Republicans. Even Marc Andreessen managed back in 2016 to tweet, "I'm with her."

The relationship between the Obama administration and the tech world was strong, with talent now flowing in both directions. Obama's campaign manager, David Plouffe, would become Uber's top lobbyist and policy advisor, and later an advisor to Mark Zuckerberg. Obama's press secretary, Jay Carney, would take a top job at Amazon and later move to Airbnb. Twitter's general counsel, Alex Macgillivray, joined those moving in the other direction when he went to Washington to serve as deputy CTO under Megan Smith.

Hillary Clinton was fifteen years older than Obama and never shared his affinity for technology. Yet it was only fitting that San Francisco would have a strong hand in trying to elect a woman to the White House. With locals Dianne Feinstein and Barbara Boxer holding the state's two US Senate seats, and a long tradition of feminist activism that extended to the working conditions of strip club dancers, the city had been at the vanguard of women's rights for decades. The traditionalist Christian dogmas that a woman's place was in the home, and that abortion was a sin, had been all but banished as acceptable views in the city. Questioning gay rights and gay marriage was similarly off the table, and that came to extend to all the letters in LGBTQ; the gay community was foundational to contemporary San Francisco, and if transgender people needed help, well, San Francisco would be there for them too. The same went for undocumented immigrants.

The Tompkins-Buell penthouse at 2500 Steiner Street would become a key fundraising hub for the Clinton campaign, with soirees featuring Clinton, Obama, Nancy Pelosi, Gavin Newsom and a predictable who's who of tech CEOs. Ron Conway's luxury spread was just a couple of blocks over and he was hosting events too, flexing political influence that was starting to extend beyond the city.

But that particular set of parties would end in abrupt and shocking fashion.

Hardly anyone in local politics foresaw Trump's victory in 2016. A man whose violent misogyny was immortalized in the phrase "grab them by the pussy," who'd kicked off his campaign calling immigrants "rapists," who'd never hidden his contempt for the poor or the different or the disabled—this man was not going to find friends in San Francisco. He didn't even believe in the technology-driven global economic system, and trade with Asia in particular, that had made San Francisco, and California, so prosperous. He was actively hostile to US higher-education and to the scientific and intellectual culture that had created Silicon Valley and made San Francisco a juggernaut.

If you were to invent a person to repel San Franciscans, you'd end up with someone pretty close to Donald Trump.

There were a few dissenters. Peter Thiel was a rare venture investor who didn't follow the herd, and he made a thousand headlines when he threw his support to Trump and spoke at the 2016 Republican convention. Thiel's firm, Founders Fund, was based in the Presidio, and he owned a home in Pacific Heights, but he was distant from the SoMa scene and would soon relocate to Los Angeles—partly to avoid the stink eye he'd get in San Francisco for being a Trump supporter. At Facebook, Palmer Luckey, a young entrepreneur who'd sold his virtual reality company to the social media giant, was enraging his colleagues with pro-Trump Internet memes, and would be pushed out of the company for making trouble.

Another contrarian was Louis Rossetto, the *Wired* founder, whose politics were always libertarian and who'd been known to complain about liberal political bias in his magazine. He never endorsed Trump publicly. But later on he'd be thoroughly bought into the Trumpian critique of Democratic politics.

It was a grim and confounding irony for San Francisco's Democratic Internet industry that of all the politicians out there, it was Trump who proved the master of social media, and especially Twitter. A medium that was supposed to empower the people had instead become a megaphone for a man openly dismissive of the Constitution and determined to rule as a king. Twitter's efforts to manage that contradiction would help bring it to ruin.

CHAPTER 9

The Late Empire

Even amid the mounting strife over gentrification, Mayor Ed Lee made a lot of friends with his cheery, modest, man-of-the-people style and appreciation of the city's quirks. He championed the heartwarming adventures of Batkid, a nine-year-old with leukemia who was made a San Francisco superhero for a day in 2013, an event that inspired national headlines and a documentary film. He was always ready to cut a ribbon, the oversize scissors kept close at hand. John Law, the urban adventurer and Burning Man cofounder, recalls Lee's earnest efforts to help rescue the city's last "Doggie Head"—an amusing mascot that once adorned dozens of outlets of a now-defunct Bay Area restaurant chain, and had become an icon of sorts for underground artists. A longtime North Beach resident, Law had known Lee slightly since the International Hotel protests decades earlier, where the future mayor had been a part of a failed effort to stop the demolition of an SRO on Kearny Street that was a lifeline for low-income Filipinos. "He had a real interest in local history," says Law, recalling that Lee showed up for several events where the carefully restored, ten-foot-tall Doggie Heads—Law had three of them mounted on a trailer—would be brought out to schools. Only in San Francisco would there be a movement to save six-hundred-pound fiberglass sculptures of a floppy-eared hound, but Lee was a supporter. "You must love these events, it's the only time people aren't yelling at you, right?" Law teased.

He hadn't wanted to be mayor, Lee confided to Law. But he was definitely having fun with it.

In truth, being mayor carried some serious perks, and Lee's aw-shucks approach to his status belied the fact that he was quite happy to take them, especially when it came to traveling in style. In China, and in many US cities with significant Asian populations, Lee was treated as an A-list celebrity, recalls Francis Tsang, his deputy chief of staff. In San Francisco, and throughout the US, Asian Americans had been slow to accrue political power to match their demographic, economic and cultural might. Lee was an important symbol of progress on that front, and maybe the avatar of bigger breakthroughs to come. It was no small thing to be the Chinese American mayor of a major US city, especially one that in the 2010s was the envy of cities around the world, and Lee was routinely greeted with pomp and reverence. In 2013 alone, he would travel to China twice, and visit South Korea, India, Ireland and France—all on official business. San Francisco's "sister cities" abroad—including Paris, Shanghai, Haifa, Israel and Cork, Ireland—apparently needed a lot of support from their California sibling: The mayor led a 2017 delegation to Cork that counted no fewer than seventy-nine people, including three supervisors. "Globe-trotting Mayor Makes Himself Scarce in SF," read the *Chronicle* headline in September 2017, with a photo of a grinning Ed Lee at a Hong Kong opera premiere.

Lee felt he'd earned it, what with his conscription into the job. And he had plenty to brag about when it came to the business of San Francisco.

In just five years, the city's economy had almost doubled in size, at least according to inexact measures of its "gross domestic product." Local government got much bigger as head counts and wages climbed and spending on social services—often administered by nonprofits under contract with the city—rose alongside tax revenues; the city budget hit $10 billion in 2017, up from $7.3 billion five years earlier. Unemployment, the big bugaboo when Lee took office, was almost nonexistent. On Ron Conway's simple calculus of "jobs equal prosperity," the city was in great shape. The streets thrummed with commerce and culture, however unequally distributed.

The housing squeeze and conflicts over gentrification were still a major issue, but Lee was also proud of his record on housing, with some twenty thousand units built during his tenure, a third of them subsidized to be "affordable." He led a 2012 ballot initiative that created a Housing Trust Fund to prevent evictions and develop affordable housing, which proponents say kept thousands of people in their homes.

One of the beneficiaries was Chris Carlsson, the activist and Critical Mass founder, who in the early 2010s was living in an old six-unit, rent-controlled Edwardian in the Mission, along with other artists and activists.

The elderly landlord was motherly toward her tenants, and her eccentric affections for the local birds led to it being dubbed the Pigeon Palace. In 2015, she was declared incompetent by a conservator and the building put up for sale. Carlsson's friend Randy Shaw, founder of the Tenderloin Housing Clinic, suggested the building would be a good candidate for the trust fund, and after theatrical street protests against the proposed sale of the building to a developer, it was acquired by the city. Carlsson and the other tenants were spared a near-certain eviction.

Still, even with the booming economy and earnest efforts by Lee and the progressive-led Board to address the affordability crisis, all was not well in the newly crowned tech Camelot—not by a long shot.

On top of the intractable strife over gentrification and the ongoing homelessness problem, the national political warfare that had begun building in the early 2010s was now roiling San Francisco, with the Black Lives Matter movement pushing racism to the top of the city's political agenda. San Francisco may have been a civil rights bastion, but the old sins of redevelopment and police brutality still stung, and the Black population continued to shrivel in size—now just 5 percent of the population, down from a peak of 13.5 percent in 1970. Greg Suhr, who'd been captain of the Mission station during the first dot-com boom, was a well-liked police department veteran who'd built strong relationships in the Black community, and he was named police chief in 2011. But racial tensions simmered, and they exploded when racist text messages among policemen emerged as part of a federal investigation in 2015. When San Francisco police officers in 2016 shot and killed a Black person in questionable circumstances for the third time in six months, Suhr was ready when Lee asked for his resignation.

The bald, broad-shouldered Suhr looked the part of the good-ole-boy cop, and he was a City Family survivor, a comer from early on who'd been sidelined for a time over his handling of a domestic violence incident involving a female friend, but eventually rising to chief. Despite the look, he was a liberal-minded leader who'd overseen many reforms, and he remained personally popular across most of the city's factions. He'd never himself fired his gun in decades on the streets. But he couldn't overcome the shadow of the callousness and racism that persisted in the department, and there was little resistance to his departure. The national racial awakening was the priority of the moment—and the political strife that would come along with it was about to enter a new and darker phase, both locally and nationally, with the election of Donald Trump.

Trump's shocking win in the 2016 election abruptly put Ed Lee and the rest of the city government shoulder to shoulder with the Black Lives Matter movement and what came to be called the resistance. The immediate fight was over immigration; San Francisco was a "sanctuary city," meaning local law enforcement would not cooperate with federal immigration agents or turn criminal suspects without papers over to the feds for deportation. Trump was determined to break through that impediment, and the city quickly engaged the fight, becoming the first city to sue the president over his efforts to force cooperation by withholding funds.

Trump's ascendance energized the local left—a clear enemy was now in their sights—and San Francisco progressives capitalized on the mood to strengthen their position over the moderates. Aaron Peskin of North Beach, their master tactician, was now back in the arena, having been elected to the Board again in 2015; the city's two-term limit only barred consecutive terms. In 2018, and especially in 2020, voters backed progressives for the Board. The left also gained control of the DCCC, the Democratic Party committee that had the power to grant candidates the big D endorsement—a key lever in more obscure local races. Democrats were united in their resistance to Trump, though the leftward shift of the party that came along with that would have a variety of consequences in the years to come.

The tech industry was ready to rumble when it came to attacks on immigrants, who had been central to its success for generations. After Trump announced a ban on all travelers from Muslim countries as one of his first acts as president, the statements of solidarity from tech CEOs came fast and furious. "We should also keep our doors open to refugees and those who need help," wrote Mark Zuckerberg, citing his own immigrant in-laws. "That's who we are." Uber's Kalanick, while insisting the company would be "standing up for what's right," nonetheless accepted an invitation to Trump's corporate advisory council, only to back out after an employee uproar. Lyft, seeking the high ground, declared: "We stand firmly against these actions, and will not be silent on issues that threaten the values of our community."

Trump's election, though, had produced another more complex and far-reaching dynamic, one that would prove far more consequential than the industry's stance on immigration policy: It shattered the national social consensus on the benefits of technology.

Trump's ascent was so contrary to what political professionals thought they knew about the American electorate that it begged for a hidden explanation, something other than the policy differences or debate performances or personal characteristics that typically led people to choose one candidate over the other. Backlash against the burgeoning social justice movement wasn't yet obvious. The deep despair in many corners of the country over an economy and culture that didn't include them remained invisible from the coasts. There had to be something else driving the electorate. An obvious candidate—and scapegoat—was social media.

In the four short years since the previous election, social media activity, now happening mostly via smartphones, had exploded. Everyone had a Facebook account, and the social network's "news feed" had become a roiling stew of real-time information—much of it of uncertain provenance. Twitter had seized a powerful niche as the place where journalists kept up on what was happening and collectively decided what was important. *The New York Times* had long dictated the mainstream news agenda: At the *Los Angeles Times* back in early 1990s, we'd get a faxed copy of the front page of the first *New York Times* edition, which came out the night before and ahead of our West Coast deadlines, to see if there was anything we'd missed. Now it was Twitter that prompted the questions from top editors: *Did you see that tweet? We need to chase that!*

Trump was a Twitter maestro, constantly seizing the news cycle with shouting, all-caps missives that in their rawness tapped the live nerve of a network whose design rewarded argument and volume. His campaign team was shrewd in understanding the power and potential of Facebook, and eagerly agreed when the company offered to send staff to help. Clinton's campaign, foolishly, declined the invitation.

Still, better social media tactics didn't seem like an adequate explanation by themselves for Trump's win. In the spring of 2018, two British newspapers and *The New York Times* seemed to fill in some blanks when they reported that a political consulting firm called Cambridge Analytica had gotten its hands on the Facebook data of more than fifty million people and used it to help the Trump campaign. Cambridge Analytica was partly owned by Robert Mercer, a right-wing hedge-fund billionaire who was pouring money into Trump's bid, and suspicions swirled that it was part of a Russian effort to tip the election. Federal prosecutors in Washington, led by Robert Mueller, the former US attorney in San Francisco, were already investigating the Russian influence campaign and whether Trump and his team were part of it. Cambridge Analytica could be a smoking gun.

The blowback against Facebook was immediate, with the company losing some $100 billion in market value as a "delete Facebook" campaign gained traction. Congressional leaders demanded explanations from Mark Zuckerberg as the negative press coverage mounted.

To many in the tech world—and certainly in the executive ranks of Facebook—the attacks on the company were misguided at best, conflating sloppy handling of data and routine electioneering with a grand Russian conspiracy. Facebook had inadvertently allowed an unscrupulous academic to get access to data about user actions on the site, which then made its way to Cambridge Analytica and the Trump campaign. But there was never much evidence that any of it played a meaningful role in the election.

The turn against the tech industry writ large was more gradual, but there was a palpable shift in sentiment. Big Tech was just *too* big. It was marginalizing the independent Web with what were being called "walled gardens"—places like Facebook, where you never had to venture beyond their pages. These platforms were monitoring and manipulating our behavior and selling our information. Critics called it surveillance capitalism.

The blogging revolution that had brought so many new voices and styles to the media conversation was withering by the mid-2010s as Facebook and Google vacuumed up all the advertising dollars. Federated Media, John Battelle's blog advertising network, had struggled through the ad slump that came with the Great Recession, but as the business began to pick up, the more existential threat emerged. "Our sales force kept losing deals—to Facebook," recalls Battelle. His ad reps related that when they took clients out and got drunk, all they wanted to talk about was Facebook. "They're all saying, it's just so easy, and you can get exactly the audience you want. And, you know, it's cheaper than you guys." The dramatic rise of smartphones as the main means for surfing the Web also hurt. Federated diversified into automated "programmatic" advertising, and the blog advertising network was sold to a TV broadcaster. "The independent Web was starting to falter and Federated was the canary in the coal mine," says Battelle.

San Francisco's local media was suffering too, as advertising dollars moved to Google and Facebook. *San Francisco* magazine, a large-format glossy, had a good run of journalistic excellence before converting to a more lucrative catalog-like title focused on high-end luxuries. The *Examiner*, the *Bay Guardian* and the *SF Weekly* were barely husks of their former selves, and the *Chronicle* continued to suffer from shrinking print ad dollars and a weak digital strategy. *The Bay Citizen* had died along with its patron, Warren Hellman. Innovative efforts like *SFist* and *Hoodline* weren't getting much traction. Kara Swisher and Walt Mossberg at Recode, the company they'd spun out of *The Wall Street Journal*, were doing very well with their high-powered Code conferences, but the Web publishing business was another matter. "Google and Facebook were eating all the juicy bits," Swisher lamented later. "We couldn't figure out how to make money on the journalism side."

The smartphone economy was now having a big impact on many aspects of urban life, some of it for the better and a lot of it for the worse. Amazon and all the new delivery services were beginning to undermine brick-and-mortar retail and sit-down restaurants, though many of the impacts would only be visible later. The streets were clogged with Uber and Lyft drivers, who often came from out of town to compete for rides and slept in their cars. The tech industry had driven up the price of everything—especially real estate—and pushed a whole generation of San Francisco residents to seek cheaper locales in the East Bay and elsewhere.

Tech's promise of personal empowerment and social connection, in the meantime, had instead yielded political strife, culture wars and Internet oligopolies that were impossible to escape. There were disquieting changes in day-to-day interactions as everyone looked at their phones. The resentment could be very personal.

"One of the curious things about the crisis in San Francisco—precipitated by a huge influx of well-paid tech workers driving up housing costs and causing evictions, gentrification and cultural change—is that they seem unable to understand why many locals don't love them," wrote the essayist Rebecca Solnit, a fierce critic of tech's impact on the city, in 2014. Some years later she lamented the social isolation that had arisen, paradoxically, in the connected world. "I think of what has come to my city as 'the great withdrawal.' People on the street often seem to have their eyes elsewhere, usually on their phones: they might video a crime, but they might also not notice it's happening."

The sorrow of natives like Solnit could easily be dismissed as gauzy nostalgia for an imagined good-old-days. But she was far from alone in her sadness for a city that seemed less and less about people, and more and more about money. Between the anxiety over housing, the growing cynicism about the city government and the tech industry, and the culture war being waged by Trump, San Francisco's Internet era was becoming a lot less fun.

The change-the-world optimism about the wonders of tech and a new style of company was fading fast by the late 2010s. John Battelle, after leaving Federated Media, launched a venture called NewCo to celebrate mission-driven companies. It put on events in cities around the world, where attendees could visit with various businesses and see how they did things differently. The festivals, as they were called, were quite popular. But the only way to make it work as a business would be to tap the dollars that big companies were committing to diversity, equity and inclusion initiatives and other corporate responsibility programs as the social justice conversation gained traction. That was not what Battelle had in mind.

"In the 2017, 2018 period, there were a lot of large organizations who wanted to tell a story of being mission-driven, and they would spend millions of dollars, like, setting up elaborate proofs of their 'wokeness' at Davos," recalls Battelle. "The only way to save [NewCo] was to tack hard into the budgets that existed, to follow the money. But it was all window dressing." The ideals of the early Internet industry had been fully co-opted into corporate sloganeering. Subsequent events, alas, would make it quite clear that tech company commitments to a purpose beyond financial success were almost always a transitory matter of convenience.

For Ed Lee and the City Family, Trump's election presented a unique kind of threat, far less visible than his antics on immigration but in their own way more ominous. A Republican in the White House meant a new attorney general atop the Justice Department, which might lead to new public-corruption investigations of the local political machine. Trump had made a campaign issue of the accidental shooting of a tourist on the Embarcadero by an undocumented immigrant, and he rarely missed a chance to bash the city. He'd seem to have every reason to go after its Democratic politicians.

Federal prosecutors had spent many years investigating Willie Brown, whom they viewed as a master practitioner of pay-to-play politics, but had never been able to make a case. Brown was out of office now, though, and a less appealing target for prosecutors. But other skeletons loomed.

An FBI investigation into a notorious Chinatown gangster known as Shrimp Boy had turned up an outlandish scheme involving Leland Yee, then a state senator, to funnel money into his campaign through the sale of guns imported from a Philippine rebel group. Yee would plead guilty to racketeering and fraud in 2015, and one thing tends to lead to another in a criminal conspiracy investigation. A Bayview political operative named Keith Jackson had also pleaded guilty for his involvement in Yee's machinations, and the probe then revealed separate crimes by Jackson—a Brown protégé who'd gotten elected to the school board—as well as an associate of his named Zula Jones. She would eventually plead no contest to extorting $20,000 from a developer, to be used as a campaign contribution to Ed Lee. "You pay to play here," Jones allegedly told an FBI informant.

That's just what everyone feared—that getting things done in the city required the greasing of palms.

Keith Jackson had been part of a Bayview network of nonprofit groups that got a lot of money from City Hall, but often didn't seem to be spending it in the community. All over the city, social service providers that depended on government funding engaged in legally questionable lobbying and favor trading, essentially spending taxpayer money in order to get more of it. The Department of Building Inspection, a relatively new city department, appeared to be in the pocket of the construction companies and developers it was supposed to police.

There was a lot of corruption out there, if anyone really started looking.

Lee, though reliably chipper in public, was feeling the pressure of a job he hadn't asked for in an endlessly contentious town. He was often in serious pain from gout, those close to him say, dimming his enthusiasm for making the rounds. More worryingly, the corruption investigation in all likelihood would be coming for him and his closest associates. He was again looking forward to retirement.

Lee was shopping at his local Safeway late one evening in December 2017, a regular guy despite all, when his heart stopped beating. He was revived briefly as an ambulance raced him to San Francisco General, the sprawling old city hospital on the edge of the Mission, where his wife and a few aides would arrive minutes later. Doctors worked frantically to revive him again in the emergency room, but it was for naught. Lee died a bit after one o'clock in the morning as a crowd gathered outside the hospital, and soon the whole city would be sharing in the sadness. Ed Lee was a good man, most people agreed, dedicated to his city and trying to do the right thing.

His friends, and his wife, Anita, were sure that being mayor had killed him. He was only sixty-six.

London Breed, the president of the Board of Supervisors and a loyal member of the City Family, automatically became mayor upon Lee's death, which was convenient for Willie Brown and his allies. Brown had passed the reins to Newsom, who had then handed off to Lee. Many senior staffers—Steve Kawa, Tony Winnicker, Phil Ginsburg, Rudy Nothenberg and Mike Farrah, to name just a few—had served in multiple administrations. There'd been continuity in the mayor's office for more than twenty years. Now the torch would be passed to the next generation of the Family, without even the bother of a difficult election—so long as everyone played their cards right.

Breed had gotten her start as an intern in Brown's office after a friend had recommended her as a promising up-and-comer, and she'd been on the fast track ever since. Brown put her in charge of the African American Art & Culture Complex, a city-controlled nonprofit, which she was credited with revitalizing. She was sent out to Treasure Island to front for the development efforts there, and added posts on the Redevelopment Agency Commission and later the Fire Commission—oversight bodies that were well-used tools for political advancement.

Breed had an inspiring personal story of growing up in the Western Addition, one of the city's poorest and most violent neighborhoods, and a depressing remnant of the vibrant Black community that once existed in the adjacent Fillmore. She never knew her father and was raised by a stern and loving grandmother, along with three siblings. Her sister would die of a drug overdose, and one of her brothers would land in prison for murder, a poignant and sometimes-complicated aspect of her political career.

"I'm from the 'hood," she declared as she ramped up her first campaign for the Board of Supervisors, in 2012. "I spent more than half my life in what was considered the worst public housing in the city. Taxis wouldn't come there. People wouldn't come to my house. I saw my first homicide when I was twelve."

Greg Suhr, then an SFPD officer, remembers Breed as the little girl who lived in a second-floor apartment in the projects, which he frequented in search of her brother.

"There was a lot of violence in the Western Addition in the late 1980s, early '90s, so we were assigned out there a lot," recalls Suhr. "Her brother Napoleon, Sonny Boy, he was a really good crook back in the day, a little smaller and a little smarter than his brother . . . there were so many chases, fights, whatever. I arrested Napoleon so many times." Breed's grandmother was "a great lady," Suhr recalled, trying to keep her family safe in very tough circumstances.

Tom O'Connor, then a rookie fireman in a Western Addition station, remembers Breed from those days too. She had an aunt with special needs who liked to hang around the fire station, and every afternoon a teenage London Breed would come by to pick her up. O'Connor says it was clear she was a special person in the neighborhood, protected even by the gangbangers. If somebody had a chance to get out, to make it in the big world, everyone was in for that.

"I always felt that there were people trying to protect me, whether it was my grandmother or my brothers or folks in the community," Breed said later in an interview. "So there was a bit of, 'You can't hang out over here. You can't be here.' It was pretty much everybody, and it was interesting to think about it. I did feel protected in the community where there was genuine love and respect and support."

Despite impeccable City Family credentials, Breed had been an underdog when she ran for supervisor in the district that included the Western Addition and the Haight. It was among the most left-leaning parts of the city, and the incumbent, Christina Olague, a housing-rights activist appointed by Lee, appeared more aligned with the electorate than the centrist Breed. But Breed prevailed after a nasty battle. The most memorable moment of her campaign was an expletive-filled rant about how she wasn't a pawn of Willie Brown.

Having by then been tapped as Board president by her peers, Breed was reelected in 2016 over Dean Preston, a wealthy housing-rights lawyer who was endorsed by the Democratic Socialists of America. Her ascension to the interim mayor's post after Lee's death needed to be endorsed by a majority of the Board, though that looked like a formality.

But Aaron Peskin had other plans.

As in 2011, whoever was interim mayor would have a big advantage in the special election scheduled for June. Peskin did not want that person to be London Breed. He was holding out hope that a progressive board colleague—maybe Jane Kim, or Mark Leno—might be the next mayor. But neither of them had enough support to win the six Board votes needed to secure the interim mayor slot. Determined to deny Breed and the Brown machine any advantages, Peskin then made a shameless deal, throwing his backing, and that of his progressive allies, behind conservative Supervisor Mark Farrell. Hillary Ronen, who represented the Mission, passionately denounced Breed as the candidate of "white, rich men, billionaires in this city who have steered the policies of the past two mayoral administrations, if not more," a direct reference to Ron Conway. But it was an odd argument to make in support of Farrell, a rich, white venture capitalist, if not a billionaire.

Farrell was duly sworn in as interim mayor, to furious protests from the Black community, though he'd promised not to run for the full term, and in this case the pledge would stick. There would be an election in five months, and Willie Brown mobilized the Family: Elected officials, senior city employees, union leaders and assorted operatives were summoned to a January meeting at the Far East Cafe, a venerable banquet hall in Chinatown. "There were like eighteen of us in the room," recalls one participant. "Willie said, 'Okay, we're going to do what we've been doing since 1992. London Breed's our pick. Now all of you need to make a commitment right now.' And he went around the room and all of a sudden we had like $1.4 million." That would be used for an independent expenditure committee supporting Breed, a group that was not allowed to coordinate with candidates—and whose donors did not need to be disclosed.

Breed would narrowly defeat Mark Leno in the special election. She'd have to wait six more years for a direct shot at Farrell.

Homelessness remained the city's most visible and insoluble problem, and a constant reminder of the class divide. The number of people living on the streets or in shelters had been ebbing and flowing since it first became an issue in the 1980s—growing worse during the first dot-com bubble, then declining during the 2000s, under Newsom. But it had never sunk below about five thousand, at least according to the annual "point-in-time" count that many believed missed a lot of people. Now, with the relentless rent increases and population growth that began in the early 2010s, the numbers were climbing again, with a rising percentage of people living in the alleys and on the sidewalks rather than in shelters. With so much tech money in town, it was shameful to see so much suffering in the streets. Maybe the moment was finally right for a big swing at the problem.

Jennifer Friedenbach, the veteran advocate who led the city's Coalition on Homelessness, was constantly on the lookout for sources of money, and now the stars seemed to be aligning for a significant new tax. The City Charter required that taxes be approved by a two-thirds majority of voters, but there was some wiggle room in the language, and after a lot of consultations—with allied nonprofits and politicians, but also business groups—she led the creation of what would be known as Prop C, a new tax on large corporations that would generate $300 million a year for homelessness programs. Trump's big corporate tax cut in 2017 turned out to be an excellent talking point for the 2018 Prop C campaign. "We put this forward and said look, this is just replacing the taxes that the corporations are not having to pay anymore," Friedenbach recalled.

Ballot measures had long since become a defining feature of San Francisco's governmental system, giving residents direct decision-making authority over many big issues, and many very small ones too. It wasn't uncommon for an election to feature a dozen or more such measures, some of which were simply new laws, but some of which amended the City Charter, which defines how the city is organized and run. The mayor, or a group of supervisors, could put measures on the ballot, and they often used it as a way to sidestep opposition or promote their own priorities. Prop C, though, came about the old-fashioned way, or at least the way reformers had envisioned more than a century earlier when they made direct democracy a key feature of California's political system: Friedenbach and her allies pounded the pavement with a petition. They needed signatures representing at least 5 percent of the votes cast in the most recent mayoral race, which meant about ten thousand.

Friedenbach, who grew up in a wealthy family in San Mateo County, had always proceeded from the view that homeless people have rights too, and the city, with its wealthy companies and taxpayers, should do more to help them. For thirty years she'd made that case, and learned to endure the brickbats coming her way from people who believed the city had grown too tolerant and that the homeless needed to get out of the way. She was maternal toward her charges, but also a fierce bureaucratic infighter who knew the system and wasn't afraid of confrontation. Her uncompromising tactics would earn her a lot of enemies over the years—even among those who agreed with the cause—but she'd spend more than a year soliciting input on the ballot measure and building the coalition she'd need to win.

Christin Evans, who owned a bookstore on Haight Street and had become involved in helping the homeless, agreed to be an ambassador for Prop C. The proposed tax would affect three to four hundred companies, and they'd be facing a hike of 25 to 35 percent in the so-called gross receipts taxes they paid to the city, which was no small thing. Evans's business background, Friedenbach figured, would be helpful in some of their more difficult conversations.

Evans had grown up in the East Bay, the daughter of an aluminum-company executive. After college at Vassar she'd gotten an MBA, and in 2006 she was working at the consulting firm A. T. Kearney's San Francisco office, where she'd met her partner Praveen Madan. Evans had done well at A. T. Kearney, but something was missing. She'd been raised in a family that taught her the value of money and investing, but also the importance of giving back.

There was something else going on too, something in the air in San Francisco in the mid-2000s. "There was this sense that anyone could be an entrepreneur and do a startup, that if you were smart and hardworking you could go and start your own thing," Evans recalled. She and Madan took a leave of absence and began to explore ideas, and landed on a concept for connecting independent bookstores with book lovers through a website. After quitting their jobs and digging further into what book people really wanted—and what startup investors found interesting—they decided a better idea was to see what they could do with a traditional, brick-and-mortar bookstore.

They bought the Booksmith, a struggling but long-established outlet on Haight Street, and a couple of years later acquired Kepler's in Menlo Park, a famous community institution that they'd eventually turn into a nonprofit. Booksmith then took over a craft-cocktail bar next door, called

Alembic. Evans started to become quite familiar with the woes of the homeless teens and twentysomethings who'd been finding their way to the Haight for decades.

The Booksmith was a neighborhood business, with readings and other events, and Evans lived just around the corner. "One of the things that was very obviously of huge importance to the Haight-Ashbury community was the topic of housing affordability and homelessness," Evans said later. "It was very clear that the boom-and-bust cycle of the city, and more people coming into the city, and the price of housing going up and up and up, was leading to more people becoming homeless." Several of her employees had been evicted and were barely hanging on. She became especially worried about the homeless teens who'd aged out of foster care and had nowhere to go; the data showed they were at very high risk of becoming permanently homeless. The social services system, though, gave priority to veterans, and to those with the most severe illnesses.

"There was no focus on going upstream, to when people become initially homeless," Evans said. "And that is actually the most effective use of our homeless dollars." Instead, waitlists for housing were based on how long someone had been on the streets. "If you were about to die, if you had full-blown AIDS or late-stage cancer, you would get bumped to the top of the list. But you had to be homeless for a very long time. There was no focus on these youths."

By the time Jennifer Friedenbach asked her to get involved in Prop C, she was more than ready. "It was also the Trump thing," says Evans. "A lot of people were saying, 'You have to get involved.' "

Evans hoped at first to build a broad coalition. She met with Jim Lazarus, the longtime head of the San Francisco Chamber of Commerce, but soon after it became clear that the Chamber would be implacably opposed.

She held out hope that London Breed, who'd just won the mayor's office, would support it, a potentially pivotal endorsement.

Evans was no fan of Breed, who'd been her local supervisor, partly due to what she regarded as an unfair attack by Breed on her favored candidate, Christina Olague, in the race for the Board seat back in 2015.

Evans was also critical of her style. "She's very much a person like, if you go against her, you're dead to her. She's not a politician that governs through building coalitions, bringing people together."

It wasn't too big a surprise to Evans, then, when Breed, along with

Supervisor Scott Wiener, dropped a letter a month before the election announcing her opposition to Prop C. She worried the corporate tax would drive businesses out of the city—it had been the Twitter tax break of 2011, after all, that had helped persuade many tech companies to make the city their home in the first place. Unbeknownst to most at the time, though, Evans just a few days earlier had hooked a supporter who would prove even more important than the mayor.

Evans was on Twitter in the fall of 2018, following a stream of comments about a conversation between Breed and Salesforce CEO Marc Benioff at the company's big Dreamforce conference. They were discussing local issues, including homelessness, and at one point the comments in the thread noted that Benioff had called San Francisco "the Four Seasons of homelessness." Critics of the city's approach routinely contended that the homeless were overly coddled.

"Is this for real?!" Evans tweeted at Benioff. "Did @Benioff just compare SF's homeless services to a luxury hotel chain? How out of touch can a billionaire be?!?!" She continued: "We can't count on the largesse of corporations to solve the homelessness crisis. Vote Yes on C in November!"

Benioff messaged her back privately after noticing her profile, defending himself as a guy who gave a lot of money to charity and supported more spending on social services. Evans mentioned her sour experience with the Chamber of Commerce, the business group that had rejected her entreaties, and Benioff seemed unhappy that his fellow CEOs weren't more invested in the issue. She exhorted him to support Prop C.

Her message struck a chord. Within days, she was talking with Benioff's advisors, who wanted to know whether a million dollars would be enough to help it not only win, but get the two-thirds needed to avoid a legal challenge. Evans had been hoping to raise $100,000 for Prop C, maybe $150,000 at the outside.

"Our city is in a crisis," Benioff said at the time in explaining his motives. "A crisis of cleanliness, a crisis of inequity and inaction . . . and we can't just sit by anymore." His support was all the more meaningful because Salesforce would pay more tax under Prop C than any other company in town.

Not all of his fellow CEOs saw it the same way. Jack Dorsey, now the CEO of not one but two major San Francisco tech companies, was furious about the proposal, and Benioff's shaming of his peers. The gross receipts tax, which had gradually replaced the payroll tax that tech companies despised, hit financial firms like Square especially hard, and Dorsey said the additional levies under Prop C would force it to leave the city. Patrick and John Collison, the Irish brothers who'd built the hugely successful online payment company Stripe, were vocal opponents too, announcing their opposition alongside Breed. So were most of the other tech executives and venture capitalists in town, including Ron Conway and Michael Moritz.

Benioff responded with fire. "He just doesn't want to give, that's all," he said of Dorsey in an interview with *The Guardian.* "And he hasn't given anything of consequence in the city." Benioff was equally contemptuous of the Collison brothers.

"[Prop C] will be a direct tax on Stripe that they don't want to have to pay," he added. "Even though they've made $20 billion in San Francisco, they're not willing to give back at scale. Isn't that amazing?"

Benioff would ultimately contribute $8 million to support Prop C—a measure that if passed would cost his company tens of millions a year. The whispering among political gossips said he must be planning a run for mayor, though there were never any concrete signs of that. He was the champion of corporate social responsibility—it was built into the credo of Salesforce—and wanted to show that it could make a difference in his hometown.

It would only be a few years before he and other proponents of that concept would beat a hasty retreat.

Mayor Breed announced her opposition to Prop C at a very inopportune moment—just a few days before Benioff's endorsement gave it a huge lift. It won easily, with more than 60 percent of the vote, though it fell short of the goal set by Benioff and the organizers to reach two-thirds. Breed had now alienated progressives without gaining much for it and would stand to take the blame if Prop C didn't make the difference it was meant to. Joe Eskenazi, a shrewd local political columnist writing for *Mission Local*, called it a "crushing and credibility-destroying decision."

Breed soon had the worst of both worlds. Anti-tax groups, as expected, sued to block the measure. Though the city ultimately prevailed, it was barred from spending any Prop C money for almost two years while the legal proceedings played out. The new tax revenue, though, was collected immediately, and Dorsey and the Collison brothers quickly followed through on their threats. Stripe, with about one thousand local employees, announced it would move its headquarters to the neighboring city of South San Francisco; Square, now called Block, would dissolve its headquarters office entirely and move staff to Oakland. By most measures the city was still booming, but the prosperity concealed an accelerating flight of businesses. The Prop C cash piled up, unspent, even as the number of unhoused people continued to grow.

Alarmingly, overdose deaths had doubled over the course of two years, to 441 in 2019, and they were still rising fast, piling a new humanitarian crisis onto the beleaguered Department of Homelessness and Supportive Services, which had been established under Ed Lee. The synthetic opioid fentanyl was sweeping the streets, and nobody had a grip on the problem. The Department of Public Health and its nonprofit contractors were committed to the philosophy of "harm reduction," which essentially meant keeping people alive until they were ready for treatment. In practice that meant distributing overdose reversal drugs widely and supplying paraphernalia like pipes and needles, so people didn't share. A lot of harm-reduction advocates—though not all—also rejected any role for law enforcement, even when it came to arresting dealers. The police regarded this approach as absurd but didn't have many ideas of their own, and overdose deaths marched higher amid a pointless ideological battle between the cops and the social workers.

A program called LEAD, for Law Enforcement Assisted Diversion, illustrated the problem well. It was supposed to bring specially trained police officers and social workers together to help the most troubled addicts, but not many officers volunteered, and those who did weren't welcomed by some of the nonprofit workers. "There are harm reduction people who were horrified when I decided to do LEAD. There are people who were anti–LEAD, because they're anti-cop," a LEAD case manager from Glide Memorial Church, which helped run the program, told David Sjostedt at *The SF Standard* in 2021. It was shut down after two years, despite promising results.

As Breed navigated the shifting political terrain, she made another serious blunder, one that would haunt her for the rest of her term.

George Gascón, who had succeeded Kamala Harris as district attorney, announced in 2018 that he wouldn't seek another term—and then quit just a month ahead of the 2019 election to run for district attorney in Los Angeles. Breed, campaigning mostly unchallenged for a full term as mayor, could have left the DA job in the hands of a career prosecutor in the office, given it was only a month until the election. Instead, she installed her preferred candidate, Suzy Loftus, as the interim DA. She may have seen it as an inconsequential appointment, but the move backfired badly, as Breed was excoriated for putting her thumb on the scales ahead of the vote.

That created an opening for a young, left-wing lawyer named Chesa Boudin. He'd been working in the city's public defender's office, which prided itself as among the best anywhere in protecting the legal rights of those without means. Now he was in the running to be San Francisco's top law enforcement officer, even though by his own admission he didn't believe in a lot of the laws, and didn't have much faith in the police.

Boudin's personal story was extraordinary, and polarizing in itself. His parents, Kathy Boudin and David Gilbert, had been left-wing militants and members of the Weather Underground, whose campaign for the revolutionary overthrow of the American government in the '60s and '70s included bombing banks and government buildings. In 1981, just after dropping a thirteen-month-old Chesa at the babysitter's, his parents participated in the group's most notorious crime: the attempted robbery of a Brink's armored car in Rockland County, New York, in which two policemen and a security guard were killed. Boudin and Gilbert, though not the shooters, were convicted of felony murder and sentenced to life in prison. Chesa was raised by Bill Ayers and Bernardine Dohrn, also leaders of the Weather Underground, who'd managed to negotiate plea deals that had kept them out of jail after the group's bombing campaigns were done.

Chesa Boudin didn't see his mother outside of prison walls until he was twenty-three, and his father was still inside as he ramped up his campaign for DA. Growing up with his parents incarcerated, in a household where hard-left orthodoxies had never died, did much to shape his views of the criminal justice system. He did not like what he saw.

Boudin attended the Lab School in Chicago, run by the University of Chicago and based on the progressive principles of philosopher John Dewey, and then Yale and Oxford, where he was a Rhodes Scholar. After a stint working as a translator in the administration of the Venezuelan strongman Hugo Chavez, whose socialist-populist policies, though ultimately a horrific failure, had made him a hero on the left for a time, Boudin returned to Yale for law school. He moved to San Francisco in 2012 to take his first job as a lawyer.

Public defenders don't usually make a lot of money or get a lot of glory, and the rewards lie in achieving justice for the less fortunate, often by fighting the state. They don't usually move to the other side of the table and become prosecutors, much less seek election to a high-profile district attorney job. Certainly the left-wing pugilist Terry Hallinan, who had served as DA for two terms before being defeated by Kamala Harris, wasn't a lot of people's idea of a prosecutor, and the recently departed George Gascón was a progressive by any measure. But Hallinan was a colorful local celebrity who'd been in the public eye for decades. Gascón had been a cop and a police chief earlier in his career, not a translator for Hugo Chavez.

Boudin, though, had a cause—decarceration—that suddenly matched the moment. Still in his thirties, he was articulate and forceful, making an eloquent case for alternative sentencing and diversion programs and for prosecuting bad cops. He supported "restorative justice," which seeks reconciliation rather than punishment, and the elimination of cash bail, both popular ideas in a criminal justice reform movement that had been super-charged first by Black Lives Matter, and then the election of Trump. Though balding already, with a gap-toothed smile, he was tall and handsome enough, with a trim beard and a gleam in his eye, and San Francisco progressives were smitten.

"For whatever reason, the people who love Chesa love him with all their heart," said University of San Francisco law professor Lara Bazelon, an ardent supporter herself.

Still, it seemed like a reach. Suzy Loftus and another veteran prosecutor named Nancy Tung were both credible and experienced contenders, and Boudin was pretty far out there, even for San Francisco. But the leftward pull was strong, both nationally and locally, and the city's ranked-choice voting system, where candidates were picked in order of preference and that was theoretically supposed to help centrist candidates, in this case helped Boudin. Loftus, wounded by Breed's ill-conceived appointment, won the most first-place votes, but Boudin would prevail in the ranked-choice calculations.

The moderate Breed was now stuck with him, just as worries about public safety were on the rise. It would prove a disastrous combination.

■ ■ ■

In January 2020, just a few months after winning her full term, Breed got news that many in City Hall had been worrying about for a long time: A federal grand jury had indicted the head of the Public Works department, Mohammed Nuru, on corruption charges. The head of the Public Utilities Commission, Harlan Kelly, and a half dozen other officials were soon charged too, with crimes ranging from straight-up bribery featuring envelopes of cash to more subtle influence-peddling and favor-trading schemes. And it wasn't clear if the prosecutors were done.

Nuru, even above and beyond his job running Public Works, had a special role at City Hall. He was the guy who could get a mess on the streets cleaned up, or organize a party, or do special jobs that required heavy equipment. He showered his staff with parties and other favors, a jovial presence who always seemed to be on the move and happy to help. That could explain why he paid a $5,600 car repair bill for Breed, even though it's illegal for a department head to do favors for an elected official. The official explanation offered by Nuru and Breed's lawyers is that they were "friends," and Breed soon acknowledged they had been romantic partners at one point.

Nuru had caught the attention of Willie Brown in 2000 after building a small nonprofit called the San Francisco League of Urban Gardeners into a hundred-strong organization supporting community gardens, though his ascent raised eyebrows even at the time. In 2001, Ed Lee, then head of Public Works, hired Nuru as his deputy, and charged him with cleaning up some of the city's dirtiest streets. He and Lee grew close, and when Lee was promoted to city administrator, Nuru stepped into his shoes at Public Works, where he continued to report to his old boss.

Nuru's crimes included trying to bribe an airport commissioner for a restaurant concession, and taking gifts including a $35,000 Rolex watch from contractors doing business with the city. He also ran a slush fund with money provided mostly by Recology, the city's privately owned trash-collection monopoly, which needed his approval for rate increases. The cash was laundered through a nonprofit, and Nuru used it to throw boozy parties and otherwise curry favor.

The scandal implicated some of the city's most entrenched institutions. Recology's franchise dated to the 1930s, when violence among warring trash companies competing for routes prompted the city to step in with a settlement, which was written into the City Charter. The company changed its name from Golden Gate Disposal to Norcal Waste Systems in the 1980s as it consolidated its local monopoly and began competing for contracts in neighboring cities and states, and then again to Recology in the 1990s to distance itself from a legacy of corruption. Against all odds, it had lately succeeded in branding itself as the friendly, neighborhood garbage collection company, employee-owned and on the cutting edge of recycling—a message it drove home by spreading the lucre around. Occasional efforts to end the monopoly at the ballot box were thwarted by the company's financial might, and its status in the City Family.

A common form of legal influence-peddling was something called a "behested payment," in which a person or company looking to win favor with the city would make a donation to a specified charity, or city project, at the "behest" of a public official. Politicians counted on charity dinners and community events and various sorts of good works to do favors for allies and court voters, and Recology was an excellent source of cash. Behested payments were legal as long as they were reported to the city ethics commission and followed certain rules, but in the view of federal prosecutors, Nuru and Recology pushed it too far with the "donations" to the slush fund; a top Recology official was indicted too.

Most damning of all, at least from the point of view of anyone who'd beaten their heads against the city's absurdly time-consuming permit approval processes, was the saga of Walter Wong. He was a "permit expediter," which sounds exactly like what it is—someone you pay to get your city permits approved more quickly. It's a job that shouldn't exist, on the face of it, since a permit system shouldn't require paying an independent middleman to navigate it. San Francisco was hardly the only city where such a function had sprung up. But its byzantine process for even minor home improvements, or building out a store, or constructing anything new was especially daunting; in many cases the rules were *designed* to slow things down, in keeping with deliberative slow-growth approaches. A veritable industry of facilitators, many of them former city officials, had arisen to grease the wheels. There were plenty of incentives for bribery.

One of Wong's favorite gifts was lavish trips to China, which he bestowed on Ed Lee, among others, and his indictment was a laundry list of small-scale pay-to-play schemes, some of which didn't even work. Bernard Curran, a building inspector, ultimately signed off on more than 150 construction and renovation projects in exchange for payments or other favors. The seemingly routine nature of it all was shocking.

Corruption, of course, has long been endemic in big cities across America. San Francisco was likely worse than most in its earlier days, but the postwar years had brought cleaner and more professional government; the Brown/Burton machine, even with all the patronage, was a different kind of operation than the criminal politics of yore. Still, Brown's style was always close to the edge, and since he'd left office his seemingly very lucrative and barely disguised lobbying activities were a constant reminder of how things were done. Investigations by the *Chronicle* and others in the 2000s—one of them titled "Willie Brown Inc.—How SF's Mayor Built a City Based on 'Juice' Politics"—showed that he had dozens of clients.

Gillian Gillett, Ed Lee's aide on transportation issues, recalls a meeting during her first week on the job with the founders of Cubic Transportation Systems, the vendor of the Clipper Card fare system that served Muni and BART. Cubic was seeking a major expansion of its contract, though Gillett and others considered the technology vastly overpriced and outdated. Willie Brown led the Cubic delegation, Gillett recounted, and did all the talking. "Then he just stared straight at me and said, 'I'm confident the city will do the right thing when decisions need to be made,'" she recalled with a shudder. The decision wasn't up to Gillett, but the message was clear enough: power before public benefit. When it came time for a Board vote on the contract, Brown was there in the doorway of the Supervisors' chambers, making sure everyone could see him. Gillett knew how things worked, but it was still startling to see it up close.

London Breed had enjoyed all the benefits proffered by the City Family, but now she found herself inheriting all its baggage too. Mohammed Nuru, Harlan Kelly, and his wife, Naomi Kelly—then the city administrator—were three of the most powerful officials in her administration; Breed had reappointed them to the same jobs they had under Ed Lee. If Lee hadn't died, many at City Hall believed, he might well have been indicted too. Mayor Breed's allies worried that prosecutors might yet be coming for her. For the City Family, it was another layer of stress and strife on top of all the tensions that had built up over the previous decade. The dam would soon burst in a most unexpected way.

PART FOUR

The End of Innocence (2020–2024)

CHAPTER 10

Covid Comes to Town

The indictment of senior city officials in January 2020 came at a moment when political tensions were coming to a boil. The economy was very strong, but the citizenry wasn't happy. Progressives thought Mayor London Breed was too friendly to business, and not doing enough on homelessness and displacement. Businesses were still bristling at the new Prop C taxes and the ever-rising cost of operating in the city. In the tech world, the "VC subsidies" for Uber rides and meal delivery and much else had gone away, along with a lot of the excitement around what the new economy might bring. The surviving innovators of Web 2.0 had gotten big and corporate, and the mega-corps down south—Google, Apple and Facebook—were devouring everything in sight.

The city's leadership wasn't in a good place. But London Breed was ready when Covid-19 came to town.

Two of the first Covid cases in the US were discovered in nearby Santa Clara County in late February 2020, and Breed quickly declared a state of emergency in the city. Two weeks later she imposed restrictions on large gatherings, but she was unconvinced the measures would stem the spread of the virus and began thinking that a lockdown might be the best option. "I felt a shutdown was potentially inevitable, and I believed it could help us, maybe we could stop something so that it never reaches us," she said later. Growing up in a poor and violent neighborhood, where bad things had a way of getting worse in a hurry, had taught Breed not to mess around.

On March 17, 2020, Breed issued a "shelter in place" order together with public health leaders in the surrounding counties; their agencies had all been through fire together during the AIDS epidemic and built a close working relationship, a big asset in the Covid fight. It was the public health officers who actually wielded the legal authority to order emergency measures like lockdowns, but Breed insisted that she would be the public face of the government response.

"There was a lot of pushback, from elected leaders and business people and others, people wanted to go to the symphony, the Warriors game," she said later. "It was not comfortable. It was not pretty. There were some threats. I remember really, truly believing this was the right thing, and somebody has to be the bad guy."

Mayor Breed became a steady presence on video as the pandemic unfolded, giving briefings a few times a week from her living room as she tried to reassure residents. San Franciscans, always more willing than most to do their civic duty and unconcerned about the Republican bugaboo of government overreach, mostly stuck to the social-distancing and stay-at-home rules, and they appreciated Breed's forthright approach. Like then–New York Governor Andrew Cuomo, she became a Covid star as isolated residents tuned in for any scrap of information and found a mayor who was both cautionary and encouraging, handling the problems but also suffering them, just like the rest of us.

The swift lockdown, and a citizenry that was diligent and informed, undoubtedly saved many lives: San Francisco ultimately had one of the lowest Covid death rates in the country, even though it's one of the most densely populated cities in the US and many of the first domestic cases had bubbled up nearby.

"Mayor London Breed's early and aggressive moves to contain the outbreak have made San Francisco a national model in fighting the pandemic," declared *The Atlantic* in April of that year. A more jaded *Wired*, after some throat-clearing about the city's chronic dysfunction, could only agree, writing in its August 2020 issue that "San Francisco executed by far the most successful initial response to Covid-19 of any major American city."

Bob Wachter, who as head of the department of medicine at UCSF was among the nation's top doctors, emerged as a steady and trusted voice on Twitter with his sober discussion of what was being learned about the virus and how to stay safe. He was proud of his city's approach, and residents' belief in the science.

"The local community was almost uniformly supportive of wanting to know the science, and valued hearing from someone who they thought had legitimate credentials," Wachter recalled. With the benefit of hindsight, he believes public health measures such as lockdowns, masking and vaccines saved many lives; with the important exception of school closings, he considers the virulent criticisms of the pandemic response that became dogma on the political right to be "revisionist history." San Francisco clearly did much better than other cities, he said, even if the exact reasons for that are difficult to parse; the population started out healthier than most, for one thing.

Breed's initial success with the stay-at-home order, though, couldn't ultimately shield the city from the many kinds of suffering—and the fevered political wars—that Covid brought to America.

At the outset of the pandemic, the city had stopped admitting people to its so-called congregate homeless shelters, where people slept in large, barracks-like rooms that were easy places for the virus to spread. Public health workers handed out tents and moved to establish "safe-sleeping" sites where people could camp at a safe distance from one another, with meals delivered by nonprofit service providers. That quickly accelerated a proliferation of tents on sidewalks, especially in SoMa, Civic Center and the Mission, that had begun a few years earlier as the housing crisis intensified. Large encampments began growing in other parts of the city too, especially in the Bayview and Hunters Point, which got much less attention from the city than downtown.

There was also an ambitious effort to get everyone inside. With the tourism industry shut down and the state providing extra money for homeless services, the city began doing deals to lease tourist hotels to house the homeless. Some twenty-five properties would ultimately be part of the "shelter-in-place" hotel program.

At first, the so-called SIP hotels were widely hailed as a success, a humane alternative that would help keep everyone in the city safe. Progressives saw it as the possible start of something bigger—maybe some of the properties could be permanently converted to subsidized housing. Dean Preston, the Democratic Socialist who'd been elected to the Board of Supervisors after Breed became mayor, viewed a lot of the pandemic-era programs in that light. "The federal support, the rent relief programs, some of the health care that was offered with no barriers, these were models of how you could do things better," he said later.

Yet the hotel program would also become a talking point for critics of the Covid response. Some of the lodgings were anything but safe havens as the problems of the streets moved inside; at the Hotel Whitcomb on Mid-Market, an elegant Art Deco structure that had once served as a temporary city hall, there would be twenty-one overdose deaths between April 2020 and August 2022. The program was also easy to exploit. A partnership led by Aby Rosen, a big-spending social climber and real estate investor based in New York, had bought the Whitcomb in 2018 to serve the hip, up-and-coming business district that city leaders had envisioned. The partners collected $89 million in rent from the city for the thirty-three months that it served as a shelter-in-place hotel and then sued for damages in the aftermath and collected another $32 million—none of which would be spent on repairing the property. All told, the SIP hotels cost the city a bit over $450 million to house about 3,700 people over two and a half years, and most of the promised federal reimbursements would never come. Reflecting the ever-widening political divides of the Covid era, there'd ultimately be only argument over whether the program was a problem or a solution.

For local tech companies, the sudden shuttering of downtown offices and the rise of a new paradigm around work-from-home came as a shock. Yet it wasn't long before the arrangements became not so much a temporary emergency measure as a revelation: Work-from-home really could be done on a mass scale, and it turned out to have all sorts of advantages. CEOs and investors fantasized about the possible cost savings. Rank-and-file office workers rejoiced, not so much because people didn't like their workplaces, but because they won back the time they had once spent traveling back and forth. The hoary dream of "telecommuting," which computer and telecom companies had been promoting as far back as the 1980s—long before the tech was good enough—was becoming an instant reality.

Working remotely was already part of the culture of the tech industry. WordPress cofounder Matt Mullenweg's startup, called Automattic, was a classic case in point: It had arisen from an open-source project with a community around the world, Mullenweg noted, so it was an obvious choice to be a "distributed" company, as he calls it, when it was founded back in 2005. That logic would now apply to much of the industry and beyond.

A lot of big companies had put contingencies in place for major emergencies, and this was a moment of truth. Del Harvey, the trust and safety head at Twitter, had a team of some three hundred staffers and several thousand contractors around the world, and though the pandemic lockdowns caused a certain amount of chaos at first, the group was already prepared for work-from-home in the event of emergencies.

"We'd set up a whole protocol," Harvey recalled later. "That was incredibly fortunate, because it meant that when the company switched to all-remote, trust and safety people already had their setup." It wouldn't take long for the rest of the company to catch up, and in May 2020, just two months after the lockdown, Jack Dorsey told Twitter employees in an email that they could work from home "forever." Dorsey himself, by many accounts increasingly detached from the company as he remained CEO of both Twitter and Square, sent the note from a redoubt in French Polynesia.

At Salesforce, the city's largest private employer, Marc Benioff said staff could work from home at least until the end of the year. He'd soon be talking about new modes of work and new ways of employees connecting, and Salesforce leased a luxurious retreat center sprawling across seventy-five acres in the redwood forests of Santa Cruz County—a temporary location, said Benioff, while the company searched for a permanent campus. Benioff himself hated working in an office anyway, and was now spending a lot of time in Hawaii, where he'd been accumulating properties on the Big Island.

The stock-market route that greeted the pandemic's arrival proved remarkably brief, with tech company shares recovering their losses by May 2020 and resuming a steep upward climb. Nobody would quite say it, but work-from-home was turning out to be very good for the tech business, trapping everyone on their screens and replacing real-world living with its virtual counterparts, in the workplace and everywhere else. Social networks saw a surge as the homebound legions turned to virtual communities. Struggling computer companies like Dell and HP, along with vendors of accessories like webcams, enjoyed a sales boom as people outfitted spare rooms for work. The price of Bitcoin and other cryptocurrencies skyrocketed as boredom merged with a vague sense that maybe the social order was collapsing and crypto would be a safe haven. The digerati flocked to Clubhouse, a new app for large group conversations, for rowdy discussions of Covid, the economy, the Internet—and, of course, politics. The 2020 election was approaching, and emotions were running high.

Yet even as the tech companies enjoyed a pandemic boomlet, and banks, law firms and other white-collar institutions barely missed a beat, downtown San Francisco was dying. There wasn't any other way to say it.

Over the course of a decade, the city center had been remade and revitalized, with major new developments like the Salesforce Tower and the reclamation of Mid-Market as a business district. Even though soaring prices and all the problems in the streets had taken some of the shine off, the city was still bustling in early 2020, and empty offices or vacant storefronts were not on anyone's list of worries. City Hall's reliance on business tax revenues to fund its social services programs was hotly debated, but it was a rational enough strategy. Until it wasn't.

Around 250,000 people were working in San Francisco offices every day before the pandemic; in January 2021, fewer than 50,000 were back at their desks. BART, the regional transit system designed to funnel commuters to downtown San Francisco, was carrying more than 400,000 riders a day in 2019. By the end of 2020, nine months into the pandemic, the number stood at barely more than a tenth of that. As cities in Florida, Texas and elsewhere began to see their office districts come back to life in 2021, San Francisco remained at the very bottom of the rankings among all cities on return-to-office metrics and would remain there throughout the pandemic and its aftermath. It turned out that tech workers, who by now accounted for 35 percent of the city's workforce, tended to like working from home.

Twitter had conceived of its Mid-Market headquarters as the anchor of an urban neighborhood and an idealistic model for how Internet companies could partner with local government to solve problems and spread the wealth. It fit seamlessly with the belief that Twitter and the rest of the city's new-wave tech companies were a positive force, not just for the economy, but for civil society and culture too.

Now the Market Square complex was all but empty, and Mid-Market, at a stroke, was back to where it had been before the tech industry arrived, the street corners frequented by drug dealers and their customers, the homeless camping in doorways and alleys—and increasingly on the sidewalks. Much of the rest of downtown, and SoMa, was starting to look like that too. Like a receding tide exposing old skeletons, the emptying of the city during the pandemic showed just how little progress had been made on the conjoined problems of homelessness, addiction, mental illness, street crime and wealth inequality.

The killing of George Floyd by a Minneapolis police officer, in May 2020, brought fresh anguish. Mayor Breed and almost all of the political leadership, reflecting the mood of the moment, were eager to align with the Black Lives Matter movement, which was seen as a natural successor to the civil rights struggles of yore. But the anti-police protests that followed Floyd's killing, on top of the Covid-induced troubles in the streets, contributed to a growing sense of lawlessness. There was quiet resentment that public health officials seemed to have a double standard on Covid precautions when it came to demonstrations. One big BLM protest concluded with a wave of looting in Union Square, while another led to the toppling of monuments in Golden Gate Park, including likenesses of Ulysses S. Grant and Francis Scott Key. Bitter over the rallying cry of "defund the police," the cops often appeared to stand down.

As summer turned to fall, major forest fires swept the state, from the Sierra foothills east of the city to the rolling fields of Sonoma County to the north, and even engulfing the damp redwood forests of Santa Cruz and Big Sur to the south. A sickly orange haze descended on San Francisco, nature's signal that all was not well: It would be the worst fire season in California history.

With the 2020 election approaching, Twitter was a maelstrom of heated Covid policy debates, absurd conspiracy theories and bizarre pronouncements from President Donald Trump. Del Harvey and her team struggled with unprecedented decisions. There was pressure from many quarters, including the federal government, to try to stop the spread of dangerous misinformation about Covid—and contrary to what was alleged later about government censorship, the company was doing it anyway, for its own reasons. To that point content moderation had mostly focused on illegal activity, abuse, and incitements to violence, but the company had been working on policies to label and down rank verifiably false claims and foreign propaganda, and link to authoritative sources. Covid prompted an expanded and accelerated rollout of the public health portions of those policies, which would draw ferocious and mostly partisan criticism later on.

It was clear from the earliest days of social media that it was in the interest of the companies—as well as advertisers and the vast majority of users—to bar hate speech, personal threats and incitements to violence, as well as pornography. Such speech was usually protected by the First Amendment, but just as a newspaper or TV station might prohibit the use of the f-word, social media companies had the right, and arguably the responsibility, to block people from, say, celebrating the Nazi Holocaust as a model for how Jews should be treated. The "Overton Window" of socially acceptable views hadn't yet undergone its radical shift to the right.

When it came to disaster response and public health, working with the authorities to provide accurate and useful information in an emergency was also a routine media company practice; when I was the West Coast bureau chief for Reuters, a global newswire, we had our own small role in the regional earthquake plan. It never occurred to me that we might be accused of nefariously taking orders from the government.

Twitter wanted to be a good source of information during the pandemic, but policing the truth or falsity of content was new territory, and public health policy presented novel and wrenching issues. For starters, what standard should be used in separating dangerous misinformation about Covid from differences of opinion on a virus that was poorly understood? And what to do when life-threatening suggestions, like drinking bleach, were coming from the president?

National political tensions were reaching a breaking point as the 2020 election approached, and political disinformation, much of it originating in Russia and China but a lot of it also coming from allies of the president, was another hazardous conundrum for Harvey and Twitter. Trump himself was peddling lies and conspiracy theories (not to mention using popular songs without permission) and at times crossing into abusive and threatening rhetoric. Should Twitter hold the president to the same rules as everyone else and block his account for repeated violations? There was also the question of how to prevent the service from being used for "hack-and-leak" operations, designed to extort or humiliate people, without stepping on freedom of expression.

CEO Jack Dorsey was a spectral presence on moderation issues, remaining closely involved in policy but often staying out of the fray on enforcement and leaving decision-making primarily to Harvey and her team, along with chief legal officer Vijaya Gadde.

Dorsey had asked for more information over the ban of right-wing provocateur Milo Yiannopoulos and conspiracy theorist Alex Jones, but he didn't intervene. He seemed to be uncomfortable with some of the moderation decisions, but his views were enigmatic and mostly filtered through Gadde, according to Harvey, so everyone "maintained a polite fiction of we have no idea where this is coming from." To many in the company, Dorsey seemed checked out, and was taking his preference for delegating to a new level.

Fissures in the local Covid consensus were now starting to appear too. Small businesses, and especially bar and restaurant owners, were getting restive about San Francisco's ever-changing rules on masking and social distancing, which were stricter than those in most places and sometimes crossed the line into nanny-state paternalism.

Doug Dalton, the Netscape alum and entrepreneur who'd gone into the nightlife business after the first dot-com bubble, now co-owned a dozen cocktail bars around town, and he was incensed by the rule that said you couldn't sell alcohol from a takeout window unless you also served food from a professional kitchen. The theory was that drunk people were less likely to follow Covid guidelines, so drinking should be discouraged. "From what they stated to us, restaurants are adults, and bars are children, and they don't trust children," recalls Dalton.

In a worrying portent for the city's much-celebrated hospitality sector, service workers were leaving in droves. Tech executives decamping for Miami or Austin tended to get the attention, but the data would show that San Francisco's pandemic population decline—it lost about 5 percent of its residents in the first year—was mostly due to the departure of the working-class wage earners who cooked in restaurant kitchens, served cocktails and cleaned up after everyone. City economist Ted Egan called it an underappreciated problem, as those workers would be much harder to bring back.

Controversies were even starting to brew over the popular restaurant "parklets," which had opened up the city to the pleasures of outdoor dining but also drew complaints, since they entailed private businesses taking over parts of public streets. Another celebrated innovation, the "slow streets" that were established to lure people outside, would come under fire too, for making the city less friendly to cars.

But the seminal pandemic-era ruptures in San Francisco would involve two of the most high-stakes functions of local government: keeping the streets safe, and running the public schools. A momentous national backlash against progressive policies on both of these issues would come a couple of years later. But it was in San Francisco, where progressive ideology had flourished more than anywhere else, that the revolt truly began.

Sunset resident Patrick Wolff, a two-time US chess champion and hedge-fund manager, was among the San Francisco parents who in the fall of 2020 began to sound an alarm about disquieting developments in the public school system.

Wolff had stepped back from his finance job after making plenty of money, and with his kids now in their teens, he'd turned his attention to a long-standing interest in education. He and his wife were both big believers in public schools, but they were losing confidence in the San Francisco Unified School District. There were so many ways that the city fell down in serving families—San Francisco has fewer children per capita than any major city in the country—and Wolff joined up with a friend to create Families for San Francisco, a political group that would advocate on a variety of issues affecting students and parents.

San Francisco Unified suffered from many ills that were common to big-city school districts. The student population was much poorer than the city as a whole and much less white—just 13 percent as of 2024, compared with 40 percent for the city. Educational performance in the schools that served the most impoverished communities were at the very bottom of the state rankings. The teachers' union was controlled by progressives who insisted on taking positions on inflammatory political issues like the wars in the Middle East. Mismanagement and corruption had plagued the district for years; a faulty payroll system that began spitting out incorrect checks and missing payments in 2022 was still messing with employees' lives in 2025.

Still, the school district's enrollment had stabilized after a steep decline in the early 2000s, even as it continued to fall precipitously in other California cities, including Los Angeles and San Diego. The number of public school students in San Francisco in 2019 was about the same as in 2007, and there were pockets of excellence across the district. The city took special pride in Lowell High School, long reputed to be among the best in the nation. It was a public school with selective admissions, similar to Stuyvesant or Bronx Science in New York, and like its East Coast counterparts was much more white and Asian than the district as a whole—a touchy issue that the Black Lives Matter movement had brought to the fore. Unlike the New York schools, though, Lowell was the largest high school in the city, and was seen by many middle-class residents and upwardly mobile immigrants as a crucial route to a good college.

The school district was funded by state and local taxes but was independent of the city government and overseen by an elected school board, which had become a stepping stone for aspiring politicians. In the fall of 2018, three up-and-coming progressives—Alison Collins, Gabriela Lopéz, and Faauuga Moliga—won official Democratic Party endorsements in the school board race, the fruit of their factions' successful takeover of the local party committee. Their aim was to promote equity and root out institutional bias in the school district, and they rode the anti-racism wave of the moment to victory. Lowell High School, from their point of view, was not a beacon of opportunity, but rather a citadel of historic discrimination that needed to be made more broadly accessible.

In October 2020, as the pandemic raged, the school board convened an emergency meeting to eliminate the test-based admissions process at Lowell for the coming year and replace it with a simple lottery. Administering the test while schools were closed for the pandemic would be impractical, Board president Collins argued, and there was really no choice.

Wolff, whose son was in his freshman year at Lowell but had yet to set foot in a classroom, was immediately suspicious. "They said they were going to do this for only this one year, but actually they wanted to get rid of merit admissions altogether," Wolff said later, and subsequent events would show that he was correct. His group got involved in a petition drive to stop the change, and though it failed, it brought a lot of attention to the issue and got more people involved. Then, just a few weeks later, Wolff read an article in the *Los Angeles Times* about the work of a committee the Board had established to rename public schools, many of which honored historical figures whose words and deeds had not always aged well.

The Board had already shown censorious instincts when it ordered the destruction of the magnificent *Life of Washington* murals on the walls of George Washington High School. Commissioned by Franklin D. Roosevelt's Works Progress Administration and painted in 1936 by the artist Victor Arnautoff, the mural's thirteen panels depicted different aspects of

George Washington's life, and two of the panels had long sparked controversy for their depictions of Native Americans and slaves. Unmoved by furious protests even from fellow progressives like former Board of Supervisors President Matt Gonzalez—or by the fact that Arnautoff was a communist whose message can easily be read as subversive—the Board voted to paint over the murals on the grounds that they created a hostile environment for students of color. Only after news of its actions went national did the Board agree to cover them instead.

The mural battle turned out to be just a preview. Wolff had gotten ahold of the spreadsheet the school renaming committee was using to formulate its recommendations, and he couldn't believe what he was reading. "They claimed that Paul Revere was a colonizer," Wolff recounted. "They decided that James Lick [one of the city's original real estate moguls] was responsible for something terrible that happened twenty years after his death. They wanted to rename Roosevelt Middle School, but they didn't know which Roosevelt it was named after, so they came up with a reason for each."

Wolff found it so over the top that he decided to go deeper. The volunteer renaming committee had been gathering for eighteen months and all the meetings were recorded. Wolff, with time on his hands, endeavored to watch every single one. "It was a trip, a complete trip," he recalled. Wolff wrote up a report, and Families for San Francisco published it. It was after *Mission Local* columnist Joe Eskenazi, generally sympathetic to progressive causes, wrote a story about how the renaming committee had indeed gone off the deep end that a furor erupted.

Undeterred, the Board voted in January 2021 to accept the committee's recommendations and expunge Washington, Roosevelt, Revere, Thomas Jefferson and even Abraham Lincoln, censored for his attitude toward American Indians. Dianne Feinstein, then still a sitting senator, would be de-honored too, for once ordering the replacement of a damaged Confederate flag. To say that this was catnip for the right-wing media doesn't come close to describing the sheer glee with which Fox News and the Trump-loving commentariat greeted this fresh evidence that left-wing lunatics had taken over San Francisco. Even locally, people were indignant at best, and shocked that the Board would even be discussing such things while schools were still closed due to Covid. Breed called it "offensive and completely unacceptable." Her chief of staff, Sean Elsbernd, convened a meeting with dozens of people inside and outside of government to talk about making the school board an appointed body, rather than an elected one.

But two recent arrivals to the city, not knowing what the local political pros considered possible, had a bigger idea.

Autumn Looijen and Siva Raj had met on Tinder in the summer of 2020, seeking companionship in the early days of the pandemic. Both were single parents with careers in the tech industry, Looijen living in Los Altos, in Silicon Valley, and Raj in Pleasanton, in the East Bay. San Francisco was a logical place for them to meet up, and Raj especially loved the city for its open and accepting culture. He'd grown up in the slums of Chennai, in India, and savored the life he'd built in the US, even if his recent startup, which made exercise machines, was failing. Looijen could see that Raj, gregarious, intellectual and proudly bisexual, loved being in the city. "You're so much more yourself here than you are anywhere else," she recalls telling him. "Why are you not living here?" When he'd moved to the Bay Area from the East Coast in 2018, navigating the city's school system for his two boys, then both in middle school, seemed impossible, and San Francisco was frightfully expensive too. Now school was remote anyway, and apartment rents were way down, at least in the center city, so maybe it was time. Looijen and Raj were walking the streets of the Mission in early November, cheered by Joe Biden's victory in the presidential election, when they decided to take the plunge together, renting a scruffy second-floor flat in an aging wooden walk-up on lower Haight Street. The five guys who'd been living there, Looijen said, had split for Hawaii, and the apartment was going for little more than half what it had rented for a year earlier.

With five kids between them, ranging in age from six to fifteen, Looijen and Raj were counting on the schools opening up again early in the new year. That's what the school board had indicated back in October, and Looijen had noted that kids were already back in the classroom in Los Altos and many other cities and towns. So Raj was shocked when he got on a Zoom call for the January school board meeting and realized that reopening was barely even on the agenda. "They did all this renaming stuff and all this other stuff, and they made us wait like eight, nine hours before they even got to reopening. And when I saw the state of preparations . . . nothing had been done." There was no testing infrastructure, no agreement with the teachers on how to proceed, no protocols on class sizes or distancing or anything else. It was mainly the teachers' union that was standing in the way of reopening, but its leadership was aligned with the progressives on the Board.

Raj was mystified as to why San Francisco seemed to be so far behind and began researching what other cities were doing. "Across the top ten metros, we had the best Covid rates and the worst reopening rates," Raj said later. The school board began getting a lot of heat on the issue: City attorney Dennis Herrera sued the district for its failure to develop a reopening plan. Meredith Dodson, a child development expert and parent of two young kids, had formed a group called "Decreasing the Distance," later renamed the SF Parent Coalition, which was laser-focused on getting schools to reopen. There was Wolff's group, too, with similar goals. But it wasn't clear if or how these efforts would move the needle.

Raj, meanwhile, had noticed that in the East Bay town of San Ramon, a group of parents had threatened to mount a recall drive against the school board if the schools weren't opened, and all of a sudden a reopening plan materialized. Maybe that would be the way to go: a recall of the school board.

In the midst of rising anger among parents and moderate politicians over reopening and renaming, the Board in early February voted to permanently eliminate merit-based admissions at Lowell. It was a move guaranteed to enrage two very important constituencies: Asian parents, who saw Lowell as the route to success for their children, and the school's extensive network of powerful alumni, who took great pride in Lowell's history and reputation.

The Lowell question wasn't an easy one. Its merit system was central to its excellence, but there was plenty of evidence supporting the argument that competitive testing effectively locked out many Black and Brown students and reinforced the massive achievement gap within the district. The Board wasn't interested in various middle-ground options, though, and instead pushed ahead with the hugely contentious change, even as reopening of the district's 120-odd schools remained on the back burner.

Looijen and Raj were intrigued with the idea of mounting a recall, but people who were savvier about local politics, including Patrick Wolff and Meredith Dodson, waved them off. It would take the signatures of 10 percent of all San Francisco voters—about fifty thousand in total—to get a recall on the ballot, and that was a huge hill to climb. The city's commitment to direct democracy meant there were ballot measures all the time requiring the collection of signatures, and the campaign consultants, who had it all down to a science, told Looijen and Raj that they didn't have enough time or enough money. They'd never make it without paid signature gatherers, they were told, and it would be hard to even hire them quickly enough to make the September deadline.

There was another problem too: They'd wanted to recall the whole school board, but based on when they'd been elected, only Collins, Lopéz and Moliga were eligible. The political optics of targeting three people of color at a racially fraught moment were terrible, and maybe needlessly divisive, with the regular election coming the following year. "There was a lot of fear among people we spoke to about it," Raj said later. Looijen said she was told "you'll never be welcome in this town again if you do something like that." On a Zoom call with school district alumni, the couple was urged to stand down, lest the effort fail and further strengthen the incumbents.

For all the resistance, Raj and Looijen were inspired by how many people shared their concerns, and frustrated that for all the conversations surrounding schools, reopening remained as distant as ever. It was on Valentine's Day of 2021 that they decided "we should just do this recall thing," as Looijen put it later. They began calling around to other individuals and groups who'd seemed interested. Nobody wanted to take the lead, so they'd have to do it themselves. "We don't care, win or lose, it doesn't matter, if there are enough people out there who want to support it, we're going to do it because it's hurting our children and we have to do something," Raj said later, reflecting on his thinking at the time. "We need our kids to know that we would always stand up for them."

The couple proved creative in building momentum. There were already active Facebook conversations among concerned parents, and one of their first moves was creating their own Facebook group, which gained seven hundred members in a week, and a mailing list, which soon had seven thousand names. A blog post in the *Chronicle* by Joel Engardio, a journalist and perennial candidate for the Board of Supervisors, helped get the word out further. Looijen said she'd message five people and ask each of them to message five people, and soon they had a volunteer corps collecting signatures on street corners and at weekend farmers' markets. They didn't have money for polling, but instead created a list of "thirty reasons to recall the school board," and then tweeted them out one by one and monitored the likes and retweets for clues on which messages were resonating. They got a big assist when a Board critic discovered Twitter posts from Board President Collins that denigrated Asian Americans for being insufficiently anti-racist, comparing them to "house slaves" whose desire to be a "model minority" led them "to use white supremacist thinking to assimilate and 'get ahead.' "

Both Raj and Looijen had been on the Internet forever, and that would prove an important asset. Looijen, back in the 1990s, had worked at Netscape's Open Directory Project, a pioneering effort to create a crowd-sourced index of the Internet that would serve as an inspiration for Wikipedia. Raj had been a soldier in India's social media wars, battling propagandists for President Narendra Modi who advocated turning the country into a Hindu nationalist state. Both remembered a time when there was joy in online connection, and they proved adept at managing the anger that was now dominating social media and even their own Facebook group. "It's amazing how much consensus you can achieve in a widely disparate group, when you give people the opportunity to debate and discuss and hold the boundaries," says Raj. They were committed to a truly grassroots campaign, with collective decision-making and a $99 limit on donations.

Everyone told them it would never work.

Raj and Looijen had formidable allies, though. Chinese American parents in particular were up in arms about what they saw as threats to their children's future. At the same time, the city's Asian population was confronting racial violence that it hadn't seen in many years, and it wasn't going to passively accept that either. The sleeping giant of city politics was awakening.

From the start of the pandemic, Trump had blamed Covid-19 on China, referring repeatedly to the "China virus" and framing it as part of a larger threat to the American way of life. It was an appeal that neatly combined nativist bigotry and America-first economic nationalism, and it set off a wave of violence against Asian Americans around the country—one that would hit especially hard in San Francisco. During the first year of the pandemic, there were more than a half dozen unprovoked attacks against Asians, mostly elderly people. Among the assaults captured on video was the January 2021 murder of an eighty-four-year-old Thai man,

Vicha Ratanapakdee, who was violently shoved to the ground by a random assailant at eight o'clock in the morning and died after hitting his head on the curb. Another showed a seated, eighty-four-year-old Rong Xin Liao being brutally kicked to the ground as he waited for a bus in the Tenderloin.

The incidents were a dark reminder of both the virulent anti-Chinese racism of San Francisco's early days, and of the ongoing conflicts between Blacks and Asians. The former sometimes seemed to resent the latter's "model minority" status, as evidenced by Alison Collins's Twitter thread. At the same time, there was often a strain of anti-Black racism in Asians' views of street crime—especially after the perpetrators of several attacks on Chinese elders turned out to be Black men.

Ed Lee had broken the ceiling on how high Asians could rise in San Francisco politics, and there was now a strong cadre in senior positions representing a new generation of leadership, including City Attorney David Chiu, State Senator Phil Ting, Supervisor Connie Chan, former Supervisor Jane Kim, City Administrator Carmen Chu and local Democratic Party Chair Nancy Tung.

Still, much of the Chinese American community, and the smaller Filipino, Korean, Japanese and Pacific Islander cohorts, hadn't been very engaged in politics, with significantly lower turnout numbers than the white and Black populations.

That was about to change.

On St. Patrick's Day in 2021, just a month after Siva Raj and Autumn Looijen had resolved to go ahead with their recall campaign, Mary Jung was enjoying some corned beef and cabbage at the house of her friend and mentor, Caryl Ito, and awaiting news on the birth of her grandson when another disturbing video made the rounds. A seventy-five-year-old woman named Xiao Zhen Xie had been punched by a young man on Market Street, leaving her face bloody and swollen, though she'd managed to turn the tables and strike a few blows herself.

Jung was a City Family veteran who had chaired the local Democratic Party Central Committee and now served as the political director for the influential San Francisco Association of Realtors. She remembers feeling a little relief that at least the perpetrator wasn't a Black person. The last thing the city needed was another wedge in the Black/Asian divide.

She did, however, believe there was a person who should be held responsible: District Attorney Chesa Boudin.

"I was just really tired of seeing all the beatings of the Asian elders," she recounted later. "You see these people beaten up, being bandaged up in the hospital, and you think, my God, this looks like my mother, my grandmother, right? It's like every little old man and little old lady that I see in Chinatown. It was just really distressing. And I thought OK, I'm going to do a little bit more the next time the opportunity comes."

Jung says she likely wouldn't have launched her campaign to recall Boudin if he had done more to acknowledge the issue of violence against Asians and shown some willingness to revisit policies that routinely left repeat offenders on the streets. Many people who'd supported decarceration and "defund the police" in 2019 were reconsidering their stance in the very different context of 2021. Chesa Boudin was not one of them.

A Republican gadfly named Richie Greenberg was already gathering signatures for a recall of Boudin, and a few months earlier Jung had considered helping out. But he wasn't getting much traction.

Mary Jung was a far more formidable force. She was blunt, even acerbic at times, brooking no patience with bleeding-heart progressives and fighting the moderate side with passion. But she was also respected as a serious, diligent person who was sincere in her beliefs and could get things done. With her mainstream Democratic history and deep ties to the Willie Brown machine, Jung was a much more comfortable figure for the city's wealthy elite to ally with than Richie Greenberg. She was also a much savvier operator.

She got a call from Brian Giraudo, an heir to the famed Boudin Bakery (no relation), whom she knew had a passionate dislike for the DA; he was on board. Andrea Shorter, a friend with deep experience in city government and social services, and a lot of organizing chops, joined in too. It wasn't a given that they could muster enough support, but Jung figured it would pay dividends either way.

"My feeling was, if we could educate the public on the difference between a public defender and a district attorney, when it came time for Chesa to run for reelection, then at least we would have a shot at electing somebody who was a true district attorney," Jung recounted.

Jung had earned her political stripes, along with deep credibility when it came to civil rights, back in the 1970s, as a young secretary in Cleveland, Ohio. Fed up with the way women were treated in the office in those days, she helped create the "9-to-5" movement for women's equality in the workplace, which was later immortalized in a documentary, and a film starring Jane Fonda. She'd then gotten married, moved to San Francisco and had a daughter. It wasn't until Willie Brown's campaign for mayor, in 1995, that she began tiptoeing back into politics, and she proved very effective at mobilizing Asian voters. Brown appointed her to an office manager job at City Hall, and she later did a stint at PG&E before landing with the Realtors group. Along the way she was appointed to the influential Democratic Central Committee and would serve as its chair for a decade, before being ousted in the progressive wave of 2016.

But in the anarchic summer of 2020, she had seen a chance to fight back. The street chaos of the early pandemic had persuaded a lot of the city's more conservative residents that the time had come to confront progressive policies they believed had gone way overboard. Jung seized the moment and rallied key members of the city's tech elite, along with a few hedge-fund managers and old-money families, behind a new political group called Neighbors for a Better San Francisco.

Founded with the avowed purpose of improving quality of life in the city, Neighbors' biggest financial supporter was an obscure hedge-fund manager named William Oberndorf, who was in fact a Republican and had given many millions nationally to his pet cause, charter schools. Other key backers who plunked down $300,000 apiece at the start were Michael Moritz, the Sequoia venture capitalist; Chris Larsen, the Ripple founder; and John Pritzker, the hotel heir. Jung installed a protégé named Jay Cheng as executive director, and the group would bridge the gap between real estate, tech and financial services interests to build a war chest of some $3 million to spend on local races in its first year.

The emergence of Neighbors marked the start of a profound political shift. Ron Conway and Marc Benioff may have been heavily involved in local politics for a decade, but with a few exceptions—notably Benioff's Prop C campaign—they mostly got involved in issues that directly affected their companies. Now Neighbors would regularly be bringing big money to the macro issues of local politics, especially public safety and land use, with the aim of making the city less permissive and more business friendly.

There were already political groups of seemingly every stripe when

Neighbors entered the fray, but they'd emerged in earlier eras, and were often organized around social justice causes, old business alliances, or ethnic or racial cohorts. Now there was a flurry of new organizations on the moderate side of the local political spectrum. A couple of generations ago, these groups might have been comfortably Republican, of the Nelson Rockefeller or Warren Hellman kind. That faction didn't exist anymore, though, and the takeover of the Republican Party by Trump's MAGA movement had made it a toxic brand up and down the West Coast. In San Francisco, it was Democrat or die if you wanted to win elections.

Michael Moritz, in addition to backing Neighbors, launched his own political operation called TogetherSF, appointing Jay Cheng's wife, Kanishka Cheng, as executive director. Onetime Twitter and Apple engineer Sachin Agarwal, together with software developer Steven Buss, created GrowSF—"rejecting the politics of scarcity"—with a weekly newsletter and regular meetups. Web software entrepreneur Zack Rosen, who had been involved in Tim O'Reilly and Jen Pahlka's Gov 2.0 efforts a decade earlier, launched Abundant SF, the first node of what he envisioned as a national network. On the lobbying side—as distinct from electoral politics—Ron Conway's sf.citi was still working it, with Alex Tourk, the onetime Newsom campaign manager and principal of Ground Floor Public Affairs, as its point person. There was also Advance SF, descended from the old Committee on Jobs and representing a cross section of the downtown business community, and of course the venerable Chamber of Commerce, boosting enterprises large and small. There was no shortage of mobilizing efforts around a business-friendly, law-and-order agenda.

It was a moment for politics. The pandemic gave a lot of tech professionals and others the time, and sometimes the tools, to dive into conversations they might once have ignored. As in the country at large, the Covid era in San Francisco would not just upend the economy and the rhythms of daily life; it would profoundly reshape politics in ways that were just starting to come into view.

CHAPTER 11

Recalls and Recriminations

Unlike most local tech moguls, Michael Moritz was well versed in city affairs, having been sporadically involved for many years. He'd bankrolled the ballot initiative on pension reform back in 2010, and with his wife he'd established Crankstart, a foundation that doled out tens of millions a year to civic causes. When people complained that tech chieftains didn't support charities in the way that the old money did, they weren't talking about Moritz.

The son of German Jews who'd escaped Nazi Germany, Moritz had grown up in Wales, and made his way to America to pursue journalism after being blocked by the apprenticeship system of the British press. After a stint covering the tech industry for *Time* magazine and penning a book about Steve Jobs and Apple, he went over to the other side, landing a job at Sequoia Capital. He'd spend forty years at the firm, much of it as the managing partner, and his investment wins made for quite a list: Google, Yahoo, PayPal, LinkedIn, Airbnb and YouTube, to name only the most prominent. Sequoia under his leadership was widely considered the top VC firm of them all.

He'd begun to pull back from Sequoia in 2012, due to a mysterious illness that he later identified in a privately published memoir as a rare blood cancer, though he remained the firm's chairman until 2023 and was still involved with some investments. He had a home in Italy and a farm in Bolinas, the erstwhile hippie community on the Marin coast whose lack of water and resulting freeze on any new home-building had made it one of the most exclusive places around; the Tompkins-Buells and Mark Pincus were among his neighbors. But his base was a comparatively modest mansion in Pacific Heights. He owned a gourmet chocolate company in the Mission too, along with some commercial real estate, where he had a partnership with the veteran contractor and investor Angus McCarthy, a leader among the "Irish builders" who had a lot of clout in town.

Like so many of the city's wealthy residents, he was fed up with progressive dogma and conditions in the streets, and had created TogetherSF to support causes and politicians he believed in. But he didn't think that was enough. Like Warren Hellman before him, he thought the city needed more and better journalism to shed light on its inner workings, build engagement in civic affairs and celebrate the glorious place that San Francisco still was and could be.

The slow-motion collapse of journalism institutions that had dominated for most of the twentieth century was ongoing, both nationally and locally. The *Examiner*, now owned by onetime mayoral candidate, political consultant and real estate investor Clint Reilly, was a shadow of its former self, with fewer than a dozen reporters. The *Bay Guardian* and *SF Weekly* were gone. Nonprofit outlets like *Mission Local,* the *San Francisco Public Press* and *The Frisc* were spirited but tiny. The *Chronicle* trudged along with a newsroom of around a hundred and fifty—about a quarter of what it was back in the mid-2000s—and branded itself "the newspaper of Northern California," with the city sometimes looking like an afterthought.

One could make a case that when it came to national political journalism, and niches including technology news, the decline of newspapers and magazines was mirrored by the rise of new media outlets. For local news though, where big-city newspapers once routinely employed many hundreds of journalists and even small towns had their own dailies, the Internet business models simply hadn't worked. San Francisco was still in better shape than most places when it came to local journalism, but that wasn't saying much.

Moritz thought it was time for something new.

He had no particular answer to the business question, but the venture capitalist in him couldn't help but think that maybe there was an opportunity there too. His political group, TogetherSF, launched Here/Say Media in early 2021, hiring a half dozen people to create what turned out to be a confusing hodgepodge of random news stories and interminable video features. Moritz then decided to get serious about it, putting the publication into a for-profit LLC and retaining a headhunter to find an editor in chief. In the spring of 2021, the headhunter found me.

I'd never gotten to know Moritz during my years covering the tech industry, though I was well aware of both his business acumen and his interest in local politics, having covered his pension reform initiative at *The Bay Citizen*. I hoped his early career in journalism might have inoculated him against the ignorant and arrogant attitude that tech executives typically brought to the news business—namely, that there was nothing wrong with it that proper Silicon Valley thinking couldn't fix. My job was to build a serious publication with top-shelf journalism, laser-focused on the city, that would somehow make money.

I'd ultimately do okay with the first part at least.

My initial task when I joined in September 2021 was to change the name. Here/Say was a comically poor choice for a news publication that aspired to be a trusted source, and as we brainstormed alternatives, I threw out *The San Francisco Standard*. It felt weird to propose something so resonant with *The Industry Standard*, and I made it clear that I wouldn't push for it, but others liked it too, and a couple of months after my arrival we rolled out the rebrand, which we'd completed with surprising alacrity. Building the newsroom would be tougher, and I'd made the hiring even more difficult by insisting on four days a week in the office at a time when most newsrooms—and in San Francisco, most white-collar organizations of every kind—were still all-remote. Creating a news organization from scratch is challenging in the best of circumstances, and so is running a high-quality, hard-hitting, fast-paced publication where accuracy and good judgment are critical. I didn't think I could do it via Zoom and Slack.

At first we occupied part of a lovely loft space at Mariposa and Hampshire Street, an inviting part of the Mission populated by restored industrial buildings, workshops and studios mixed in with small apartment buildings and modest duplexes. It was awkward though: Our office mate was TogetherSF, and they were deeply involved in the political wars that we were covering. After a few months we moved to an aging four-story building that Angus McCarthy owned at Tenth and Brannan, amid a tangle of freeway viaducts in the SoMa neighborhood known as Showplace Square.

Not long ago this had been a bustling area, with the gorgeously renovated onetime battery factory at 888 Brennan that housed Airbnb, the big Zynga building just up the street, a couple of new apartment complexes and plenty of restaurants. In the fall of 2021, though, it was something else entirely: Tent encampments stretched for blocks on the sidewalks beneath the freeways, while disheveled addicts nodded in alleys and makeshift shelters. Small fires burned, sometimes escaping to scorch nearby vehicles, or even on occasion those on the highway above. A 24 Hour Fitness was open, but otherwise commerce in the neighborhood was minimal, with most of the offices and restaurants closed, and hardly anyone about except the homeless, the junkies and the occasional gym rat. It was a daily reminder of what was at stake as the city staggered through the pandemic.

From a news standpoint, at least, *The San Francisco Standard* had picked a good time to launch. The district attorney and school board recalls were both charging ahead, and it was becoming very clear that there was going to be no snapback to the pre-pandemic boom. The media environment we entered, though, was treacherous.

The recalls came at a moment when the news and information business had taken yet another turn, after the fall of blogging and the rise of social media, with the growing popularity of chatty, informal podcasts and individual "influencers." The San Francisco startup Substack, founded in 2017, was proving a popular platform for people who wanted to launch their own email newsletters, a venerable format that was gaining fresh traction. It resembled nothing so much as Federated Media, the blog network John Battelle had launched back in 2005, though this time around it was based on subscriptions rather than ad dollars.

The inspiration for Substack, created by three friends who'd worked together at a Canadian messaging app called Kik, was to bypass the perverse incentives of social media, which rewarded attention-maximizing conflict and controversy rather than insightful writing or original journalism.

"If we could build a system where writers can get paid directly by their audiences, and especially if those payments are coming in the form of recurring subscriptions, you change the dynamics of who can win and who can lose," cofounder Hamish McKenzie explained in discussing the origins of Substack. "The guy who became president in the 2016 elections was especially adept at playing that maximized-attention game. Our starting Substack was not a response to Trump being elected, but it was a response to the culture of the time that was being fed to an outsized degree by the incentives of the social media system."

There was all sorts of writing to be found on Substack, from fashion to architecture to food to history. It proved especially attractive to a subset of political writers who were part of the move toward more ideological, personality-first punditry that was already stealing big chunks of the mainstream media audience. Social media had been rewarding contrarianism and argument for years. Substack now offered a platform that quickly became a medium of choice for a new generation of right-wing media personalities who were leading the charge *against* San Francisco values.

Among the Substackers who got into the San Francisco–bashing business was Michael Shellenberger, a Berkeley intellectual who'd made a small name for himself as a climate-change skeptic. In 2022, he'd weigh in with a book called *San Fransicko*, explicating how the progressive left had ruined the city with delusional policies on drugs and crime. (I asked him at a book reading whether he was worried that the derogatory title might make it harder to have the tough policy conversations needed to solve complicated problems. "No, that was very deliberate," he assured me. San Francisco progressives weren't merely wrong, according to Shellenberger, they were literally sick in the head.)

The Free Press, launched by disaffected *New York Times* columnist Bari Weiss, became a major national platform for moralizing against the sins of the progressive left and inevitably devoted plenty of pixels to San Francisco.

Nothing, however, exemplified the new style of commentary better than the *All-In* podcast, hosted by investors David Sacks, Chamath Palihapitiya, Jason Calacanis and David Friedberg. Sacks, a South Africa native, had been a far-right agitator since his Stanford days, where he'd bonded with Peter Thiel—who'd also grown up in South Africa—to make *The Stanford Review* a tribune for combative right-wing politics. Palihapitiya had earned a nine-figure fortune as an early Facebook employee before turning to investing, where he'd become well-known for hyping dubious vehicles known as SPACs. Calacanis had successfully started and sold a blog network called Weblogs back in the 2000s, but his big score was a $25,000 angel investment in Uber—eventually worth more than $100 million. Friedberg, also a South African by birth, had a successful career at Google before starting a weather company that he later sold to agribusiness giant Monsanto.

None of them had any experience in politics or local government—

Palihapitiya didn't even live or work in the city and was openly disdainful of it—and the podcast was nominally about business. But they had a lot of opinions about local public policy, and *All-In* and the Twitter feeds of its protagonists would become the hub for a social media shout-fest about the perfidy of Chesa Boudin and other local progressives. The growing swarm of media personalities who were less journalists or pundits than cheerleaders for Trump and his MAGA movement couldn't get enough of it, and they reinforced one another's messaging: Former Fox News host Megyn Kelly, for example, had Sacks and Calacanis on her show to talk about the evils of Boudin and progressive prosecutors in general. It was a bit much for anyone outside the MAGA bubble—a partisan interviewing other partisans with the same opinions, on a topic where none of them could claim even a modicum of expertise. Boudin actually called into an *All-In* chat at one point and quickly exposed the shallowness of their criticisms, but the facts weren't really the point.

The industry that grew up around shaming San Francisco, some of it funded by the political right, included numerous Twitter feeds devoted to photos and videos of the suffering in the streets. Ricci Wynne, a convicted drug dealer who styled himself a neighborhood activist, would become a Fox News star with his videos of homeless people and drug use—only to be arrested in 2024, with $80,000 in cash on hand, for pimping. JJ Smith, a genuine street activist, was more sincere in arguing that he was doing a public service, though the subjects of his photos and videos didn't always agree. Smartphone-equipped residents who spotted something ugly were adding to the fray. It was all extremely damaging to the city's reputation and helped drive the exodus from downtown.

A handful of local personalities were amplifying the message too. Susan Dyer Reynolds, who ran a neighborhood publication called *Marina Times* that later morphed into the nonprofit *Voice of San Francisco*, took justifiable pride in having been early in digging out some of the city corruption scandals. She was often shunned by the local journalistic establishment, and was rightly outraged when supervisors tried to cut her publication out of city advertising contracts on the grounds that it was a political operation. Her relentless and angry Twitter feed, though, rarely missed a chance for an ad hominem attack on Chesa Boudin, homeless advocate Jennifer Friedenbach or other progressives she held in contempt. *Voice of San Francisco* was also funded in part by Neighbors for a Better

San Francisco, reflecting right-wing political donors' national strategy of subsidizing like-minded blogs and podcasts to promote their views.

To be fair, Friedenbach and other top targets, including State Senator Scott Wiener, could give as good as they got, often attacking their opponents as racists or fascists who had no heart. And it wasn't only the right-wing media who found the city's troubles too delicious to ignore. CNN would weigh in with a documentary titled, *What Happened to San Francisco?*, replete with homeless young people describing how they'd come to San Francisco for the drugs. ABC, CBS, *The Washington Post* and seemingly everyone else would take their swings on the homelessness, shoplifting, car break-ins and of course the drugs. It was all a bit embarrassing as locals explained to friends and relatives that yes, we were okay, and no, it wasn't too dangerous to walk the streets. The worst was confined to just a few places, prospective visitors were assured, and most of the city was its regular, beautiful self.

But also: Never, ever leave anything in your car. The ubiquitous auto break-ins were known as bipping, local slang said to be derived from the gentle sound of special tools breaking glass. Among San Francisco's many nicknames, the least complimentary is surely Bip City.

For all the heated political drama, a big part of the crisis in the streets could be boiled down to one word: fentanyl.

The cheap opioid was ripping through the city's underclass like an out-of-control reaper, decimating everything in its path. Longtime city residents were accustomed to people living on the sidewalks, but now the homeless and indigent seemed to be in much worse shape, often stumbling about doubled-over, or laying face down on the pavement, challenging passersby to see if they were dead. The city recorded 222 overdose deaths in 2017. In 2021, there were 641, and the numbers were still rising.

At the end of the year, as fentanyl overdoses surged, Mayor Breed declared a state of emergency for the Tenderloin, including a crackdown on drug dealing. But the police weren't very interested in arresting street-level dealers, complaining that they were only going to be released anyway by the overly lenient Chesa Boudin. It was an attitude that looked like laziness, or insubordination—and often it was—but the cops also had a point. As we'd show in *The San Francisco Standard* a few months later, Boudin's office had recorded just three convictions for the crime of drug-dealing in the city in all of 2021. About half of the hundred-odd people who were charged with that crime were sent to diversion programs or saw their charges dismissed, while most of those who were convicted—forty-four in total—pled guilty to a different, lesser charge of accessory to a felony.

The reason for that, it turned out, was that the drug dealers were overwhelmingly undocumented immigrants from Honduras, whose nationals controlled the local fentanyl trade. A drug-dealing conviction meant automatic deportation, and in the view of public defenders—and Boudin's office—that was too harsh a fate for the people in question. "I can tell you that I have represented young kids from Honduras, and . . . many of the kids are victims of trafficking," said Francisco Ugarte, who worked in the immigration defense unit of the public defender's office. After the story was published, we took a hail of criticism accusing us of racism, but the reporting certainly seemed pertinent to the debates over crime and drugs. I wasn't surprised by the reaction, but to my eye the advocates had lost touch with what their program looked like to people who weren't already bought into the progressive program. The commonsense view, even in San Francisco, was that fentanyl dealers would be the people you *want* to deport, even if you were otherwise sympathetic to undocumented immigrants.

Without buy-in from the police or the DA, and facing active pushback from the advocacy community, Breed wasn't going to get very far with law enforcement solutions to the fentanyl problem. She did, though, stand up the second big piece of the emergency policy, a new drop-in social services location that would be called the Tenderloin Linkage Center.

With so many people dying, harm-reduction advocates had been clamoring for the city to establish a so-called safe-consumption site, where people could use drugs in a peaceful setting under the supervision of social workers armed with naltrexone, the overdose reversal medication. That was difficult, though, for a pretty basic reason: It would be illegal under state and federal law. Breed said later that she had explored the issue and even visited Vancouver to see how that city's safe-consumption program was working. But after Governor Newsom vetoed state legislation that would have legalized such an effort, Breed pulled back, and turned to the Linkage Center idea as an alternative. "I brought together a bunch of people, and I said, we've got to have a place for people to go," she recounted. "I didn't say we have to have a place where people go to use drugs. I said we have to have a place that when the cops pick up someone for drug use, instead of an arrest, they could take them to the Linkage Center, whose job is to help them through the process of trying to get into health treatment, support or whatever it is that they need."

The center was set up on Mid-Market, about a block down from UN Plaza, with a large tent and a tall fence surrounding it all. Matt Haney, the supervisor for the Tenderloin, echoed Breed in declaring his support. "A drop-in center where people can get off the streets and immediately [be] linked to services, placements and care, without delay or bureaucracy is something we desperately need," he said.

But the Department of Public Health, and the nonprofit HealthRIGHT 360, which had been contracted to operate the center, apparently had a different idea. They provided hot meals, showers and laundry, as well as Covid and HIV testing, to anyone who walked in, but referrals to drug treatment weren't necessarily part of the program. Instead, people were offered a chair in an outdoor area where they could use drugs. It was an unofficial safe-consumption site, and rather than getting people off the streets and into treatment, the Linkage Center became a magnet for dealers and addicts alike—and a highly visible one. Journalist-activists including Michael Shellenberger employed various techniques to document the drug use inside, rendering the Linkage Center another colorful episode in the running drama of San Francisco's decline. London Breed would do an inquiry of her own.

"I sent a few people over there, addicts, people that I grew up with that I've been trying to get into treatment, and I said, 'Tell me what's going on,'" she recounted later. "And one of my friends came back and he said, 'London, it's like they're happy we're on drugs.' I said, 'What do you mean?' And he said, 'We walk in there, and they're like, "Do you want a sandwich?" I'm like, "No, I don't want a sandwich, I'm trying to get housing." I told them, "I'm sick, and I want some help." And they said, okay, and then they were asking me all these questions, and I just couldn't do it. And then they told me to go outside. And then everybody was out there using dope, and all I wanted to do was use dope.'"

Breed said her friend ultimately got into treatment and is doing well, no thanks to the Linkage Center. "But he basically said, that place is a fraud."

By mid-year, after a weak effort at rebranding that removed the misleading term "linkage" from the name, Breed signaled that the Tenderloin

Center's days were numbered. It would close in November, but not before it had become a $22 million symbol of the city's conflicted and dysfunctional approach to the fentanyl epidemic.

Not everyone agreed that the Linkage Center was a failure, or that it should have closed. Politicians on the left, and nonprofit advocates led by HealthRIGHT 360 and the Harm Reduction Coalition, would tout the 350–400 "guest" visits a day, and the 320 overdoses that were reversed during its eleven months of operations, as indicators of its value. There were plenty of anecdotes from people who'd gotten some much-needed help.

"People have come to depend on the kind of care that they have been receiving, and without it . . . they will become unhealthy [and] they will literally die," said Sara Shortt, the housing activist who'd been involved with the Google bus protests and now worked as director of public policy and community organizing at the nonprofit HomeRise. Still, whatever support it may have provided for a small number of people, any clear-eyed public policy analysis could only conclude that running an illegal safe-consumption site in a prominent location on Market Street was a terrible idea, especially without a parallel effort to address all the problems on the surrounding blocks. The overdose-reversal numbers were a questionable success too, since the same people were often revived many times, and may have taken more risks with their dosages knowing that help was standing by.

The divide over drug policy extended far beyond the Tenderloin Center. The fentanyl crisis was killing people all over the country and testing the capacity of local governments everywhere, and especially in the Democratic-run cities on the coasts. It was good fodder for the partisan media environment—and for the rise of video as the dominant medium across the Internet. The right-wing outrage over smartphone clips of drug addicts acting like drug addicts, or the mentally ill acting mentally ill, was mainly performative: The MAGA media were much more interested in demonstrating their righteousness and winning followers than seeking out facts or digging into possible solutions. At the same time, the emergency was real, and the insistence of advocates that more money was the answer to every problem rang hollow in the face of fiascos like the Linkage Center. More and better policing was obviously needed, but that was far from a complete answer by itself. There were no simple fixes.

The ideological battles that undercut the fight against fentanyl were seriously undermining efforts to address homelessness too, even after the city won the legal battle over Prop C and had fresh cash to spend. Jennifer Friedenbach and the Coalition on Homelessness infuriated Breed and many others with a lawsuit that barred the city from dismantling tent encampments unless beds could be provided for everyone. Friedenbach also had a big role on the committee that decided on how the Prop C funds would be spent, and was uncompromising on the housing-first approach, which had become gospel among progressives. As a practical matter, that meant that drug use could not be considered when deciding who got subsidized housing, with the predictable result that there were a lot of drugs—and overdoses—in the city-funded supportive housing facilities. For many moderates, let alone the ever-more-vocal critics on the right, that didn't look like a way to solve the problems.

Sharky Laguana, who'd once been homeless himself and later had some success as a musician, was among those who soured on what he viewed as the scorched-earth political tactics of Friedenbach and her ally Christin Evans. Laguana was now the prosperous owner of a van-rental company—he'd been inspired to launch it back in 2003 after his own band's vehicle had broken down one too many times—and he'd gotten involved in city affairs, serving first on the city's Small Business Commission and then on the Homelessness Commission, which was established in 2022. He agreed with key tenets of the advocates' views. "There is no data point with a higher correlation to homelessness than the cost of housing," said Laguana. It was also true that the longer someone was homeless, the sicker they would get, a central component of the logic behind housing-first. Homelessness was too often conflated with drug use and street crime, and had become "a catch-all term for street behavior," Laguana noted, when in fact there were very different cohorts of homeless people, with impoverished seniors among the fastest-growing categories.

Still, San Francisco for many decades had attracted homeless people for all sorts of reasons, and the insistence that anyone on the street deserved a home with no questions asked, at taxpayer expense if necessary, was wildly unrealistic. The average stay in a supportive-housing unit was now ten years, and the city already had sixteen thousand such units, Laguana said. "No matter what we do, we can't help everyone." It was also extremely difficult to spread the subsidies fairly, resulting in what he called a "system of lottery tickets."

Friedenbach and Evans, he said, often seemed heedless of the complexities. "They built up a lot of political power, and gained advantage by painting opponents as inhuman monsters," Laguana said. On top of that, Laguana and many others saw the litigation against clearing encampments as "governing through lawsuit that's subverting the will of the voters." Laguana ended up breaking ties with Evans, whom he'd considered a friend.

Distrust of law enforcement was ingrained in progressive Democrats and libertarians alike, but the drug problem and the property crime were putting those attitudes to the test. Chris Larsen, the serial financial-technology entrepreneur who'd founded the crypto firm Ripple, would find himself in the thick of the argument over policing after he took it upon himself to fund surveillance cameras for the city's neighborhood business associations, as a means of stemming street crime.

Larsen's Ripple had survived the down cycles and was prospering on the back of the crypto boom, and Larsen was flush, certainly on paper. The value of the XRP tokens he owned had put him in the billionaires' club, though as a native San Franciscan from a working-class family, he didn't cut the profile of a stereotypical "crypto bro."

Larsen had started underwriting the video-camera network back in 2017, with the intended aim of helping local merchants keep an eye on their streets. Starting with a handful in Union Square, the network eventually numbered over a thousand cameras, operated by a private contractor but also accessible to the police. Like a lot of people in tech, Larsen saw cameras, license-plate readers and even drones as obvious crime-fighting solutions. He'd eventually write checks to the tune of $4 million.

Larsen was open about the program, and there were protocols in place around police access to the video and how long it would be stored. He wasn't oblivious to the potential for abuse. "Too much safety likely means too many disadvantaged citizens losing their right to justice, as we saw with 'stop and frisk' policing and 'three strikes' sentencing," he'd write in defense of the camera program. "Privacy has equally complex trade-offs; too much of it can enable criminals to hide and exploit the innocent."

Still, a lot of people saw the cameras as a dystopian invasion of privacy, the rich paying for tools of repression on behalf of the police. When cameras were proposed for the Castro, local political groups such as the Alice B. Toklas LGBTQ Democratic Club came out in furious opposition.

"Queer people and people of color have historically been the victims of targeted violence from law enforcement," the Toklas club said in a statement to the press. "We encourage everyone who values (what) makes the Castro special to speak out against this dangerous plan." The cameras were eventually rejected in the neighborhood, but that was the exception rather than the norm.

Civil libertarians remained very suspicious, and after a George Floyd protest in Union Square that had deteriorated into vandalism, the Electronic Frontier Foundation would report that the Police Department had accessed live feeds of the Larsen-funded cameras to surveil attendees. The EFF, along with the local branch of the ACLU, would sue the city for violating its own policies and would continue to fight broad use of surveillance technologies by government agencies and their proxies. Freedom from government monitoring and the right to privacy were foundational tenets of the EFF, even if they were getting harder to uphold.

Cindy Cohn, the EFF executive director, said the data didn't support surveillance cameras as a means of bringing down the crime rate. "What you usually see is an initial dip, and then it comes right back up. Just like police body-worn cameras don't actually make the police better, right? There's no demonstrable impact on police violence against anybody as a result of body-worn cameras, but yet we've spent billions of dollars because we really want a simple technology to solve what are societal problems."

Larsen was having none of it. "Shame on them," he said of the surveillance camera critics, including the EFF. "They just optimize on one thing as if there is no public safety problem, there's no criminal justice problem. It's super-arrogant. Everybody needs to be balancing out, you know, what's a good city look like."

The EFF, in fact, was struggling with some internal divisions of its own. The group prided itself on its nonpartisan commitment to principles, centered on the right to navigate the digital world free of government monitoring and coercion, that transcended the left-right divide. Classic libertarianism and the new economy "communitarianism" exemplified by Burning Man could find common ground here, alongside mainstream liberalism. Over the years though, the EFF's staff of lawyers and researchers, now more than a hundred-strong and working out of a small building the group had bought near Civic Center, had developed an unmistakably progressive Democratic hue. There were differences over policy issues like whether the EFF should back social media company moves to take down a mass murderer's manifesto. Cofounder and longtime board member John Gilmore was particularly upset with the EFF weighing in to support the removal of Free Software Foundation founder Richard Stallman from the board of his own nonprofit, after he'd alienated many people at the EFF and elsewhere with his behavior, especially toward women. There was friction over internal issues, including a staff unionization drive. Remote work made it all even more difficult to manage.

Tensions would come to a boil in 2023 when Gilmore, in a board meeting on Zoom, made a barbed joke about progressive speech policing, displaying his pronouns on the screen as "Pro/Noun." A lot of people didn't get it, and the incident was something of a last straw for those who thought Gilmore was no longer a good team player—including his old friend Cindy Cohn, whom he'd brought into the organization back in the 1990s. Gilmore had a long history in the sex-positive community and wasn't at all opposed to the transgender cause, but he declined to apologize, and was rebuffed by Cohn in his efforts to talk with those who might have been offended. Soon after, he was voted off the EFF board, where he had served since its founding more than thirty years earlier.

Gilmore says he never got a clear explanation of why he was ousted, but said "the staff's adoption of intolerant progressive politics" had created a lot of strife. Cohn, for her part, said it was Gilmore who had failed to see how his actions, as a revered founder, were weighing the organization down. The split was ultimately generational as much as it was political. The interpersonal norms of the Internet's earliest days, where rough argument among idiosyncratic and mostly male technologists were a routine part of the culture, felt alien to a lot of the younger generation. Times had changed.

The dynamics at the EFF reflected one aspect of the tensions that had simmered below the surface of the city's tech industry from the start. The change-the-world ethos of the Internet economy had emerged from the counterculture-inflected, quasi-libertarian idea that companies and organizations, seamlessly connected with technology, could create both a more humane present and a more promising future. There was a natural affinity with Democratic liberalism in the egalitarian- and human-rights–friendly dimensions of that ethos, even if there were differences on the government's role.

The disillusionment that came with the "techlash," and the realization that the "new" economy was just a slightly different type of profit-maximizing machine, had undermined many of those ideals. The delicate ideological balance that had kept even the super-rich leaders of San Francisco's Internet industry, and much of Silicon Valley, in the left-liberal political lane was coming undone. The leftward lurch of Democrats on social justice and economic issues during the pandemic was partly a cause and partly a consequence of the breakdown of what one might call the Burning Man consensus. So was the far-right turn of key figures in the tech elite.

For many of the young and ambitious, doing well and also doing good was falling away as their highest aspiration. Greed is Good 2.0 was now at hand.

If there was a rightward shift taking place citywide, part of it stemmed from this generational change in the Internet industry.

The city was much wealthier, and older, than it had been in the 1990s and the 2000s, with a lot of young creatives and activists—who were also a big part of the service industry—driven out by soaring rents. Many first- and second-generation San Francisco tech entrepreneurs were now well-to-do middle-aged parents; they might not have minded—or even noticed—stepping over passed-out junkies outside their offices in SoMa back in 1999 or 2010, but it was a different matter when their kids confronted it at the playground. The dysfunctional school system, the never-ending car break-ins, the squalor plaguing the Tenderloin and much of SoMa, the broken process for home repair or renovation permits, the steep taxes, the crappy public transit, the corruption . . . it was all just too much.

Jonathan Nelson, the Organic Online founder, had been away for a few months with his family in 2022 when his house in the upscale Dolores Heights district, not far from where Mark Zuckerberg once lived, was ransacked and then occupied. "I have a junkie living in my basement," he recalled. "We call the cops and we say, 'We have cameras, we have photos, they grabbed quite a bit of stuff, here's the list.'" The police did little. Nelson's wife then discovered some identifying information the squatters had left behind in the trash. They presented it to the police. "They say, 'What would you like us to do?' And I say, 'Arrest her!'" But it was for naught.

For Nelson, the city's careless attitude about the break-in at his house wasn't the only issue; he was also fed up with what he saw as its arrogance toward the business community. He recalled a conversation he had with Aaron Peskin back in the mid-2000s, after Nelson had been named head of the Committee on Jobs. "He goes 'Look, Jonathan, here's the deal. This is a really beautiful city, and you're always going to want to live here, and tourists are always going to want to come here. And so it doesn't really matter, a lot of this stuff.' I'm like, are you crazy? It's not true. I sit at Omnicom and we've got two hundred operating companies. Here you've got this gross receipts tax, you've got Healthy San Francisco [the city's universal health-care plan], you've got all this stuff, and if I moved to San Mateo I don't have any of this. Or I could just double down in New York where we actually have public transport and infrastructure, which we don't have in San Francisco."

Public safety and business taxes weren't the only issues underlying the new moderate wave. A backlash against the slow-growth and anti-gentrification sentiments that often stood in the way of development had been bubbling for years in the form of the YIMBY movement, spearheaded by an Oakland teacher named Sonja Trauss, and it was now a statewide force. The YIMBYs won enthusiastic support from techies, who were baffled and dismayed by how hard it was to build housing in a city that needed it so desperately.

Scott Wiener emerged as a YIMBY champion and successfully shepherded major pro-housing laws through the state legislature in 2021 and 2022. A new, "moderate" platform was emerging, powered by tech money and Asian voters: build more housing, bring down the hammer on drugs and crime, recall the school board and recall Chesa Boudin.

School board recall leaders Autumn Looijen and Siva Raj had done extremely well with their volunteer signature gathering and $99 contribution limit—a true grassroots campaign. But as the fall deadline for producing more than fifty thousand valid signatures came into view, it was clear they weren't going to make it. They asked the Facebook group if they should scrap the limit and raise money for paid signature gathering, and after much hand-wringing it was agreed. The donations quickly flowed—the first big slug of $50,000 came from entrepreneur and investor Garry Tan, who was soon to be named head of Y Combinator—and the group quickly had a war chest of a few hundred thousand dollars. There was a problem, though: They were very late in the game to hire a signature-gathering firm, and if they could find one at all they'd likely be stuck paying a huge premium for each valid signature. It was a big break, then, when they were able to piggyback off the recall of Chesa Boudin.

Mary Jung's campaign had been well-oiled from the start, and there were people collecting signatures all over town. Looijen would see them at the farmers' markets, and since it was a piecework business—an agreed amount of money for each valid autograph—she began asking if they might solicit signatures for her petitions, too, for some extra change. It turned out to be a good deal all around. It was actually easier to get people to sign the school board petitions than the one for Boudin, Looijen recounted, and there were three of them (one for each school board member), so the signature gatherers were motivated to take it on. The campaign would also get a fresh burst of support from the tech world when another issue bubbled up, one that had nothing to do with the pandemic but was very close to the hearts of scientists and engineers: the district's approach to math education.

In the early 2010s, in an effort to promote equity, SFUSD had revamped its math curriculum and no longer offered algebra for eighth graders, even those who were ready for it. If all students moved along at the same pace, proponents argued, the weaker students would do much better and the stronger students would be just fine, and the data showed that's just what was happening. But some parents of high-performing students, who hated the idea of their children being "held back" in the name of equity, were dubious of the data. They did their own analysis, and approached Patrick Wolff's Families for San Francisco to publish it, which they did in the fall of 2021.

The math issue quickly became part of the debate. How could the tech capital of the world fail to appreciate the importance of math? The practice of "tracking," or putting the smarter or more educationally advanced kids together in the same classes, had been controversial for years. It had obvious appeal for high achievers, but there was plenty of evidence that it hurt weaker students by branding them as dumb. Arguments over public school curricula, in fact, are as old as the country itself: Horace Mann, America's original advocate of universal public education, fought heatedly against the idea that people needed to learn the alphabet in order to read.

But the San Francisco "math wars" that began in 2021 didn't allow for much nuance. SFUSD was standing in the way of children learning faster, critics charged, and in our hypercompetitive age, that just wasn't going to cut it. That the district allegedly massaged the assessment results to make the math program look effective made it all the more outrageous.

With all of that, the recall campaign had the wind at its back. Progressive politicians and others opposed it on the grounds that recalls were a blunt weapon that moneyed interests could use to target people they didn't like, and this one was also a waste of time and money because the regular election was only a year away. Tim Redmond, the former *Bay Guardian* editor, had founded a nonprofit news site called *48 Hills* to carry the progressive flame, and he backed the arguments against test-based admission to Lowell and supported the renaming discussion too. "This is part of a national conversation, one that we need to have, about the nation's history of racism and colonialism," he wrote. "San Francisco ought to be at the front of that discussion—not playing into Fox News–style narratives about how silly it is to criticize Washington and Lincoln—or 'even Dianne Feinstein.' "

By the time 2022 rolled around, though, defenders of the Board were increasingly hard to find. Collins, Lopéz and Moliga were recalled in a landslide, with 76, 72 and 69 percent of the vote against them, respectively.

Boudin also faced an uphill battle. He was fighting for his political life when he and I met for coffee ahead of a live video interview we'd be doing at the offices of *The San Francisco Standard*. I told him that I would start out tough, going straight at the allegations that he was soft on crime, and then we'd move into a broader discussion of his record and accomplishments, and finally to a more philosophical conversation about law and justice. "I love talking philosophy," he joked, and our preliminary conversation was easy. I noted that one goal of the interview was to be interesting—and that turned out to be the only thing he took from our chat. We'd barely gotten started when he ripped into me, accusing me of lying and carrying water for the right, and then proceeded to litigate every question into the ground. I don't think it helped his cause.

Boudin was recalled in June 2022 on a 55 to 45 percent vote and replaced by Brooke Jenkins, who had worked for him in the DA's office but turned against him in part because of how he'd handled a case involving a relative of hers. Jenkins gave the recall a huge push when she quit her job and did an interview with the *Chronicle* accusing Boudin of dereliction of duty. Overnight, she'd gone from being his employee to becoming the chief spokesperson for his ouster. Breed named Jenkins DA following the recall, and Jenkins vowed to take a much harder line than her predecessor on drug dealing, property crime and much else.

Awkwardly, as we reported in *The Standard*, Jenkins claimed to have been a volunteer for the recall campaign, but that turned out to be a lie: She was in fact paid some $175,000 by an affiliate of Neighbors. There'd be no sanction for the deception though; it was just Family business.

The school board and Boudin recalls brought tech industry money, and personalities, into the local political fray like never before. It joined the newly rich, and a newly mobilized Asian American community, with the city's dynastic family fortunes in a common front against progressive policies. It papered over for a minute the gap separating far-right ideologues like David Sacks, or even establishment Republicans like Bill Oberndorf, from middle-of-the-road Democrats like Mike Moritz and Chris Larsen.

Progressives saw the new alliance as the rich circling up to prevent pandemic-era social programs from becoming the new normal, and didn't always see much of a distinction between the hard-right Sackses and the centrist Moritzes of local politics. What neither side envisioned in 2022, though, was that just two years later San Francisco's small coterie of far-right tech overlords would ascend to power in Washington.

Elon Musk had been a loud and prolific voice on Twitter for years when he began joking about buying the company in 2021. But the future of the business was no laughing matter. The company's financial results and user growth were uneven, and the product strategy was fuzzy, with missteps like the acquisition and subsequent shutdown of the video app Vine. CEO Jack Dorsey seemed to have lost interest in running the company. He was still revered among swaths of the rank-and-file—building two industry-changing startups is exceedingly rare—but Twitter was drifting, and the cultural dysfunction that traced to the early power struggles ran deep. Dorsey didn't seem eager to reengage. It was hard to argue that the company didn't need new leadership.

The prospect of Elon Musk as the owner, unsurprisingly, was viewed with horror in many quarters of the company, even if he wasn't yet the political extremist he'd become soon after. An immigrant from South Africa whose father had been a fervent supporter of white rule, Musk believed that Twitter was the vector for a "woke mind virus" that would destroy civilization if it wasn't stopped—and unlike most peddlers of ideological fantasies, he had the money to act on his theory. It seemed quite possible that his intention was to convert the "digital town square" that company's founders imagined into a megaphone for his own idiosyncratic but increasingly right-wing politics.

But there was a cohort at the company that believed a bold innovator like Musk, who intuitively understood Twitter, could be just what the company needed—and that cohort included Jack Dorsey. When Musk made a $44 billion offer in the spring of 2022, the company's board of directors, led by an entrepreneur and former Salesforce executive named Bret Taylor, accepted the bid on the grounds that it was the best deal for shareholders, whatever Musk might have in mind. Dorsey, or even Ev Williams, probably could have blocked the deal if they wanted to, but no one had a better idea.

There was an odd bit of drama where Musk said he'd changed his mind and withdrew his binding offer, only to have a Delaware court force him to go through with it. All of that, though, was almost certainly a gambit to try to get a lower price. Musk also apparently violated federal securities law by failing to disclose that he'd bought almost 10 percent of the company's shares before launching his takeover bid. When the deal closed in late October 2022, he immediately sacked all the top executives and brought in managers from his other companies, Tesla and SpaceX, to carry out swift and merciless mass firings. Within a few months, about two-thirds of Twitter's seven thousand employees were gone.

It was a devastating turn of events for those who'd given their hearts to the company. For Del Harvey and her content moderation team, there was the extra sting of Musk denigrating their work and the very idea of what they were doing. He even funded anti-woke journalists Matt Taibbi and Bari Weiss—and gave them selective access to internal emails—to conduct an "investigation" of supposed bias and political influence in Twitter's moderation practices, a project that was disingenuous at best given that the conclusions were predetermined. It's true that the government leaned on Twitter and Facebook to do certain things, but mostly those things—such as blocking dangerous Covid disinformation—were consistent with what the companies were trying to do anyway. They didn't take orders from the government; that just isn't how it worked. But Musk's enthusiasms included racist, far-right ideology and rhetoric that was often violent and abusive—the kind of talk that had been considered out-of-bounds by almost every media outlet of every type until the rise of Trump. Normalizing such discourse required discrediting the whole idea of content moderation, even though it couldn't be abandoned entirely, even by Musk.

"I just watched as my baby got murdered in front of me," Del Harvey said later. She'd left the company in 2021 after her other babies—her three- and five-year-old daughters—invented a game called "work," which was a little too spot-on. "They would say, 'Oh, I gotta go, I gotta go hop on a call. But after this call, I'll be free for about half an hour,'" Harvey recounted with a sigh.

But having been gone before the storm was small comfort. "I don't know if it was easier being on the outside, or harder in a lot of ways," Harvey said. Among his many cruel acts, Musk had gone out of his way to humiliate Harvey's successor, Yoel Roth, who'd made a point of being open-minded about Musk, who in turn praised him—before deciding he was pro-pedophile, based on nothing more than lurid conspiracy theories. Roth was targeted by the online mob and forced to go into hiding.

"There were some really great people doing some really groundbreaking work, and had given up so much to try to make things better," said Harvey. "They got treated like they were just the scum of the earth. It was just brutal. How could you not mourn for that?"

It was a grim denouement for San Francisco's Internet star, and a metaphor for the failed promise of social media. The Internet town square that Ev Williams and Jack Dorsey had envisioned as a forum for free expression and social connection would instead become a vector for propaganda and hate speech, often ruled by virtual mobs that rained vitriol on anyone with a different opinion and humiliated people for fun. In many places around the globe, those mobs translated into physical-world violence too.

The basic mechanisms of social media rewarded conflict and loud voices, and Facebook pushed to make money on that dynamic at every opportunity. Twitter, by contrast, had retained more of its public-service spirit, and Del Harvey and her team had struggled for years with how to tamp down abuse and uphold the liberal values that had shaped the company.

Now that work was being vilified as the spawn of an imagined evil conspiracy. Twitter's new owner was a strange and unpredictable man who reveled in no-holds-barred online argument and was a veritable factory of disinformation himself. It had become clear that he aimed to use the platform not merely to advance his own radical views, but to destroy the conversations of the journalists, activists, academics and other liberals that he viewed as the carriers of the "woke mind virus."

It was a spectacularly successful campaign. Musk was now spearheading the onslaught of propaganda being orchestrated by ultra-wealthy supporters of Donald Trump. And San Francisco's turn as the chief villain of their piece wasn't done just yet.

CHAPTER 12

Doom Loop

The April 4, 2023, murder of Bob Lee, a friendly and gregarious forty-three-year-old who'd made a name for himself with a payment app, was the sort of crime that makes any city dweller shudder. Stabbed three times with a kitchen knife at 2:30 in the morning, Lee was left to stagger along the deserted sidewalks of Rincon Hill, a tony district between SoMa and the waterfront, bleeding profusely and trying to summon help.

Grainy security videos, and a trail of blood, revealed a harrowing scene. Attacked beneath the soaring Bay Bridge skyway, Lee managed to walk a block to Harrison Street and approached an idling car, showing his open wounds. But the vehicle quickly pulled away. He crossed the intersection and reached the entrance of the Portside, a luxury condo building, where he pawed at the intercom and the glass doors in desperation. No one was there. He crumpled to the ground, then after a few moments got to his knees and fumbled with his cell phone. A second car passed, and he stumbled after it, waving. But it was too late.

He was found by the police a few yards away and died in the hospital soon after.

The killing came at a moment when worries about the future of San Francisco, steadily building for years, were starting to tip over into panic.

Bob Lee had worked for Jack Dorsey at Block, and for the growing right-wing tech cohort, his murder was the latest piece of evidence that it was left-wing politics, and especially permissive policies on drugs and crime, that were ruining San Francisco. David Sacks posted a video less than two days after Lee's murder, asserting that the killing was doubtless the work of a "psychotic homeless person." Elon Musk responded: "Absolutely." The comments piled up, with now-familiar rhetoric about the hellscape the city had become, thanks to the left.

The determined effort by Sacks and Musk to make Lee's murder a political talking point was exceptionally cynical, given that the basic facts of the situation made it all but certain that it *wasn't* a random street crime. Lee was killed in a quiet neighborhood of luxury high-rises where violent crime was rare, and he had no evident reason to be there alone in the middle of the night. He was viciously stabbed but wasn't robbed, common indicators that a killing is personal. And sure enough, the police soon arrested Nima Momeni, the brother of a married woman named Khazar Momeni with whom Lee had had a relationship, one that revolved heavily around drugs.

Lee's creation and calling card had been Block's Cash App, which unlike PayPal or Venmo allowed for anonymous transactions. That won it a big following in the crypto world and assorted privacy-conscious subcultures, drug commerce prominent among them. Lee was a Burner, and part of the sex party scene, which sometimes featured drugs like ketamine, ecstasy, cocaine and LSD. By all accounts he was a nice guy, with a lot of friends, known to be generous, gentle and fun.

Like a number of footloose young professionals who'd made quick money on startups or crypto during the roaring 2010s, Lee, divorced and living in Marin County, had moved to Miami during the pandemic. It was the fashionable new hot spot for those who'd grown weary of San Francisco, and along with Austin, it looked like it could become an important new tech hub. With San Francisco mostly closed down and so much disorder in the streets, sunny South Florida, or even Texas, looked very appealing by comparison—especially when you factored in California's steep personal income tax. Miami Mayor Francis Suarez rented a billboard on a San Francisco highway exit near the Twitter building, with his own social media handle emblazoned, aiming to lure the disgruntled: "Thinking of Moving to Miami? DM me."

Lee still had plenty of ties to San Francisco, though, and he was back in town for business, and some fun.

Nima Momeni's trial would show that the fun involved cocaine, ketamine and even a liquid bottle of GHB (known as the date-rape drug), as Lee, Khazar Momeni and a few other friends tooled about the city. They partied at the Battery, the private club favored by the tech set; they partied at the luxury condo of a drug dealer to the elite, not far from the Twitter building; they partied at the Millennium Tower, the city's fanciest high-rise, where Khazar and her husband, a prominent plastic surgeon, each had their own apartments. Nima Momeni ran a struggling East Bay computer service business and had a history of complaints about his abusive behavior, and he also liked his drugs. Apparently enraged because he believed his sister had been sexually assaulted by the drug dealer, he stabbed Lee after the two of them had left Millennium Tower together, leaving him to bleed out on the sidewalk.

Momeni pleaded innocent to the murder, and his family hired high-priced criminal defense lawyers from Miami for the trial. They put forth an implausible story about how Lee had attacked Momeni first, contending that Lee was so addled from the drugs that he'd tried to kill Momeni for no real reason. The jury didn't buy it, and Momeni was sentenced to thirty years.

Once all the theatrical elements were stripped away, the murder was a very sad human tragedy, as such cases usually are, even if Nima Momeni was less than a sympathetic character. But the campaign by Musk and Sacks to tie Lee's killing to city politics had inadvertently underscored an awkward truth. Critics on the right were eager to denigrate the people who came to San Francisco to buy fentanyl, meth or cocaine on the streets, and denounce local officials for turning a blind eye to the drug trade. But plenty of wealthier, healthier, whiter people—people like Bob Lee—came to San Francisco for the drugs too. Musk himself was a frequent user of ketamine, and a Burner, and even smoked a joint on set with Joe Rogan. Lee's drug supplier lived in the kind of new, upscale building that was supposed to help push out the notorious "open-air drug markets" around Civic Center. It was certainly true that drug parties in the Millennium Tower didn't tear at public safety in the same way as fentanyl use on the sidewalks of SoMa. Yet both were made possible by the city's tolerance and history. *Drugs for me, but not for thee* was a peculiar message. The libertarianism—and libertinism—that infused the local tech world cut many ways.

It was just a few weeks after Lee's murder that another shocking death rattled the city. Banko Brown, a young Black trans man, was stealing candy from a Walgreens at the San Francisco Centre mall, a routine occurrence amid the city's shoplifting epidemic. The drugstore, like so many others, had many of its goods locked in plexiglass boxes, but only recently had it told its security guards to carry their own guns if they wanted to confront thieves. Early in the evening on April 27, a guard named Michael Earl-Wayne Anthony saw Brown stuffing candy into a paper bag and tried to stop him. Security video shows them wrestling on the ground. Brown gets up and starts to leave the store, then turns back toward Anthony, who shoots him dead on the sidewalk outside, just a few hundred feet from the tourists at the Powell Street cable car.

Banko Brown lived as far as could be from the lux lifestyle of Bob Lee. He'd grown up in foster homes in the East Bay, and been in and out of juvenile hall, but he'd found a family in the city's Black trans community. At twenty-four, and several years into his transition, he was working as an advocate for young people in foster care and was a regular presence at political rallies. He died just blocks from the nation's only transgender cultural district, which sits in a rugged corner of the Tenderloin. The city had proudly made itself a haven for trans people, but the welcoming posture often wasn't matched by services: There were only forty-five subsidized housing units for LGBTQ youth in the city, while an estimated four hundred minors and young adults remained outside.

Brown was one of them. Broke and homeless, he'd been trying for many months to find affordable housing, to no avail. Shoplifting is a minor crime—and following a 2014 state ballot measure, one that wouldn't land you in jail in California. But social media videos of gangs ransacking clothing and drugstores while guards looked on cast "shoplifting" in a very different light. It now looked less like a trivial nuisance than an element of the bedlam that was making life a nightmare for many businesses and driving people out of the city.

Brown's death quickly became a political football too. Depending on your politics, he was either the victim of a merciless capitalist state, "killed over $14 in candy," as the *Chronicle* headline put it, or simply an unlucky thief who was taking his chances. Anthony, the guard who shot Brown, was also a Black man and had a plausible claim of self-defense. He was a sympathetic figure too, a guy struggling to make a living at the bottom of the food chain, forced into a violent confrontation by the demands of his low-wage job. Brooke Jenkins, the new DA, declined to charge him—a decision that caused a brief uproar among Brown's friends and a few politicians, but soon faded from public view.

The two high-profile murders contributed to a growing sense that the social compact in San Francisco was unwinding. Everyone had stories of painful encounters, or worse, with the nodding junkies and people suffering from mental illness who seemed to be everywhere, and in worse shape than ever. The extreme situations that fed the social media doomscrolling were mostly confined to a handful of blocks in the Tenderloin and along Sixth Street, with the outer neighborhoods largely spared. But downtown was in a free fall, with a collapsing commercial real estate market, a withering arts economy, a declining population and nervous talk of becoming "the next Detroit."

The office market certainly had that look as 2023 unfolded. The vacancy rate was 35 percent and still rising, and the numbers didn't even reflect the many companies that still had leases but whose offices were empty. Most properties had lost 60 to 75 percent of their worth at a stroke.

Even more alarming was the collapse of the downtown retail economy. The San Francisco Centre mall where Banko Brown was shot, on a busy part of Market Street, was suffering a slow, public death as its brand-name tenants fled one by one, sometimes filing lawsuits on their way out the door over the impact of crime and drug use. First went the mainstay chains—Adidas, J.Crew, American Eagle, Old Navy, Hollister and LEGO would all close their stores. The Cinemark multiplex went dark. Then the anchor department stores turned the lights off: The shuttering of Nordstrom, the largest outlet in the chain and a Bay Area destination since it opened in 1988, was an especially harsh blow; the parking lot where an apartment tower was once envisioned remained empty. The mall itself then fell into receivership. Bloomingdale's eventually gave up too, along with a Burke Williams spa that had held out alone on the ghostly fifth floor until early 2025. "We can no longer operate in a location where the city fails to provide for the safety and well-being of its people and businesses," the spa's president said. Mayor Breed talked about razing the building in favor of a soccer stadium, an impractical idea that showed just how dire the situation had become.

A few blocks away, in Union Square, the news was just as bad, maybe worse. The big storefronts selling T-shirts and tourist trinkets along Powell Street were almost all empty, while brand names like Disney, Banana Republic, Uniqlo, Athleta, H&M and the Gap all closed their stores. At designer boutiques, including Louis Vuitton and Dior, the talk was about installing bollards to stop the thieves who'd taken to driving cars into shop windows, an especially dramatic version of the smash-and-grab robberies that plagued businesses all over town. The massive Macy's that occupied a full block and had long been Union Square's flagship occupant was still open, but with the parent company making noises about closing stores, it wasn't clear for how long.

A new term encapsulated the worst fears: "Doom Loop." Fewer office workers meant less economic activity and lower commercial property values, and that in turn would reduce tax revenues and lead to cuts in city services, which would make downtown even less attractive. Without enough customers, bars and restaurants would shutter. Empty streets could be spooky, which would make them even emptier.

The golden brand that was San Francisco was in tatters. Nobody seemed to have any answers.

The helplessness of city leaders in the face of the spiraling problems would show itself with rare clarity on a sunny May afternoon at UN Plaza, the faded redbrick square that commemorated the 1948 founding of the United Nations in San Francisco. Just down the street from the Twitter building, and in direct view of City Hall, the Plaza was now a chaotic crossroads for the problems that seemed to be strangling the city—namely, drug addiction, property crime and troubled people living outside.

That made it a good stage for a political stunt.

And so Aaron Peskin stood on a stoop at UN Plaza, stern and stiff in a blue suit, presiding over what was, technically, an outdoor meeting of the Board of Supervisors. Official city linens bedecked the folding tables and a high-school honor guard stood by, half-hearted gestures at dignity for an occasion whose real purpose was to embarrass the mayor, London Breed.

The topic was "open-air drug markets" and the fentanyl crisis. Peskin had only recently come around to the view that more police action was needed to address drug dealing in particular. Progressives were committed to harm-reduction approaches, and drug policy activists on the left were staunchly opposed to law enforcement crackdowns. But street conditions were obviously out of control and had become a national embarrassment. Now Peskin was demanding that Breed solve the problem within ninety days.

The mayor arrived, towering over the diminutive Peskin, and quickly launched into a short, angry speech. Her sister had died of an overdose and her brother was in prison for murder. She didn't need to be instructed on the human toll of addiction, mental illness and street violence. She was done with "all the bullshit that's destroyed our city," as she'd summed it up pungently a few months earlier, referring to the left's all-carrot, no-stick approach to crime and drug addiction.

"We're willing to let people get away with murder. We cannot allow things to continue in the way that they have for far too long," she said, her voice rising, shouts of approval emerging from the few dozen people who had gathered, a diverse assortment of neighborhood denizens, curious passersby and people who had something to say. "When you know what it's like to grow up in chaos, you want nothing more than change."

"Some people are gonna like it and some people aren't," Breed added, referring to her new policy on drugs. Just that morning, she'd announced that people using fentanyl or meth out in the open might be arrested—a big move in a town where drug tolerance ran deep. She was clearly irritated that Peskin, of all people, was grandstanding on the issue. What about everything he'd done over the years to block commonsense law enforcement in the name of compassion, or prevent the housing development that might have eased the homelessness problem?

It was Peskin's turn at the mic now, ostensibly to ask Breed questions, but when he started with a ritual recitation of data showing that crime in the city was actually down, and then droned on about interagency coordination, the crowd was having none of it.

"I work in a shelter, not a funeral home," a middle-aged Black woman shouted, over and over, holding a sign with the same words, demanding that drugs be barred from homeless shelters. "Resign" others called. "Bullshit!" A burly young white woman led the most organized chant: "No more police."

Peskin, flustered amid the cacophony, soon ended the show, which lasted all of eleven minutes. The crowd briefly grew restless. The white woman threw a brick at the honor guard and was quickly wrestled to the ground, while Breed, Peskin and the rest of the politicians beat a hasty retreat to City Hall. The TV camera crews packed up their gear; a few small bands of cops, and a captain in stripes, milled about distractedly. UN Plaza, spiffed up for the occasion, quickly reverted to its regular state—the commemorative stone columns serving as leaning posts, weary people with nowhere to live sitting along a low wall with their bedrolls, and cereal boxes, and small dogs, waiting for the next thing to happen.

■ ■ ■

The city's arts economy was reeling too, with a Doom Loop of its own. The big boom of the 2010s had been a mixed bag for the arts, bringing in new money but also making it all but impossible for many artists to afford apartments and studio space; a lot of the local action had moved to Oakland. The art schools that provided a crucial foundation for the community were under serious pressure, with soaring tuition and the high cost of living putting them out of reach for most prospective students. The San Francisco Art Institute and the California College of the Arts, both long-entrenched local institutions, took on ambitious facility projects in the hopes of boosting enrollment, but both would end in disaster.

The Art Institute's heritage included the 1920s-vintage Russian Hill campus designed in Spanish colonial style by the architects Bakewell and Brown, who'd also done City Hall. It was an elegant oasis in a jewel of a neighborhood, with sweeping views, outdoor spaces and rich angles of light. One of its soaring galleries featured a 1931 Diego Rivera masterpiece titled *The Making of a Fresco Showing the Building of a City*, an extraordinary work that celebrated artistic labor and San Francisco's creation in equal measure. It was estimated to be worth $40 to $50 million all by itself.

Those assets did not pay the salaries, though. The school didn't own the land under the building and didn't have much of an endowment, so it depended on tuition, and for some years now it hadn't been adding up. All over the country, private liberal arts colleges and art schools that weren't among an exclusive elite were struggling, with ever-rising costs making them ever-less affordable, while students were being pushed toward supposedly more practical degrees in science or engineering. SFAI had more cost pressures than most, in part because San Francisco was so expensive and in part because studio-based arts instruction is especially costly, and its financial management was never up to the standards of its artists. Still, the school appeared to have a viable future in 2015 when it launched a $50 million project to convert a pier at nearby Fort Mason into studio space and a waterfront art hub. There wasn't much room on the main campus, and many students had to travel all the way across town to studios on Third Street, in the Dogpatch. The Fort Mason facility would be a big lure for graduate students especially, or so the school thought.

The new studios opened in 2017, but it didn't lead to an immediate jump in enrollment. Then came Covid. With everything shut down, tuition revenue evaporated, and the Fort Mason project had left the school without reserves, or access to credit. SFAI began to explore a merger with the University of San Francisco and announced a preliminary agreement in 2022, but the deal fell apart. There was talk of selling the Diego Rivera fresco, but there were differing opinions on whether it was even possible to move it, and given the very local and specific subject matter, it seemed artistically wrong. Aaron Peskin settled the argument, if not SFAI's woes, by successfully urging the Board of Supervisors to make the campus a historic landmark that couldn't readily be altered.

In the spring of 2023, with operations already suspended and no recourse, the San Francisco Art Institute would close its doors.

Gordon Knox, the school's last president, attributes the shutdown to the "broken model" of higher education, especially when it comes to expensive disciplines that require studio space, materials and a lot of hands-on instruction. "The cost of higher ed is so outrageous that without a huge endowment, you end up just saddling kids with completely obscene debt," Knox said later. "We're the only mammal on the planet that extracts value out of their offspring in the form of obligations."

SFAI played a unique and powerful role in nurturing the city's creative spirits, across many generations. It wouldn't be easy to replace.

California College of the Arts, for its part, undertook an ambitious expansion of its campus on the edge of SoMa but also saw enrollment stagnate. It appeared to have been saved by a $20 million state grant and a big gift from Nvidia's Jensen Huang in 2026, but that wasn't enough. The next year the school would abruptly announce that it was selling its campus to Vanderbilt University and shutting down.

The tech industry's pandemic boom hadn't lasted very long. By the beginning of 2022, inflation had spiked to levels not seen in years as Covid disruptions squeezed supplies, while the government relief programs put cash in people's hands and stoked demand. The Federal Reserve responded by ramping up interest rates, which drove the stock market down and reversed the global "trade" that had sent so much money pouring into venture capital funds.

Bitcoin had peaked at almost $65,000 in late 2020—nearly ten times its price at the start of the year—but would slide through most of 2022, along with the rest of the crypto economy. The bust was punctuated in dramatic fashion by the collapse of the crypto exchange FTX in November of that year. FTX's founder, the mopheaded Stanford wunderkind Sam Bankman-Fried, had leveraged what appeared to be an instant fortune into fame and political influence, not to mention plush living in a sprawling penthouse apartment in the Bahamas. But now it all seemed to have been an illusion. Just weeks before the collapse, the venerable Sequoia Capital published a breathless story celebrating its $200 million investment in FTX and hailing Bankman-Fried as a once-in-a-generation genius. He was sentenced to thirty years for fraud.

The crypto-skeptical Biden administration sued many of the major crypto firms, including Ripple, Coinbase and another San Francisco crypto exchange called Kraken, for selling unregistered securities. Crypto remained extraordinarily polarizing: Skeptics saw it as straight-up grift, a Ponzi scheme in fancy digital clothing. The promotors were still unmovable in their view that it was one of the most significant digital technologies of them all, though after the SBF affair, the case for that was getting weaker by the day.

Even beyond the inscrutable dynamics of Bitcoin, many of the hot city startups of the pandemic period were seeing their value plummet in 2022 and 2023. Even more significantly for the local economy, the established stars of the Internet era were retrenching too, and the hiring frenzy of the early pandemic went into reverse. Mark Zuckerberg in February declared it Meta's "year of efficiency" after laying off about 11 percent of the workforce, and the company soon abandoned the thirty-four floors it had leased in the glistening new tower at 181 Fremont Street. Zuckerberg's declaration would prove contagious.

Salesforce was among those pulling back. CEO Marc Benioff, once an enthusiastic civic booster and champion of the 2018 homelessness initiative, had mostly retreated from local politics, other than threatening to move the big Dreamforce conference out of town if the city couldn't address street safety. He'd zigzagged on the work-from-home issue, suggesting at first that offices were passe, then eventually reversing himself and mandating that most employees come in four days a week. At the same time, he was furiously trying to reposition the company for the AI era—and slashing the payroll. Salesforce axed about eight thousand people globally in 2023, amounting to 10 percent of its staff. It pulled out of a thirty-story tower on Mission Street known as Salesforce East, abandoning about half a million square feet of office space. It was scaling back at other downtown properties too.

The Twitter building was mostly empty. Shutting down the Twitter "Neighbor Nest," a flagship of the Mid-Market redevelopment effort, was among Elon Musk's first acts after he closed on his deal to buy the company in October 2022. He'd fired almost three-quarters of the employees, refused to honor severance agreements and initially even stiffed the Shorenstein Company—no longer the owner of the building but still the manager—on the rent. He'd thumbed his nose at the city by installing a large, illuminated X sign on the roof without the requisite permits, drawing scads of complaints and several visits from city building inspectors over several days before he took it down. It seemed like a matter of time before he abandoned the headquarters building entirely.

All the money and sweat that had gone into the Mid-Market revival a decade earlier now seemed for naught. Uber had moved to Mission Bay. Block had decamped for Oakland. One Kings Lane had fizzled, while Yammer had been absorbed into the Microsoft borg and abandoned its floor at 1355 Market Street. The Yotel in the old Grant Building, where Chris Carlsson had dreamed up Critical Mass, went bankrupt. A Whole Foods Market, opened with much fanfare in 2022 in the new Trinity Plaza apartment complex, not far from the Twitter building, lasted less than a year, with rampant theft and drug use in the bathrooms spoiling the show.

Layoffs swept the city's tech sector. Lyft cut fully a quarter of its staff, about 1,100 people in total. Dropbox, Twitch, Pinterest and the social network Nextdoor all shed workers. An even grimmer fate was in store for the self-driving car startup Cruise, cofounded by Kyle Vogt, who'd earlier been part of the "Y-scraper" gang behind the video streaming company Twitch. Cruise had raised more than $7 billion before being fully acquired by General Motors in 2022, and had begun testing autonomous vehicles on San Francisco streets. Then, in October 2023, a Cruise vehicle hit a woman and dragged her under the vehicle for twenty feet as it pulled over.

Instead of immediately sharing all the video and other information about the accident with California regulators, Vogt and other Cruise executives—ignoring the frantic pleas of their political consultants—

handed over only what it had been asked for specifically, which was the video of the collision, and not the videos showing the woman being dragged. California DMV officials, livid over the attempted concealment, suspended the company's operating permit. After first replacing the executive team and slashing half of the 2,100-strong workforce, General Motors would eventually dissolve Cruise entirely and fold the remnants into the parent company. Breaking things sometimes had consequences.

London Breed didn't have much chance of persuading Elon Musk of anything—no one seemed to be able to do that—but doing what mayors do, she'd reached out to his team after he'd bought Twitter to welcome them to town and offer her help. Eventually Alex Spiro, Musk's lawyer and all-purpose fixer, got in touch and paid a visit to City Hall. "He was very friendly, very nice, and said, 'If you need something from Elon Musk, give me a call,'" recalls Sean Elsbernd, then Breed's chief of staff. The mayor reiterated that she'd love to meet Musk and tell him she wanted to be a good partner. A few weeks later, in the summer of 2023, she and Elsbernd were sitting in the lounge area above the soaring concourse of 1355 Market Street, waiting for the world's richest man.

"We're looking down, and all of a sudden we see this scrum of people come through a door," Elsbernd recounted later. "Emerging from this scrum is this very large person with dumbbells in both arms, and as he's coming up the stairs, he's curling the dumbbells, and he's wearing cowboy boots so we can hear as he takes each step, click, click. He comes up and the mayor and I are standing there and he says, 'Hey what's up?'" After brief pleasantries, Elsbernd continued, Musk said, "'Let's talk, follow me,' and he starts charging down this hallway." Breed, in her heels, couldn't quite keep up, and Musk began doing lunges with the dumbbells as he waited in front of the meeting room door. Musk's young son X was with him, attended by a clutch of what Elsbernd described as "former Navy Seal nannies."

They sat at a table, Musk huffing and puffing, dripping sweat through his dark T-shirt. "So you're getting your workout in?" Breed offered politely.

"I'm getting ready for Zuckerberg!" Musk exclaimed. The two moguls had been engaging in a surreal public sparring match about a cage fight, and Musk was apparently quite serious. "I'm going to take him down!"

Musk kept on about it, recounted Elsbernd. "He said he was talking to his friend the prime minister of Italy because he was trying to nail down the Coliseum [as a venue], and then he wanted to know from the mayor, if that fell through could he get the yard at Alcatraz." The legendary island prison, long since converted to a tourist destination, didn't belong to the city, but the mayor didn't say no.

After that, says Elsbernd, the conversation was very cordial. "He said, 'Everyone wants to leave San Francisco, but I'm committed to this city and I want to work with you, I want to work with everybody.'" That night, as it happened, there was a dinner for business leaders associated with the upcoming Asia-Pacific Economic Cooperation—a big event that Breed hoped would showcase the city's fledgling recovery—and Musk was again very solicitous. "He got a little emotional about how San Francisco was so important to him, he gave me the impression that he really cared," Breed recalled. She and Musk struck up a text dialogue, and he was grateful when she reached out to congratulate him on the birth of his next child. "We're not here to pick on you," she assured him. "We want to work with you, help us understand what you need." Musk seemed to back off on his attacks on the city, a small victory at least.

Then one day, some months later, Musk posted that he was "trapped" in the garage of 1355 Market Street by a crowd of drug users outside. Breed was sure that the area had been looking good after many months of effort, and she made the short trip over from City Hall to see for herself. "That whole alleyway was completely clean and clear," she says, and she texted as much to Musk. "He was not responsive to that."

In the spring of 2024, the company formerly known as Twitter moved to Texas.

The memorial service for Dianne Feinstein, on an exceptionally hot October afternoon in 2023, was a chance for the great and the good of San Francisco to forget about doom loops for a moment and extoll the achievements of a politician who could lay a claim to being among the lead architects of the modern city.

She had died a week earlier, having served thirty years in the US Senate, and, before that, a decade as a mayor. She'd led the city with grace and compassion in the dark days after the murders of George Moscone and Harvey Milk, and rose to the occasion again when the AIDS epidemic arrived, earning the gratitude even of some who vehemently opposed the downtown development program she'd advocated as mayor. In the Senate, Feinstein had championed gun control, broken glass ceilings and brought home federal bacon like the spectacular new sunken freeway and Tunnel Tops park on the northern waterfront. She still wasn't much loved by many progressives, let alone the libertarians, but longevity had helped her transcend some of the divides, and on this day the remembrances seemed to offer a cathartic moment for the city's political class.

Kamala Harris offered the requisite hagiography: an "icon of California . . . an American patriot . . . a giant of the Senate," she rhapsodized. Nancy Pelosi, no longer Speaker of the House but still the nation's leading symbol of "San Francisco values," celebrated Feinstein's honorary status as the "forever mayor," a tribute to the lasting impact of her time in City Hall. Mayor London Breed, welcoming the chance to talk about anything other than crime, homelessness and fentanyl, called out her predecessor's influence as a role model and a women's rights pioneer: "She created a world where girls like me could be tough, where we could lead."

The men were there too: Gavin Newsom, the governor, and Jerry Brown, the former governor, and Willie Brown, who even more than Feinstein could properly be dubbed the forever mayor. The Navy's Blue Angels aerobatics squadron, in town for the annual Fleet Week festivities that Feinstein herself had brought to life, roared overhead, and the golden dome atop City Hall twinkled in the sun.

If only it weren't so damn hot.

Feinstein had earned her spurs as a leader who'd stepped up in a huge way when it mattered most. But even though she was a business-friendly Democrat, Feinstein was no fan of the Internet and had shown little interest in the extraordinarily important global industry that was headquartered in her city. On top of that, she was positively hostile to the privacy and security concerns of the Electronic Frontier Foundation and many of the more libertarian techies; as a leader of the Senate Intelligence Committee, eventually becoming its chair, she tended to view the Internet more as a potential threat than a boon. Back in the 1990s, while the hackers of SoMa were inventing the modern Internet, Feinstein was leading an effort to require tech companies to build "back doors" into their products to allow for government eavesdropping—an idea that horrified most in the industry and ultimately failed. More recently, in 2017, she'd demanded that Apple break into the iPhone of the perpetrator of a mass shooting in San Bernardino, California. That would have required Apple to take extraordinary steps that would compromise its own security protocols, and CEO Tim Cook refused. Feinstein traveled to Cupertino and the two had a heated, hour-long exchange, but she failed to change his mind. She did succeed, though, in further convincing the tech-savvy that she didn't have a clue.

It was truly, unnaturally hot on the day of Feinstein's memorial. People were passing out—including Barbara Boxer, Feinstein's former Senate colleague, who'd had the good sense to stand down for reelection in 2016 after three terms, at age seventy-six. The biggest discomfort, though, was not the heat, but the truth that nobody could say out loud: Feinstein's final years were an excruciating embarrassment for all involved, and a depressing symbol of the troubles confronting both her political party and her hometown.

Feinstein was a very wealthy woman who was so detached from reality, or so drunk with power, or both, that she'd insisted on running for a six-year Senate term at age eighty-four, even after she failed to win the state Democratic Party endorsement. Within a couple of years, her memory and cognition problems were obvious enough to draw regular notice in the press. Yet she persevered, rarely able to make it to the Senate floor in her final year, and barely coherent in her occasional public comments. She still sat on the Judiciary Committee, though, which forced her colleagues into painful discussions about how to get around her to approve judicial nominations, a crucial party priority.

A tawdry family drama also spilled into public view. Feinstein had been removed as the executor of her family trust in 2022, and her daughters and the children of her third husband, the late private equity mogul Dick Blum, were now fighting over the fate of a $9 million weekend house in Stinson Beach, on the Marin coast. Blum and Feinstein together had left a fortune in the hundreds of millions: At Feinstein's death, the real estate portfolio alone included a twenty-five-acre ranch in Aspen, a huge spread in Tahoe and a condo in Hawaii, in addition to the enormous Gold Coast mansion and the beach house. Seeing the two families' heirs bicker over their luxury real estate was an unpleasant reminder that the senator had for many decades lived a life far removed from the concerns of most of her constituents.

Feinstein's insistence on serving well past her time—and the inability or unwillingness of her friends and political allies to persuade her to stand down, or to oppose her when she refused to do so—would foreshadow the catastrophic decision by Joe Biden and those around him (and Ruth Bader Ginsburg before him) to ignore the fact that he was clearly too old for high office. Feinstein hadn't even had the courtesy, or wherewithal, to prepare any instructions for the formalities of her passing, leaving her staff scrambling to organize the memorial.

Feinstein had served her city and her state honorably for a long time. But in her final years, she proved not only a liability for the Democratic Party, but an embodiment of what had gone wrong with it—and a prelude to the political Armageddon that was to come.

If anyone needed another symbol of how far San Francisco had fallen, Burning Man 2023 more than fit the bill.

The event was canceled in 2020 and 2021, though in the first pandemic year a "rogue burn" drew a couple of thousand people, and the next year around fifteen thousand people made the trip to the Black Rock Desert for the unofficial version; it was public land, after all, and very remote, and there was no easy way to keep people away. Old-timers who attended raved about how it was like the old days, but of course there was no going back. With the help of government pandemic assistance and a handful of big donors, the Burning Man Project came through the pandemic in decent financial shape. The official Burn was on again for 2022, but it proved an especially tough year, with relentless heat reaching 115 degrees and frequent dust storms. Word on the playa was that a lot of older Burners would be calling it quits: Burning Man had been happening for more than thirty years now, and maybe it was time.

As the 2023 Burn approached, tickets were easy to come by for the first time in fifteen years. It was still sold out, but in the months leading up to the late-summer event there were plenty of people looking to unload their $550 tickets. I decided to go myself. I'd passed up invitations back in the late 1990s, when I was just too busy, and had never managed to make it since, but I had always been intrigued. It was also the type of thing that I didn't think you could say much about unless you'd been there—which would never be truer than in 2023.

I'd planned to camp out of my old Jeep Grand Cherokee, but a friend hooked me up with an established camp, and I arrived on Monday evening after an easy seven-hour drive. I was mesmerized from the start. The city was a festival of humanity of all ages, many bedecked in colorful fashion, most open to any conversation, and there were amusing diversions around every corner: Bars and bands were a staple, but only the beginning. Bikes were the basic way to get around, and I spent an afternoon with a small group touring the land art installations: a giant fish rising from the flats; a mechanical Pegasus, twenty feet tall, flying along with flames sprouting from his wings; a series of wooden posts and markers that made up an elaborate sundial. There was a thirty-foot-tall clitoris, a project celebrating female empowerment and challenging the suppression of medical information about women, conceived by artist Melissa Barron, the wife of Burning Man lawyer Terry Gross.

I got my first taste of a dust storm the next night, with gusting winds creating billowing white clouds of fine powder that at times made it impossible to see more than a few feet. But when you're ready for it—face mask, goggles, a hat with a chinstrap—it somehow doesn't seem that terrible. Rain clouds began to gather the next day, but I didn't think too much of it. Surely there wouldn't be more than a drizzle—the Black Rock Desert was one of the driest places on earth. Yet soon enough some serious showers began to come down, and it became apparent that my old, borrowed tent, though well sealed against the dust, was no match for the rain. I implemented my backup plan, and beat a retreat to the Jeep.

When I awoke around seven o'clock the next morning, the vast expanse of tent compounds, RVs and shade structures that made up Black Rock City was as quiet as it had ever been, fully stilled by the rainstorm. Burning Man was now quite literally a giant puddle of mud. I wrapped plastic bags around my feet and headed for the porta-potties, trudging slowly through the viscous soup that the alkaline desert floor had become, the misty mountains looming in vivid pastel on the horizon. There wasn't a soul in sight. After finding the door emblazoned with "You're Doing It Right," where an anonymous angel had transformed one stall in the row of smelly outhouses into a South Pacific–themed oasis—recently cleaned and scented, with tinny island music playing through a small speaker and vintage photos decorating the walls—I did my business and was heading back to camp when I heard someone shouting my way.

"Hey, come over here, we're open!"

Across the road there were two weathered guys in baseball caps sitting at a small table in the mud, in front of a camp called Hair of the Dog.

"C'mon, get over here. Have a beer."

I couldn't really say I was in the mood for a beer. I was in a bit of a daze, truthfully: There'd been a lot to take in, even before the rains intervened. On the other hand, they were open—the bar was the oldest at Burning Man, I'd learn later, and its calling card was "always open." It seemed churlish to refuse in the circumstances. We drank beer and laughed uproariously for an hour or so, though I have no idea about what, and for the next few days we saw a lot of each other. The Mud Burn, as it would be known, kept everyone from traveling too far from camp.

A friend of mine was close with one of the Burning Man Project executives, so I made my way over to see her at First Camp, where the organizers had their own compound—and Internet access. I got a note off to my wife, and we all traded stories about the weird night that was, and the weather forecast. There'd been enough rain to render the desert impassable to all but the most robust four-wheel-drive vehicles, and the road to the highway had been closed, so no one could get in or out. The good news was that it wouldn't take long to dry out. The bad news was that two more storms were in the forecast, and if both happened as scheduled it could be three or four days before anything could move. That would include a crucial piece of the infrastructure: the trucks that pumped out the porta-potties, which were already pretty full.

I thought there'd be some frantic contingency planning underway at First Camp, but nobody seemed too concerned—contending with the elements and the unexpected had always been part of the drill. A few Wi-Fi hot spots were set up to help people get the word out that they were okay but might be home a little late. In the event, the second storm sideswiped Black Rock City, leaving a window for the porta-potty trucks, and they made their rounds to applause. The climactic moment—the burning of the Man—was delayed by two days, but eventually went off in spectacular fashion. The "exodus," where the temporary city of seventy thousand empties out onto a single, two-lane state highway, was particularly slow.

The biggest surprise for me came when I got into cell-phone range and my phone exploded with messages from seemingly everyone I knew, asking if I was okay. The global media had found an irresistible story in wealthy, hedonistic San Francisco techies being stranded in a remote desert, with food supplies running low and the threat of disease lurking—a veritable life-and-death situation, it seemed. The comedian Chris Rock had done much to feed the narrative with his account of slogging through the mud to get out; he had a schedule, and didn't want to be delayed. For me, though, it was a shock to my journalistic sensibilities to see the vast gulf between what I had experienced and the way it was portrayed in the press.

"Burning Man's reputation collapsed faster than a tent in a mudslide," declared the website Vox, supposedly devoted to sober explanations of the news, citing not only the rain but also a small climate protest at the start of the event that had gone unnoticed by most attendees. "The rains turned Black Rock City and the surrounding desert, or the playa, into a sludgy wasteland and left tens of thousands of wealthy sojourners literally stuck in the mud," the story continued. "Burning Man organizers closed routes in and out of the area, forcing attendees to stay behind or ration their food, or else find alternative routes out."

The trouble with this account, for one, is that the road was open by Monday, the day the event ends, so only those who planned to leave early were "forced to wait." There was plenty of food. The tell comes a few sentences later, when the author acknowledges that a lot of people had a great time despite the weather. "But," she continues, "across social media, onlookers reacted with glee at the thought of a bunch of desert glampers who'd paid between $200 and $2,800 a ticket having to rough it."

Burning Man's dramatic rise, and its bumps along the way, had paralleled the spectacular growth of San Francisco's Internet economy. Now, as with the city itself, the national discussion of its doppelganger in the desert was all about the *schadenfreude*.

CHAPTER 13

Regime Change

Manny Yekutiel, the proprietor of an eponymous Mission district café, had already established himself as the new master of ceremonies for city politics as the pivotal 2024 election year dawned. Now he seemed to be everywhere: organizing a series of downtown street parties, moderating mayoral debates, leading trash cleanups and promoting the city's prospects at every opportunity. For all the troubles that hung over San Francisco, the new artificial intelligence industry was beginning to revive the tech economy, and there was hope that the worst might be over. And nobody embodied positive civic spirit more fully than Manny.

Yekutiel was a man of the Castro, a gay kid who'd been disowned by his Orthodox Jewish father in Los Angeles and then found a home, and a community, in San Francisco. He'd arrived in the city in 2012, after graduating from Williams College, and cut his political teeth on Hillary Clinton's 2016 presidential campaign. After her crushing loss to Trump, he'd worked briefly at a coffee shop, and then begun to formulate a plan.

"I had this idea of opening up a coffee shop, but I was interested in politics. And I thought, what would a political coffee shop look like?" he recounted. He'd been traveling with his then-partner, a touring ballet dancer, and seen different versions of civic and political gathering spaces, including Busboys and Poets in Washington, DC. "After Trump won, it became very clear to me that it was necessary to build a physical space for people to go to, to become better citizens."

Yekutiel took a course from the Small Business Administration on how to write a business plan, and then pitched it to three hundred people—and got fifteen bites from potential investors. After many months looking for the right spot, in the summer of 2018 he leased a restaurant space on the corner of Sixteenth Street and Valencia, not far from the sloping lawns of

Dolores Park and the bustle of the Castro, but also close to the gritty blocks around the Sixteenth Street BART station and the outer parts of SoMa.

He'd soon find himself negotiating with a coalition of Mission nonprofits on what he'd be doing for the community; despite his civic mission, local activists decided his business was presumptively a negative, a "gentrifier." He formally agreed to keep prices low, hire locally and offer free space for some events—mostly things he'd been planning to do anyway—though his good faith wouldn't ultimately buy him peace.

After a mad scramble on the renovations, Manny's opened on election night in November 2018.

"There was a line wrapped around the block," Yekutiel recalled incredulously, some five hundred people eager to watch the results of an election that would see the Democrats take back the House, and serve as a rebuke to Donald Trump. "In front of my eyes my premise was proven right, that people wanted a physical space to gather politically. It was beautiful. I gave a speech, I was very drunk. My mother and my sisters were here and when they left the next morning in a taxicab I started sobbing. I'd been so scared that I made a huge mistake and didn't know what I was doing and no one was going to come and my idea was going to be a flop."

People kept coming as Yekutiel ramped up an impressive slate of events. Hillary Clinton would appear there, and Kamala Harris. Sam Altman, cofounder and CEO of the artificial intelligence juggernaut OpenAI, gave a talk there before he was famous. Jane Fonda's visit, for the Kamala Harris event, was especially memorable for Yekutiel: She had a stomach bug and threw up on his shoes.

Mostly the focus was local, with politics at the center but encompassing the arts and business too. After surviving the pandemic, Manny's would emerge as a go-to place for civic discussions, with functions most evenings, sometimes more than one in a day. Yekutiel began hosting a podcast with Sharky Laguana, the musician turned businessman, and his profile grew: He was a worker, and very smart, with equal doses of empathy and common sense, brimming with ambition but wearing his emotions on his sleeve. He cut an impressive figure as a buff, handsome and charming man who knew how to handle a room, enjoyed appearing in drag and had a crucial sixth sense for how to avoid alienating people without being a milquetoast. The wags were whispering that Yekutiel himself would be a pretty good candidate for mayor. He certainly wasn't lacking in drive.

He might have taken a shot, too, but the events of October 7, 2023, put that idea to rest, for this election cycle anyway. Yekutiel was in Israel on that day, in Jerusalem for the Bat Mitzvah of a niece, when Hamas militants in Gaza breached border defenses and massacred some 1,200 people in southern Israel and kidnapped 250 more. It was a searing event for all Jews, and especially traumatic to be on the ground when it happened, scrambling to an air-raid shelter and wondering what might come next. On top of that, many on the political left immediately deemed the attacks to be the fault not of Hamas, but of Israel and the Jews. For many left-leaning American Jews, it was a double tragedy, as erstwhile allies in the progressive movement joined arms with Israel's enemies. Yekutiel, whose father had immigrated to Israel as a refugee from Afghanistan, decided it would not be his year to run for office.

Even before the October 7 attacks, Manny's had been targeted by anti-Israel activists, who'd decided that its proprietor's familial ties to the country made him a suitable representative of the evils of Zionism. Yekutiel was born in Los Angeles and was a "Zionist" mainly in the sense that most American Jews were Zionists: He believed that Jews had a right to a homeland, even if he didn't support the policies of the Israeli Prime Minister Benjamin Netanyahu and his government. Tagging anyone with those views as an evil oppressor often made "anti-Zionism" indistinguishable from anti-Semitism.

At first, pro-Palestinian activists picketed Manny's every week, calling for a boycott and scaring away some potential patrons who'd been persuaded that the café had a right-wing agenda. Even before opening, Yekutiel had dealt with protests from a group called Gay Shame, which also objected to what they believed to be his politics. Now he was trying to organize a discussion of the Israel–Palestine conflict, but those on the Palestinian side wouldn't speak to him; there was a boycott. Committed to building bridges, Yekutiel eventually persuaded the protest leader to have a conversation.

"It was pretty mean, what they were saying," Yekutiel recalled later as he discussed the issue in an open forum at his café. "But it was a perfect example of how when you don't know someone and you've never really had an interaction with them, it's so easy to create this picture in your mind of who they are and what they think." After a long discussion, the protest leader acknowledged she'd been wrong about Manny and agreed to call off the weekly picket line.

That wouldn't be the end of it, though. Much of the city's activist left had grown extremely censorious, demanding adherence to a long list of positions, opposition to Israel prominent among them. A Gaza war protest by artists at the Yerba Buena Center for the Arts led to plainly anti-Semitic attacks on the center's director, Sara Fenske Bahat, who got little support from the group's board, or the city, and resigned. "Cancel culture" had been running strong in the city for a decade, and Manny's would be targeted repeatedly with anti-Semitic graffiti, and sometimes have its windows smashed, for more than a year after the October 7 attacks.

The mayor's race kicked off in earnest in April 2024, with Yekutiel moderating a candidate debate at the stately Sydney Goldstein Theater, originally built in the 1920s as a high school auditorium. More than two thousand people turned out, with only a single, brief protest to disrupt the proceedings.

It was a respectable field of candidates, if collectively lacking in inspirational ideas. Mayor London Breed was the moderate centrist, backed by the City Family and holding all the advantages—and in this case disadvantages—of incumbency: The city was a mess, and she'd face an uphill climb. Mark Farrell, the onetime supervisor and interim mayor, would run to her right as the law-and-order candidate. Aaron Peskin, still president of the Board of Supervisors, would represent the progressive left, whose continued strength was in question but not necessarily exhausted. Supervisor Ahsha Safaí was the clear laggard, with some labor support but not much else.

The big wild card was Daniel Lurie, a centrist like Breed and an heir to the Levi's fortune. He ran a housing nonprofit called Tipping Point and had been in charge of the 2016 Super Bowl festivities in the city, but he didn't have any experience in government or business. Raising money for good works is always easier when your mother's a billionaire.

"The guy's never had a job in his life," groused one top Breed aide, and that wasn't too far from the truth.

The debate itself, and most of those that followed, were notable for how much agreement there was on the main issues: public safety, housing and the troubles of downtown. Scratch a little bit, though, and you could find some subtle but significant discord. Peskin, who'd spent twenty-five years fighting development, insisted that he supported more housing, though he still wasn't in favor of encouraging "market rate" housing, nor did he support rezoning that would bring taller buildings to the more suburban Western neighborhoods. Farrell, too, tacked away from citywide rezoning—he'd need to win big in the residential, NIMBY-dominated Sunset and Richmond districts—and had the most aggressive plans for police crackdowns on drug use and sidewalk tent encampments. Breed was the most hawkish on housing and the clear choice of the YIMBY crowd. Lurie stayed vague on the touchy zoning issue and stressed the need for fresh eyes on the problems and better coordination among departments.

Truth be told, the debate was pretty dull. The politics of the race, though, were just the opposite.

The political operatives, Asian American activists, tech executives and investors who'd organized the defenestration of Chesa Boudin were anxious to build on that success and usher one of their own into City Hall. Farrell was the obvious choice; he'd even been a venture capitalist himself, albeit not a very successful one. He ticked a number of other boxes—city native, politically experienced, pleasant enough personally, an ally of the police. But as countless businessmen turned politicians had learned over the years, money and endorsements usually aren't enough to win races, and that was especially true in the intricate and sophisticated world of San Francisco elections.

The first problem for Farrell was that some of his most important natural allies among the tech crowd, notably Chris Larsen and Ron Conway, were sticking with Breed. Both of them had been deeply engaged in local politics for more than a decade now, and had built close ties not only with Breed but with many of the power brokers in national Democratic politics. Conway boasted of his frequent text exchanges with Nancy Pelosi. Larsen, who'd funded the controversial crime-prevention surveillance cameras around town on his own dime, was talking regularly with the Police Department and the DA's office about tactics for bringing order to the streets, while ramping up his profile as a national Democratic donor. Breed was a known quantity, with the experience and the relationships to work the system, and a compelling personal story. They were loyal to her, and could fairly expect that she'd be loyal back.

Another issue for Farrell was the bumbling by his most important supporter, Michael Moritz's Together SF. A key part of their election strategy was a ballot measure designed to reduce the number and blunt the authority of the city's many oversight commissions and give the mayor more power. But the group fumbled its first pass and ultimately ended up with a proposition widely criticized as overly simplistic. It was common for candidates to ally with a proposition and establish their own fundraising committee to support it, effectively giving them two ways to raise money while sidestepping limits on direct donations. What would be called Prop D, though, would contribute to Farrell's undoing.

As for Lurie, the risk of electing an outsider was that the machine would chew him up and spit him out before he could make a dent. San Francisco technically has what political scientists call a "strong mayor" system, where the city's top elected official enjoys executive authority. That's in contrast to a city manager system, common elsewhere in California, where the elected city council appoints a professional executive and the mayor's role is secondary. Still, the idiosyncrasies of San Francisco's endlessly amended City Charter meant that the mayor didn't have anything resembling straight-line authority over much of the bureaucracy. Willie Brown had spent his mayoral career getting the right people in the right places to overcome these obstacles. Aaron Peskin had spent much of *his* career erecting them. Whichever way you came at it, though, the City Family ran on relationships, and governing effectively required deep knowledge of the system. Lurie, at the very least, would have a lot of catching up to do.

Crucially, though, while Lurie wasn't a machine insider, he was extraordinarily well connected in other ways, even beyond what came with being rich. His father, Brian Lurie, was a rabbi and the longtime head of the Jewish Community Federation of San Francisco. Brian Lurie had married a young woman named Miriam Ruchwarger whom he'd met on a trip to Israel, and the couple had two children, including Daniel. They would split when he was five, and Mimi, as she was known, would then marry Peter Haas Sr., patriarch of the clan that founded Levi's and CEO of the apparel icon for many years. The city's successful Jewish merchant families had made a big imprint on San Francisco from its earliest days, and Temple Emanu-El, with congregants like the Hellmans, the Fishers, the Shorensteins, and the Haases, was a network in its own right. There was some inevitable quiet grumbling that Mimi, almost thirty years younger than Peter, wasn't *really* a Haas, and that she didn't fully deserve the immense fortune—estimated at $1.4 billion—that he left her when he died in 2005. Regardless, Daniel Lurie grew up in the most rarefied of realms, and now enjoyed luxuries that included the obligatory Pacific Heights mansion and a $15 million Malibu beach house.

By refusing big donations and funding his own campaign, Lurie wouldn't owe anyone anything, as he liked to stress on the stump. That approach also meant that nobody would owe *him* anything, and City Hall could be a cold place if no one was interested in doing you a favor. Lurie was easily tarred as a trust-funder using his mother's money to buy an office—an uncomfortable truth, especially in a year where the wealthy elite were pouring unprecedented sums into both local and national politics. His campaign and the associated PAC would ultimately spend some $17 million—around $9 million of it his own money—far more than all the other candidates combined.

Manny Yekutiel was working overtime as the city's hype-man in an uncertain moment. The Civic Joy Fund, as he'd named it, was sponsoring free concerts and street parties with the goal of "activating" public spaces, which mainly meant getting people out to spend money at bars and restaurants. Yekutiel was especially proud of hanging a disco ball on a crane in an alley off Second Street, where three blocks were now closed off once a month for food, games, music and dancing, courtesy of the city. His mass emails, with subject lines like "Appreciating being alive," exhorted San Franciscans to "help get [San Francisco's] groove back even better than before."

The streets were indeed looking a little cleaner as the mayoral campaign wore on, the white pickup trucks of the Public Works crews—often accompanied by police cruisers—much more visible alongside the roving street-safety workers from the nonprofit Urban Alchemy. Phil Ginsburg at the Parks Department had what turned out to be the inspired idea of turning UN Plaza into a skate park, and during the daylight hours at least, the drug activity that was still rampant in the Tenderloin and Mid-Market was supplanted by cleaner fun.

Breed, who'd taken flak for painting a too-gloomy picture of her city, was now telling anyone who'd listen that San Francisco was back. Drug overdose deaths were starting to tick downward for the first time in a decade, and crime of all kinds was falling—even the infamous car break-ins.

Especially among wealthy tech donors, the conventional wisdom was that there wasn't much ailing the city that a little law and order wouldn't fix: One document circulating among them featured a legally vetted eighteen-point plan for eliminating outdoor drug dealing and sidewalk camping. A Supreme Court decision had freed the city to sweep homeless encampments, and those operations were scaling up.

The empty-office problem still loomed over downtown—there could be no recovery without more people around—but in September came a shard of better news there too: the reopening of the Transamerica Pyramid, an unlikely city treasure that sat between the financial district and Chinatown, and had been a dominant feature of the skyline since it was completed in 1972.

The building had been greeted with horror by the city's rising slow-growth movement when it was first proposed in 1969, foreshadowing decades of development fights to come. Neighborhood activists were scandalized by the proposed destruction of a landmark called the Montgomery Block, home of the Bank Exchange, a much-loved watering hole. Critics denounced the "Manhattanization" of San Francisco, arguing the project would blight the skyline with a futuristic design and steal the visual thunder of iconic monuments like the Golden Gate Bridge. But stopping development was a lot harder back then, and three years later it was done. The Pyramid would prove worthy of its status in the mind's eye of the world, an appropriately daring building for a city on the edge.

Being featured in postcards doesn't pay the rent, though, and the Pyramid had lost a lot of its glamour in recent years. Transamerica Corp., an insurance conglomerate originally founded by A. P. Giannini, the Italian American entrepreneur behind Bank of America, had long since been sold and given up all its office space in the Pyramid, even if it still used the image in its logo. The upper floors with the best views were quite small by the standards of more recent top-of-the-line office towers, and the location was a little off too, a couple of blocks northwest of the financial district. Newer high-rises like the Bank of America tower on California Street, a beveled, brown rectangle nearly as tall as the Pyramid, and One Market Plaza, on the Embarcadero, were the destinations of choice for companies that could afford the highest rents.

There was no small amount of head-scratching, then, when an Israeli American real estate impresario named Michael Shvo, backed mainly by a sleepy German pension fund, plunked down $650 million for the Transamerica Pyramid in the fall of 2020, while Covid was raging. The depth and breadth of the pandemic-induced office meltdown hadn't yet become entirely clear, but the timing still seemed strange, even with a reported $61 million discount from the originally agreed-upon price. Undeterred by the parlous state of downtown, Shvo was promising a $350 million upgrade, including a swank club occupying several lower floors. Architectural legend Norman Foster, at eighty-nine, was himself brought in as a consultant on the new design.

Cracks in Shvo's story started to appear soon enough, with the operators of the envisioned club suing him for breach of contract and problems popping up with his properties in Los Angeles and Miami. The show would go on though, and on a pleasant September morning in 2024, beneath redwood trees that had been planted fifty years earlier and now shaded the tranquil private park at the base of the building, Shvo welcomed several hundred VIPs to celebrate the upgraded tower. Willie Brown kicked off the speechmaking, noting the old controversies over the "radical" design and the neighborhood protests, which he said were actually funded by competing building owners. He then handed off to Breed, who enthused about "infusing new, exciting creativity and talent in our downtown neighborhood." Peskin followed, and compared the "reimagining of this structure to the reimagining of San Francisco." Lurie, a candidate but not an officeholder, diligently worked the crowd—earnest, well dressed and unfailingly polite, a couple of advisors in tow. His diligence in showing up everywhere, and the consistency of his manner, would prove to be significant assets.

Shvo eventually took the stage and told a charming story, backed up by a vintage crayon drawing, about how he'd been enamored with the Pyramid since he was a kid. The bulk of the renovation was done, he said, and leasing was going strong. It was hard not to be wowed by the setting, the interlocked, latticed beams of the tower's base framing an open, ground-floor lobby that spilled out through glass walls and doors to the surrounding gardens and the redwood grove—a sublime design even without the soaring pyramid in view. Shvo skeptics smirked to themselves when a sign plane, evidently hired by the erstwhile club operators, flew overhead with a message: "SHVO MUST GO LAWSUITS CLAIM FRAUD." But he wasn't going anywhere, at least not yet.

Shvo definitely had good instincts about the location. The Pyramid sat at the northern edge of Jackson Square, a historic part of the city dotted with old redbrick warehouses and factory buildings that was now emerging as a chic neighborhood of elegant low-rise offices, expensive condos and designer boutiques. Jackson Square wasn't undiscovered, exactly: The Battery social club had opened in the old marble-cutting factory in 2014, and even back in the late 1990s *The Industry Standard* was among the clutch of new-economy companies occupying century-old masonry buildings that had been freshly retrofitted to survive an earthquake. There were good restaurants, with upscale stalwarts Bix and Kokkari chugging along and the newer fine-dining favorite Cotogna regularly booked solid. Modest noodle shops and cafés served the employees of the ad agencies, design firms and law offices of a neighborhood that had now gained fresh cache among wealthy professionals, in part because of its proximity to their homes in Pacific Heights and Russian Hill.

Owing to one of those PacHeights residents, it was now becoming the hottest office spot in town.

Sir Jony Ive, designer of the iPhone, was a London native who'd lived in San Francisco since joining Apple in 1992, making the long commute to Cupertino for more than twenty-five years as he rose to become the company's chief designer, and together with Steve Jobs set the standard for elegance, simplicity and aesthetic beauty in tech products. Since leaving Apple in 2019, he'd established his own design firm, LoveFrom, and during the pandemic he'd started buying property in Jackson Square, acquiring five buildings along Jackson and Montgomery Streets between 2021 and 2023 for a little over $75 million. In 2024, he spent $60 million for another property on the same block, a former movie palace called the Little Fox Theatre that was now a modest office building owned by the onetime political consultant and *San Francisco Examiner* proprietor Clint Reilly.

Ive was the kind of mogul that San Franciscans could appreciate, an extraordinary creative professional who seemed to have the right politics. Even Aaron Peskin, not known for applauding the real estate escapades of the super-rich, had nothing but praise for Ive's plans, declaring him a "responsible" investor who truly loved Jackson Square. The chatter was that Ive had a lot more in mind for his new urban campus than high-end design. He was also working with Sam Altman on a secret project involving—what else?—artificial intelligence.

■ ■ ■

OpenAI had gotten its start in 2015 as a nonprofit devoted to developing artificial intelligence "for the good of humanity," building on ideas pioneered back in the 1980s by the British Canadian computer scientist Geoffrey Hinton. He'd developed computer models, called neural networks, that could be trained with vast amounts of data to seemingly think like humans, and now there were computer chips and data centers powerful enough to put the ideas into action. The computing resources required were very costly, though, well beyond the means of even the richest nonprofits, and in 2019 OpenAI created a for-profit subsidiary and proceeded to raise a billion dollars from Microsoft. That was just the start.

Sam Altman, OpenAI's charismatic leader, was a perfectly distilled exemplar of the modern, venture-backed San Francisco tech entrepreneur. Each of the city's tech waves had catapulted off the one that came before, incorporating the inventions, the lessons and the profits that would make the next one far bigger than the last. Altman, already worth hundreds of millions, was a human embodiment of this dynamic.

A gay man with a nice-guy persona, Altman had punched his ticket as a Y Combinator entrepreneur with a modestly successful startup, and then in 2014 was handpicked by founder Paul Graham to lead the prestigious incubator, though he was only twenty-eight at the time. Altman was a smooth media operator and very good at schmoozing the powerful, and had just enough technical chops to manage the computer science geniuses who were making AI happen. He also had deft instincts around money, with numerous startup investments and commitments even outside of his day job—something that would have been frowned upon, at the least, in earlier days.

The AI technology based on Hinton's work had been producing impressive results at OpenAI—and also at Google, where researchers in 2017 had developed a breakthrough advance called transformers. Still, what was now being called generative AI made a lot of mistakes. It didn't seem quite ready for prime time. Then at the end of 2022, with little fanfare, OpenAI had released what it called a "research preview" of a product dubbed ChatGPT, a Web-based chatbot that could answer questions like an intelligent human. More people would use it more quickly than any other product in the history of technology. Sam Altman, for his part, would burst upon the world stage as a marketer for the ages, with a gift for selling both the virtues of artificial intelligence and his own sincerity as a man who could be trusted with the future.

Fittingly for a San Francisco enterprise, OpenAI had an idealistic mission written into its charter. Equally fittingly, given how such aspirations had worked out in the past when the money got very big, those ideals mostly fell by the wayside as Altman raced to corral the tens of billions that would be needed to build AI data centers. It was an awkward tension. Several top OpenAI researchers grew uncomfortable with Altman's approach and left to form a rival San Francisco company, Anthropic, with the pledge to proceed more thoughtfully. In the fall of 2023, the nonprofit OpenAI board actually fired Altman after they lost trust in him; he could be slippery with his earnest charm. But that wasn't going to fly. OpenAI was valued at $80 billion by then and might be worth only a fraction of that without Altman, and the natural laws of American capitalism did not allow for a nonprofit board controlled by academics to vaporize $80 billion. Altman was reinstated after five days, more famous and powerful than ever, and the Board was changed instead.

The inevitable unwinding of the nonprofit structure would ultimately prove very arduous, especially because Elon Musk, who'd donated about $45 million to OpenAI at the start and was now a direct competitor in the AI business, was determined to make trouble. But that was a problem for another day. Sam Altman had established himself as the public face of AI, and of San Francisco's next tech revolution. He'd bought a newly built mansion in Russian Hill, not far from the famous squiggles of Lombard Street, for $27 million, and would later buy several adjacent properties for another $38 million. The new mega-rich liked to have their compounds, and Altman was settling in for a long run.

For all its much-publicized woes, San Francisco was almost effortlessly emerging as the undisputed global capital of AI, reclaiming a leadership perch that had seemed to be slipping away. The city was capturing around 40 percent of all VC dollars going into the technology worldwide, and was the home not only to industry leaders OpenAI and Anthropic but also a clutch of other rocket-ship upstarts including Scale AI, Databricks, Snowflake and Grammarly.

Jackson Square was far from the only hot spot. At the Ferry Building, a short eight blocks away, a Norwegian entrepreneur named Jørn Lyseggen had opened a coworking space and social club called Shack15 on the eve of the pandemic. After a few rough years it had become a go-to spot for AI-obsessed twentysomethings; in the evenings, EDM DJs, often with a Burning Man pedigree, were livening up the soaring third-floor space, the city and the Bay glistening outside the floor-to-ceiling windows.

A few miles to the south, past Market Street and SoMa and across the old steel drawbridge next to the baseball stadium, the new commercial district of Mission Bay was humming too. OpenAI, which had occupied a converted trunk factory in the Mission, in 2024 subleased half a million square feet from Uber in what was starting to look like a second downtown, with newer buildings and far fewer homeless people. Mission Bay was a destination now, some forty years after it was first conceived, anchored by the Chase Center arena and the sprawling UCSF hospital campus, with biotech labs, office buildings and six thousand units of housing.

A little farther south was a neighborhood called Dogpatch, which boasted an impressive array of renovated industrial buildings and a colorful arts community. There had been a burst of apartment construction in the 2010s, and there was a lot more to come with the repurposing of an old power plant. It was here that Garry Tan, after taking over as president of the Y Combinator in January 2023, decided the incubator would make its new home, even as many Internet-era companies were heading in the other direction. The startup world was all about AI now, and for Tan it was hardly even a choice. "Most of the founders wanted to be in San Francisco," Tan recalled. "OpenAI was just down the street in the Mission. This is the town that birthed [the AI industry]. Anytime a new technology comes out there's a bunch of knowledge that gets spread very quickly, and San Francisco always has been the mecca for people who love technology and want to create something."

The rhetoric around how AI was about to change things was almost mystical at times. At an AI event at the sparkling San Francisco Jazz Center in Hayes Valley in the fall of 2023, Vinod Khosla, the tireless venture capitalist who'd hired John Gilmore at Sun forty years earlier, blithely assured the crowd that in ten years we'd all have "free doctors, free lawyers, free tutors"—so long as the Chinese didn't win the AI race and enslave us all. He humble-bragged about being the first venture investor in OpenAI, which he thought might make him a cool $5 billion. Reid Hoffman, the LinkedIn founder and angel-investing savant, was in character as the friendly, reasonable, humane industry voice, the counterpoint to the rude right-wingers who were now dominating the conversation about tech. "Tech can be amazing. Let's be intentional about building," he implored. He couldn't help mentioning, though, that he, not Khosla, was the first money into OpenAI, through his foundation. And whatever his liberal credentials as a major Democratic Party donor, he sat on the board of Microsoft and was all-in on the virtues of Big Tech.

All of the action in Jackson Square and Mission Bay and Dogpatch, and the promise of the AI startups, looked like palpable signs of the city's revival. Certainly London Breed was spinning it that way. Yet the buzz in those neighborhoods was less an indicator of San Francisco roaring back than a marker of how the city's economic geography was changing. The 1980s, under Feinstein, had seen a renewal and expansion of the financial district, with major new office towers like 101 California and the completion of the Moscone Center. In the second half of the 1990s and into the 2000s, the dot-com bubble and the new baseball stadium would supercharge the transformation of various patches of SoMa, as developers like Dan Kingsley's SKS remade old industrial buildings into enticing new offices or tore them down to make way for new construction. The second tech boom, beginning around 2010, brought the Salesforce Tower and transit center, a flood of new buildings in the financial district and SoMa, the Mid-Market reclamation project, and the rise of Mission Bay.

Like a neutron bomb, the pandemic had scoured downtown of people, leaving empty office towers as monuments to the urban wreckage. When the city's economy began to stir again, it took a different shape. A secular shift had overtaken the technology business, with the "efficiency" mantra of 2023 turning out to be more than a mere correction. Elon Musk's success in slashing Twitter's workforce by almost two-thirds without bringing the site down had gotten the attention of his fellow bosses. So had an essay by Paul Graham, the father of Y Combinator, celebrating the concept of "founder mode," where the entrepreneur stays hands-on and manages the company for the long term as if it were a struggling startup. Cost cutting was suddenly a point of pride, and Wall Street expected nothing less.

It was also becoming clear that the first jobs to be taken by generative AI would be in software development—a field that just a few years earlier was considered a surefire route to a good job. In the boom times of the 2010s, an entire new class of businesses dubbed coding academies, or boot camps, had arisen to train programmers, promising tech riches. Now the lower-level coding tasks that might have employed boot-camp graduates were being automated, and only the best software engineers, and those with expertise in AI, could readily find lucrative jobs. The new AI tools, it turned out, were especially good at writing software.

The modern office complexes of the center city had been built by the Internet economy, employing two generations of technical, financial, marketing and design talent and feeding a robust service economy. San Francisco's Internet industry had produced many a brand-name company, and many a fortune, and its cultural influence was unparalleled. The Covid pandemic, paradoxically, had showcased the triumph of the Internet as a defining force in modern society: It was the nerve center of economic, political and social life in America and around the world, with the capacity and sophistication to keep everything running while the physical world was all but closed. Now it was also the platform for the next great innovation. ChatGPT lived on a website, and was available around the world from the day it launched. The global Internet was built out now, and the next boom would harness its power in ways yet to be imagined.

As 2024 wore on, most companies were starting to come around to the view that they still needed people working in an office, if not every day. But the city's vacancy rate still sat at 35 percent and many companies were still pulling back. Google gave up its choice space in the Landmark Building, and consolidated at the old Hills Bros. coffee factory on the Embarcadero. LinkedIn (now owned by Microsoft) and Slack (now owned by Salesforce) were laying people off and dumping office space. At 520 Third Street, where *Wired* magazine once sprawled across several floors and Organic Online and Vivid Studios pioneered the web-hosting business, the building sat half empty, the magazine now occupying just a corner of the third story. The "Organic" sign, hung by Burning Man founders John Law and Michael Mikel back in 1999, remained a visual signpost along the Bay Bridge skyway even though the company had been gone for more than twenty years. It was a forlorn reminder of the glory days of Multimedia Gulch.

A giant foundation hole and some rusting steel girders were all that had come of the Oceanwide Center, on First and Mission, which was supposed to be one of the city's tallest towers. The massive Flower Mart in central SoMa, where generations of wholesalers anchored a busy and close-knit horticulture community, had been emptied to make way for a major office development, and the new Flower Mart had opened in Potrero Hill. The old site now sat idle, the big development project on indefinite hold.

Around South Park in SoMa, where the consumer Internet first came to life back in 1994, and where a generation of major global companies was born starting in the late 2000s, the streets were still quiet in 2024, the lawns and benches of the oval, once crowded with enterprising twenty-somethings, now hosting only dog walkers on most days. A few blocks away, Salesforce Park, the pride of the new transit center development, was similarly deserted. Salesforce Tower had been a big success for developer Paul Paradis and Hines, but the transit center suffered cost overruns and construction problems that led to it being closed for a year of repairs shortly after it opened in 2018. Worse, the high-speed train to Los Angeles that was supposed to be departing from beneath the terminal remained a mirage, as did the tunnel (estimated cost: $10 billion) that would extend the rail tracks from the Caltrain station at Fourth Street. With bus traffic way down too, the city now had a shiny and massive new transit terminal with almost no passengers.

San Francisco's population had mostly recovered from the pandemic exodus, with many of the tech professionals who'd left for Austin or Miami starting to drift back as the AI boom took hold. But service workers, who were often part of the creative economy too, weren't coming back so quickly. In mid-2024, some two years after the city had fully reopened, retail employment was still more than 20 percent below pre-pandemic levels, according to city economist Ted Egan. Leisure and hospitality, including restaurants, wasn't doing much better. The center city still felt deserted except on Tuesday, Wednesday and Thursday—three days a week in the office was becoming a new normal.

It was now inescapably clear: There was going to be no "return" to the pre-pandemic economy in San Francisco. The Internet era was over. What artificial intelligence would bring was the question of the moment, for San Francisco and the world.

As summer turned to fall and the mayor's race entered the homestretch, Manny Yekutiel's frenetic schedule was further crowded with a couple of new activities: phone banking for Kamala Harris and leading bus trips to Reno to door-knock on her behalf. The city was facing a remarkable convergence of events: a secular shift in its dominant industry, a pivotal mayoral race and a national election pitting a hometown candidate against a dark and dangerous figure who despised everything that made San Francisco great.

National politics had been shadowing the local election from the start, even if there was never any doubt that San Francisco, and most of California, would be all-in for Team Blue. The jolt of June 2024, when Joe Biden took the debate stage and disqualified himself from the presidential race, had left the three national powerhouses of San Francisco politics—Kamala Harris, Nancy Pelosi and Gavin Newsom—in a tricky spot. Pelosi would write in her memoir that she quickly began pushing Biden to stand down, though she said little publicly at the time. Harris, as vice president, went the obvious route of embracing her boss until the end. Newsom had more leeway, but decided to lash himself to Biden too, sticking with the party elders that had gotten him so far and making appearances on the flailing president's behalf, insisting he was fine. It was left unsaid that a quick Harris ascent to the White House would probably end Newsom's own presidential dreams. They'd been wary rivals for twenty years, unique personalities with their own priorities, but they had broadly similar policies and were both City Family.

Harris had grown up in Berkeley, advertised herself on the national campaign trail as a daughter of Oakland, and she made her California home in Los Angeles. But she was a San Francisco politician through and through. She'd won her spurs by defeating Terry Hallinan in the district attorney race back in 2003, and after two unexceptional terms she was elected state attorney general, thanks in large part to the endorsement of the state Democratic Party, which was dominated by the Brown-Burton machine.

In 2016, leveraging the same advantages, Harris won the race to succeed Barbara Boxer in the US Senate and immediately had her sights set on the White House. She was widely panned as too lawyerly and lacking in conviction during the 2020 Democratic primary campaign and pulled out early, but Biden picked her as his running mate after promising to choose a woman. She was smart, charming, attractive and a good debater, but most importantly she was a walking, talking example of American diversity and economic mobility at its best: the daughter of a Black father and an Indian mother who'd made it big on nothing but her own talent and hard work. She was an identity politics candidate fit for the era—or so the party bosses believed.

Kamala Harris in the White House would clearly be a boon for San Francisco, and especially the local political class. London Breed in particular stood to benefit from having backed Harris in the 2020 primary, and she held a rally on the steps of City Hall the day after Biden's withdrawal announcement, casting Harris's ascension to the top of the ticket as the city's fight. "We're going to have to defend her record because right now, as we speak, there are people that are trying to tear her down and will consistently go after her every step of the way, but we know the real Kamala Harris," she said. "It's our job to support and defend her record." If Breed lost the mayor's race, she'd at least be in line for a plum spot in a Harris administration.

The Pacific Heights donor network, mobilized for Biden but then paralyzed by his debate performance, was delighted with the new candidate too. Mark Buell had been a Harris supporter since her run for DA, and his circle would pour tens of millions into her campaign. The Democratic tech moguls, including Chris Larsen, Ron Conway, Reid Hoffman, Marc Benioff and Mike Moritz, were on board, with Larsen underwriting an exclusive party for San Francisco politicos at the Democratic convention in Chicago. Harris was much more than simply an alternative to the dreaded scenario of Trump 2.0. She was seemingly a powerful advocate for San Francisco values and all that they meant in the national culture wars, a happy warrior who could legitimately stand for opportunity and upward mobility, and for tolerance, compassion, anti-racism and multiculturalism. As her critics often forgot, Harris was also quite good at winning elections. She'd been demonized to the point of caricature by the right, so it wasn't a surprise when she won a quick burst of support around the country as voters discovered her smarts and charisma.

Still, Harris conjured a certain ambivalence in many corners of the San Francisco Democratic coalition. She was impressive in intimate settings—Tom O'Connor, the firefighters' union operative, was among the many who were drawn by her kind, intelligent and attentive manner in private conversations and small groups. But in larger gatherings she could seem unfocused and evasive. As a local prosecutor, she'd alienated many on the left by backing the death penalty in a cop-killing case and not pushing hard enough on criminal justice reform. But she seemed to weave back and forth on whether to be "tough on crime." As state attorney general she'd infuriated Internet-rights advocates with her aggressive prosecution of Backpage.com for its prostitution listings. John Law, the Burning Man cofounder whose art sometimes involved crimes like trespassing, had only a string of unprintable things to say about Harris. John Gilmore wouldn't be voting for her—he hadn't voted for a mainstream presidential candidate in decades. For San Francisco's anarcho-libertarians, it wasn't a given that she was any better than Trump.

Rank-and-file tech workers had remained steadfastly anti-Trump: In 2020, he'd won just 13 percent of the vote in San Francisco, and 25 percent in Santa Clara County, which includes much of what is generally considered Silicon Valley. His attempts to overturn the 2020 election and his determined efforts to create a fact-free alternate reality sat especially poorly with scientists and engineers. Even Doug Leone, who'd succeeded Michael Moritz atop Sequoia Capital and was among the few public Trump supporters in Silicon Valley, had renounced his backing after the January 6 Capitol riot.

But a lot had changed since the pandemic.

Peter Thiel's Founders Fund was still based in the Presidio—it was federal land, and companies there were exempt from city business taxes—and it had evolved into a haven for right-leaning financiers with a special interest in military startups. He'd hired a young writer named Mike Solana, who created an engaging news-and-opinion operation called Pirate Wires that was heavily oriented toward florid attacks on the left. Another Founders Fund partner, Trae Stephens, was a cofounder of Anduril, a weapons company launched by scorned former Meta executive Palmer Luckey. Stephens and Y Combinator's Garry Tan were among the founding members of a new church in the city that was proving popular with a new, more culturally conservative tech cohort and was growing in size and influence. Tan had also become an energetic Internet troll, with shrill denunciations of anything that smelled of progressive.

Thiel and many in his circle were not conservatives at all in the traditional sense, as evidenced by their enthusiasm for self-styled Berkeley intellectual Curtis Yarvin, who preached that democracy was a fool's errand and we'd be better off with an appointed CEO who would rule the country like a king. Somebody like Steve Jobs or Marc Andreessen, he once suggested. Andreessen, it almost goes without saying, was a Yarvin fan too. Indeed, for more than a decade, Yarvin's extraordinary radicalism had been drawing followers in Silicon Valley, where it sometimes merged with the view that unfettered techno-capitalism was the solution to every problem.

Andreessen's "Techno-Optimist Manifesto," published in the fall of 2023, channeled this worldview with savage denunciations of liberal ideas like sustainability and social responsibility, citing a range of far-right sources and even quoting the German philosopher Friedrich Nietzsche to bolster his points. It's a strange document, and Andreessen seemed unaware that his characterization of the very bad people in the "ivory tower" who were ruining the world was a pretty good self-description: "disconnected from the real world, delusional, unelected, and unaccountable—playing God with everyone else's lives, with total insulation from the consequences."

As long as it was only a tiny coterie of oddballs advancing these theories, it could all be dismissed as unserious provocation, no matter how wealthy and high-volume its proponents might be. But with the rise of Trump and his MAGA followers, Yarvin's ideas weren't looking so unserious anymore. Steve Bannon, another fan, had been a key strategist and White House advisor during Trump's first term, and drew heavily on Yarvin in his rage against the "administrative state." As Trump ramped up his 2024 campaign, key planks—and especially the soon-to-be-notorious Project 2025—appeared to be influenced by Yarvin.

There was now something of a movement developing around it all, emerging less from San Francisco than from the wealthiest precincts of Silicon Valley. Trump, from this perspective, was the vehicle that was needed to smash the power of the state, restructure society and release the animal spirits of the innovation economy. The shift crystallized when Andreessen and his partner Ben Horowitz dropped a video in July 2024 announcing that they were backing Trump.

Most San Francisco tech professionals remained in Harris's camp, and were dubious of the motives of the Silicon Valley financiers. Andreessen Horowitz, for one, was heavily invested in crypto, and that seemed like a more logical explanation for its turn to Trump than the cultural sins of the left. Tim O'Reilly, the champion of Web 2.0, believed that crypto was mostly a grift, and that stock-based compensation combined with the huge amounts of capital at the disposal of big VCs had warped the incentives of the startup economy beyond recognition. Crypto, though it might have been created with libertarian ideas in mind, was now the best example of that.

"A bunch of people figured out you can get really rich on this thing, and the [Biden administration] was saying, no, you can't, right?" says O'Reilly. "They want their grift back."

Andreessen Horowitz in particular, it turned out, was very dependent on crypto. Closely guarded investment return data that was obtained and published in 2025 by Eric Newcomer's Substack (where I was working part-time as an editor) showed that crypto funds had been the firm's main moneymakers for a decade. Even beyond crypto, though, the tech world's allegiance to the Democratic Party was fracturing under the weight of an out-of-touch Biden administration and a brutal post-pandemic backlash against social justice activism.

An especially surprising turn came from Mark Pincus, who was among the early San Francisco Web entrepreneurs who saw the Internet as a means of empowerment and connection that could elevate politics and government. He was a Democrat by background and disposition, an enthusiastic Burner and counted Reid Hoffman among his closest friends; he didn't buy the view that unfettered techno-capitalism would create something close to utopia all by itself. Still, he'd be voting for Trump.

Pincus was long gone from Zynga, which had faded in relevance and been sold several times, but he was still active in the city. He'd bought the building on Eighteenth Street in the Mission that had been home to his failed startup Tribe back in the early 2000s, and it now housed a few small startups, and his own tinkerings. He was backing his friend Daniel Lurie for mayor, but was also fine with Mark Farrell or even London Breed—like most of the city's elite, his preference was ultimately anyone-but-Peskin. In his case, though, frustrations with the city government were enmeshed with a wholesale rejection of the national Democratic Party.

"They don't like billionaires, and you know what, we don't like them either. So let's not pretend that we're friends anymore," he said shortly before the election. "I do not like the Democratic Party. I don't like Trump. I dislike the Democrats more than I dislike Trump. I think they're full of shit. I think they're disingenuous. I think they're willing to lie, cheat and steal in service of their bigger ideals."

Like a lot of his peers, Pincus was tired of being dissed for being rich, sometimes by the very politicians to whom he was donating. He didn't understand why Democrats so often seemed anti-tech, determined to tax and regulate the industry of the future, even when they barely seemed to comprehend it. He was offended by the thought-policing and ideological conformity of the left, sharing a story about how his young daughter's lines in a school play, at a private Montessori school in the Mission, had her say "Alexandria Ocasio-Cortez is the new Davy Crockett, fighting for us all."

"I don't think politics belongs there," Pincus recounted. "But I told the school, 'As long as you're okay with her saying Trump instead of AOC, we're fine.' And they said, 'Oh Mark, you're right, we think ten percent of our parents are Trump supporters, like you.' And I said, I hate Trump, but I hate what you're doing even more."

Pincus was also among the many Jewish Americans who were scalded by the progressive left's support for Hamas in the aftermath of the October 7 attacks and Democrats' ambivalence about Israel's unsparing counter-attack. Progressive ideology had come to represent, for Pincus, the very opposite of what he thought the city was about. "My San Francisco, where it runs deep for me, is I believe in people who want to speak truths when those truths are not popular. I believe San Francisco is a melting pot for anyone . . . that is what our higher consciousness is about. How can we accept gay people and not accept Trump supporters?"

As for Trump's contempt for the Constitution, and his determination to burn down a system that had benefited him and those around him so enormously, Pincus remained unconcerned. Intellectual consistency was not a feature of the new tech right.

As the mayoral race entered its final weeks, Peskin's attacks on "the billionaires" didn't seem to be getting a lot of traction. The city's population was much different from when Peskin was first elected to the Board of Supervisors in 2000. Class-based appeals didn't have quite the same resonance, and Peskin's poll numbers were looking surprisingly weak.

For Breed and Farrell, a different set of problems would emerge at the eleventh hour.

Farrell had violated campaign finance rules by steering Prop D donations to his own campaign, and his opponents made hay with it. Lurie trumpeted the issue in accusatory mailers and TV ads. Worse, Farrell had enraged Willie Brown, London Breed and many of their allies years earlier by elbowing Breed out of the interim mayor post after Ed Lee died, and they hadn't forgotten. In October came their kill shot: a letter to the state attorney general (and the media) from three former mayors—Brown, Frank Jordan and Art Agnos—calling for an investigation of Farrell. With the centrist Brown, the progressive Agnos and the conservative Jordan all lining up, voters could be forgiven for thinking that a San Francisco version of a bipartisan consensus had concluded that Farrell was unfit. In fact, each of the former mayors was backing a different horse in the race and had their own reasons for going after Farrell. But the damage was done, and Farrell was soon sinking in the polls.

Breed was facing a much more serious scandal in her own house. A city department head named Sheryl Davis, a personal friend of Breed's, had been found to be granting money to a nonprofit run by a man with whom she lived and shared a car, though she insisted he was only a roommate. That turned out to be just one of a string of cases involving a city program Breed had advocated called the Dream Keeper Initiative, conceived in the heat of the Black Lives Matter protests and intended to funnel some $180 million over several years into projects that would help the Black community. It was a quintessential San Francisco effort to use the tools of government to address injustice, and now it was exhibit number one for those who regarded such efforts as little more than a scam for the politically connected, and illegal discrimination to boot. Dream Keeper funds appeared to have underwritten everything from vacation rentals on Martha's Vineyard to four-figure restaurant bills to appearance fees for favored city officials. Suddenly, the specter of corruption that had dogged Breed since her first days in office was front and center again. Her defenders claimed the allegations of fraud were rooted in racism, but the facts were damning, and the timing was devastating for the mayor.

Just a few weeks before the vote, at a wood-paneled private club on Sansome Street, Breed, her chances looking dimmer by the day, would join an intimate campaign event alongside Aaron Peskin, moderated by former Supervisor Jane Kim. Breed and Peskin didn't like one another and were on opposite sides of the local ideological spectrum, but as the election approached they were making a joint appeal, asking their supporters to list the other candidate second in the ranked-choice voting. For all their differences, they were both City Family, closer to one another in an ineffable way than they were to some of their ideological allies. Willie Brown stood at the back of the room, wineglass in hand, admiring his handiwork for another day. He wouldn't have many more chances.

Manny's was set up for a big party on election night, with Valencia Street partly closed off to make way for a giant video screen where people might watch the returns. The mayor's race had offered plenty of drama, and the stakes were high, but it still seemed like a normal election, where the candidates would shake hands afterward and the losers would pledge to help their victorious rivals, even as they plotted to fight another day.

The Trump-Harris showdown was a different matter. It was San Francisco in a death match against the world.

The national election, alas, was over before the big screen outside Manny's was even operational. An anxious crowd of a few dozen scanned the TVs inside, looking for signs of hope, but the red wave was easy to see from early in the evening, and the mood quickly turned somber.

Just up the street, in a former church-turned-event-venue, Daniel Lurie's party was just getting started, and no one seemed too worried about Trump. The dapper, shoulder-to-shoulder crowd, almost entirely white with a smattering of Asians, was decidedly more Marina than Mission, and seemed confident of an impending win. When the first local returns showed Lurie ahead, the army of campaign workers and consultants and friends of the candidate tittered enthusiastically over their drinks. Lurie appeared soon enough, his angular face all smiles as he took the stage and celebrated his apparent win, pledging to make the city run better and give more to its residents and businesses. It would be three days before he could officially claim victory—no one could quite remember why it took so long for the city to count votes—but the win looked assured.

Across town, at Bimbo's in North Beach, Aaron Peskin and his supporters—a scruffier, more colorful lot—vowed to fight on. Peskin rejected the idea that the campaign had shown a turn to the right. "People say, 'Is San Francisco becoming a more conservative or moderate city?' Absolutely not."

The election results certainly suggested otherwise.

■ ■ ■

Lurie's win wasn't a big surprise after all that had transpired. In most ways it followed a pattern familiar to any democratic system: Voters thought the city was heading in the wrong direction and were frustrated with the incumbent, so they voted for an outsider who represented change.

Still, Lurie's election marked a radical shift in city politics, the biggest in almost fifty years. It brought an end to the dominance of the political machine that had anointed city leaders and run much of local government—and its accompanying network of contractors and nonprofit service providers—ever since the Feinstein days. Willie Brown's extraordinary run as the top dog of municipal politics was over. The progressive left was suddenly as powerless as it had been in decades.

The governmental complex Willie Brown inherited when he was elected in 1995 had grown enormously along with the tech industry, through his two terms and the reigns of his three successors. Top politicians—Harris, Newsom, Pelosi—had used the city's glamour and success as a springboard to higher office, or a means of expanding their influence once they were there. San Francisco prospered mightily during that time, at least by conventional economic measures. The city government had many talented and creative employees, a lot of ambition and set the gold standard in areas including parks, environmental protection and support for the arts. Its airport and water system were among the best in the world. It opened its wallet to back a commitment to compassion and provided a social safety net—including universal health care—that was rare in America.

But at the same time, too many politicians and city workers seemed to have lost the plot over the years, failing on life-and-death problems like fentanyl and homelessness, and more pedestrian ones like permitting. Ideological commitments to policies like harm reduction and housing first got in the way of sober assessment of what was working. The hidden costs of environmental and safety rules, and measures aimed at demonstrating the city's commitment on things like LGBTQ rights, weren't properly taken into account. There was far too much self-dealing by public employee unions; in 2024, some 580 city employees, the majority in public safety, were taking down more than $400,000 a year with overtime, and most of the jobs also came with lavish health-care and retirement benefits. There were more than thirty different city employee unions and not everyone was so well compensated, but many salary and benefit deals were far out of whack with peer cities and the private sector. At the same time, spending on nonprofit contractors had leaped to $1.5 billion annually by 2024, but there was little auditing of the contracts and thus little accountability on where the money was going. In the tech capital of the world, there'd only been a few modest efforts, such as SFGov, to bring contemporary software and innovation to city government.

Willie Brown had gotten a lot done during his eight years as mayor and was now the object of great affection; the vast majority of the people I spoke to for this book said they considered him the best mayor since Feinstein. It was a remarkable turnaround from his last days in office, when voters had soured on his imperial style and what the *Chronicle* labeled "juice politics." His activities since leaving City Hall—namely, a lot of lucrative and often undeclared lobbying work—weren't of the sort that would naturally diffuse the criticism.

Yet the nostalgia for Willie Brown was easy to explain. Despite the transactional, favor-trading tactics that some people found dirty—or more likely because of them—he was the Man Who Got Things Done. Like Robert Moses, the twentieth-century builder of New York whom the fabled biographer Robert Caro crowned with that moniker, Brown was a master at working the system to accrue power and get infrastructure built. Unlike Moses, Brown wasn't abusive to those who stood in his way, and he never had anything close to the unilateral authority that Moses long enjoyed; Brown's genius was in attending to the people side of government. But both men had lately seen their legacies revised for the better by the failure of too many Democrats to Get Things Done.

Gavin Newsom scored with gay marriage, but beyond that his headline-hunting approach and lack of follow-through limited his effectiveness. He'd clearly had his sights set on Sacramento from early in his term, and in hindsight missed a big opportunity for the city to get ahead of the housing crisis in the 2000s. It was fine to be a fountain of ideas and a great communicator, but initiatives like generating electricity from the currents under the Golden Gate Bridge, or planting organic gardens, or requiring stringent environmental certifications for new buildings, now looked like the sort of policies that were either unproductive distractions or costly mandates with a lot of unintended consequences, or both. When the city was spiraling in the wake of the pandemic, Newsom as governor had kept state funds flowing but rarely showed up in his old hometown. He had his eye on the White House, and there wasn't much to be had in getting too close to San Francisco's troubles. A surfeit of ambition and a deficit of accomplishment is a criticism that could be made of many politicians, but it applied especially well to Newsom.

Ed Lee wasn't his own man, and the city's interest groups had their way with him. He presided over a mammoth economic boom and smoothed the way for the tech companies, and was proud of initiatives like the Housing Trust Fund. Yet he often appeared to function more as a chief of protocol—or the city administrator he once was—than as the leader of a major global city. His background as a City Family loyalist, and the means of his ascension to the mayor's office, all but guaranteed a go-along, get-along approach—including a blind eye to corruption—with none of the bold measures needed on intractable problems like homelessness.

London Breed was fatally hobbled by Covid and the weighty baggage of her predecessors. She never seemed to get her arms around the system, even though she was an insider, and too often failed to tend to the Family in the way that was expected. Unlike Willie Brown, who had no permanent enemies and was always open to a deal, Breed demanded loyalty but had no time for those who crossed her, and didn't make the phone calls or offer the gestures that would cement her support among allies. It left her without enough friends when she needed them. Breed herself, and many of her backers, often referred to her background rather than her achievements in arguing why she was the right person for the mayor's office.

Judging by their votes over the years, San Franciscans were mostly okay with the high sales and business taxes, ubiquitous fees and other extra costs faced by residents and companies alike. The city was exceptionally democratic, not just in party affiliation but in the small-d sense of the word, with voters afforded numerous opportunities to weigh in on policy directly via ballot initiatives. But the result was a City Charter that grew extraordinarily complex and unwieldy, and a lot of laws that didn't work as planned, or were never properly implemented before political leaders moved on. City government spending almost tripled over thirty years after adjusting for inflation, yet City Hall still had trouble delivering on basics such as street safety. Like the state government in Sacramento, it was good at legislating virtuous things and bad at making them happen. It always seemed to be two steps behind the boom-and-bust cycles that had come to define the city. Lurie's ability to steer the lumbering ship of city government in a better direction would be facing some very stiff tests.

That Daniel Lurie's win coincided with the election of Trump, and the rise to power of the city's right-wing tech moguls, was easy to overinterpret; the new mayor and his key backers were solidly in the Democratic mainstream, even if he also drew support from the right. He could just as readily have been elected in a year when presidential politics wasn't on the ballot. But with the radical rightward shift nationally, his profile as a scion of a dynastic fortune who got elected on a public safety platform was hard to miss.

The national media never saw it as such, but San Francisco had in fact been ahead of the curve in the rebellion against the excesses of progressive Democratic policies. The recall of three school board members, followed by the recall of District Attorney Chesa Boudin, both got their start in 2021, and both were animated by some of the same Covid-era rage that would lead to the second election of Donald Trump. While both campaigns enjoyed crucial financial support from tech executives and wealthy investors, neither effort was launched or led by them. It was Autumn Looijen and Siva Raj, newcomers to the city and political naifs who were animated mainly by a desire to get their own kids back in the classroom, who made the school board recall happen. Chesa Boudin's ouster was instigated by Mary Jung, a veteran local political operative whose immediate motivations lay in seeing repeated attacks on Asian elders that didn't seem to be taken very seriously by prosecutors. The pushback against progressive politics wasn't only coming from the rich.

The resurrection of Trump, indeed, read as nothing less than a sweeping repudiation of the ideas, and the tactics, that had made the small city of San Francisco a powerhouse on the national political stage. The platform that San Francisco Democrats had championed for decades and pushed to the top of the national party's agenda—social justice for women, minorities, immigrants and LGBTQ people; government as a vehicle for uplifting the less fortunate and protecting the environment; tolerance for social experimentation and alternative lifestyles—had been rejected by more than half the country in favor of its moral opposite. Identity politics, based on building coalitions around race, ethnicity, class and gender, rather than ideology, was almost second nature in San Francisco's one-party system, the very foundation of the electoral brilliance of the Brown-Burton machine. Harris's campaign proved a case study in its limitations.

She had learned very well the blocking and tackling of fundraising and coalition building and the endless parsing of names and voter lists that Willie Brown and his allies had mastered. She'd wielded those skills, and her personal attributes, to great effect in getting elected district attorney, and then state attorney general, and then senator, and then vice president. It wasn't a surprise that she relied on tried-and-true tactics when running for the top job. It was an analytical, almost clinical approach: stick to the message, evade uncomfortable questions, convey a vague positivity with few policy commitments, stay loyal to those who got you there and work the media and the ground game. But people wanted more than a clever tactician as commander in chief.

Harris was playing a tough hand in the wake of Joe Biden's fall, and as a Black woman from a city that had become a national punching bag for its dysfunction, she had an extraordinarily high bar in trying to convince America she was a leader they could trust. It's quite likely she would have lost no matter how she ran the campaign; it's still not clear that any woman can get elected president of the US, and the damage from Biden's catastrophic refusal to step aside much earlier was incalculable. But her unwillingness to distance herself from any of her predecessors' policies while remaining guarded, even evasive, about her passions and opinions was not a winning approach.

But then, vision was never really a City Family thing.

The San Francisco Internet industry had failed on the vision front too. The change-the-world mantra and the counterculture-infused ideals of a New Economy workplace that was less hierarchical, more freewheeling and more fun had been revealed as empty promises. Companies like Twitter, Salesforce, Lyft, Airbnb, as they got bigger, also got more conventional in their pursuit of financial interests above all else. Corporate support for local political priorities such as diversity and defense of immigrants, and workplace cultures that fostered employee empowerment and flexible lifestyles, had once looked like a new "stakeholder capitalism" paradigm in the making and a victory for the values of the early Internet. Instead, such policies turned out to be entirely situational, at least at the bigger companies, emerging not from any belief system but rather a simple desire to placate restive employees at a moment when good people were hard to find. Corporate efforts to help the city, such as the community benefit agreements that were part of the Twitter tax break, or Marc Benioff's support of the Prop C homelessness initiative, proved to be the exceptions and not the rule.

In another era, companies like Stripe and Coinbase that had thousands of local employees might have been part of something like the old Committee on Jobs, where big businesses worked with the city in a way that went beyond lobbying for their own interests. Instead, two of the most successful startups in city history grew resentful of the extra costs and hassles of doing business in San Francisco, and the hostile attitude of the local left, and moved their headquarters out of town. Internet companies were less tied to the city than their old-economy forbearers; top executives and investors, with their multiple homes and private jets, especially so.

Ev Williams, the Twitter cofounder, was as frustrated as anyone with the city government. "The politics in San Francisco drive me crazy," he said in an interview before the election. But he faulted the companies for their greed too. "I think that companies should do more, but tech is a very selfish industry," Williams said. "I think that's bad and it's also one of the tragedies of San Francisco." Benioff, too, had criticized his fellow moguls for not doing more for the place that had been so central to the success of their enterprises, though his own commitments would soon come into question too. Ron Conway's approach notwithstanding, tech companies were not interested in being part of a City Family.

Donald Trump's hostility to all that made San Francisco great had rendered him a pariah in tech circles back in 2016. In 2024, he and his movement were more extreme than ever, but a lot of investors and entrepreneurs had gotten used to the idea and were warming to the merits of lower taxes and deregulation. Chris Larsen, a staunch backer of Kamala Harris, was among those who believed it was the anti-tech attitude of the Biden administration that had pushed so many in the industry toward Trump. "The Dems totally had tech under Clinton. Obama policies were always pro-innovation," Larsen said shortly after the election. "The Biden people didn't understand any of that. They didn't care. They vilified the whole industry, and I think they lost the election over it."

In truth, the two main political strains of the local tech industry since its emergence in the 1990s had been the libertarian-infused "communitarianism" that was exemplified by Burning Man and had largely favored Democrats over the years, and a more conventional liberal-democratic outlook focused on civil rights and social welfare. Both those worldviews were at odds with the priorities of many atop what had become the most important and lucrative industry in modern history. The country's broad cultural and political turn against government—brewing since the Reagan years, and supercharged by the anger over Covid-era public health protocols—fit neatly with the desire of many technologists and investors to do their thing without worrying about taxes and regulation, much less social problems like income inequality and homelessness. For erstwhile left-libertarians of the Burning Man lineage, it wasn't always a big leap.

Still, the rise to power of the long-marginalized right-libertarians of the local tech industry was jarring. David Sacks, the PayPal mafia alum and *All-In* podcaster who lived in Pacific Heights, had spent the campaign posting vituperative attacks on Kamala Harris along with pro-Putin talking points about the war in Ukraine, asserting that it was Ukrainian President Volodymyr Zelenskyy who was to blame for the conflict and leaping to Russia's defense at every opportunity. "The smart people know the Ukraine War is a grift and they are in on it," he wrote in a typical missive on Elon Musk's X. "The dumb people think it's about 'democracy.'" Sacks liked to sort people into "smart" or "dumb," or sometimes "low-IQ," according to whether they agreed with him. But with Trump's election, Sacks's nasty online persona and far-right views were suddenly an asset; he'd soon be appointed the White House's AI and cryptocurrency czar, even as he kept his day job as a VC. The Biden-era lawsuits against crypto companies, including Coinbase, Ripple and Kraken, were quickly dropped as the new president and his family set about enriching themselves with crypto schemes on a scale that appalled even many of the industry's boosters.

It had been less than a decade since Google employees had pressured the company into pulling out of a Pentagon contract for ethical reasons. Now, military technology was one of the hottest sectors for VC investment, with Andreessen and others joining hands with the Trump administration under the banner of "American Dynamism." Twitter was now X, and while it remained the main forum for public discussions among venture capitalists and others in tech, it was unavoidably infused with Musk and his MAGA allies ranting against Democrats, alongside countless spammers and scammers taking advantage of Musk's retreat from policing the platform. Liberals tried promoting alternatives like Threads or Bluesky, but they ultimately had nowhere to go.

The notion of the Internet as a medium for a new kind of politics and more functional government, which had drawn Jen Pahlka, Brian Behlendorf and Megan Smith to Washington during the Obama administration, hadn't made much of a dent in San Francisco. For all of Gavin Newsom's periodic enthusiasms for tech and innovation, and Ed Lee's coziness with the startup stars of the 2010s, and the promise of civic initiatives like Code for America, it's hard to point to government-industry tech partnerships in San Francisco of any scale. The city of the future didn't spend much time on how tech might help it get there. San Francisco had a status-quo city government through a period of relentless change, and an economy married to companies and wealthy individuals whose concern for the health of the city was often fleeting.

Boom-and-bust cycles are literally built into San Francisco's bones, and it's not obvious even in hindsight how the unprecedented influx of capital that inundated San Francisco in the Internet era could have been navigated without a lot of dislocation. Still, the failure of the City Family and the liberal order it represented to better manage one of America's wealthiest and most iconic cities would have monumental repercussions. San Francisco values, however much they might have been parodied over the years, had shaped global culture in extraordinary ways. The Internet economy made its home in San Francisco because the city welcomed new ways of thinking, and reflected and reinforced a belief in a future where technology, prosperity, personal empowerment and social connection might evolve in unison for the greater good. Yet this nexus of ideas and the political apparatus that supported them had grown exhausted, and wasn't delivering on its promises. The door was now left open to something very different—a new politics for the new era of artificial intelligence.

I ran into John Law, the Burning Man cofounder, as I was traveling between parties on election night. He was unfazed by the results. The national government was a hopeless cause anyway, and the city was going to be fine, he insisted. "The kids are still out there, still doing it," he assured me, recalling a recent event at City Lights in North Beach, where an impressive pair of creative young activists who led a group called Indecline told tales of their insurrectionist exploits. Law had grown his beard out into a thick white mash, and he had a bit of a limp; more surgery was awaiting on his ravaged joints, though he waved it off as a small thing. It had been almost fifty years since he'd arrived in town, and he'd ridden San Francisco's ups and downs with gusto. He wasn't done quite yet.

Epilogue

Daniel Lurie's inauguration was blessed with the best weather the city had seen in months, a rare January day that was cloudless and hot. The stage was mounted at Civic Center Plaza, with City Hall framing the background and the city's worthies assembled in the folding chairs in front. Chris Larsen, Mark Pincus and Laurene Powell Jobs were there, carrying the flag for tech.

Warriors coach Steve Kerr, an unimpeachable representative of what San Francisco thought of as its best, was the most prominent guest speaker. He'd led the Warriors to a remarkable four NBA titles over eight years, but even beyond being a winner he was an exceedingly thoughtful and progressive-minded man, shaped by a childhood in Lebanon that ended with his father, president of the American University in Beirut, being murdered by terrorists. Like Bill Walsh, mastermind of the football 49ers in their golden years of the 1980s, Kerr was a hypercompetitive sports coach with values San Francisco could love. Lurie made a fine speech replete with the ritual invocation of "greatest city in the world." It was easy to be optimistic that the worst of San Francisco's troubles were over, and a new dawn was at hand.

Yet the inauguration also showed just how deeply the city had been humbled.

Back in 1996, Willie Brown's inauguration had featured a phalanx of national political power-players and everyone who was anyone in city and state affairs, culminating in a live phone call with President Bill Clinton. On Lurie's big day, former Mayors Willie Brown, Frank Jordan and Art Agnos were on hand, along with outgoing Mayor London Breed, but Governor Gavin Newsom was notably absent—tending to the fires that had broken out in Los Angeles, his office said—as was Nancy Pelosi, and both of California's US senators. There were around two thousand people in attendance, less than a third of the number that had turned out for Brown;

the most-senior sitting elected official was State Senator Scott Wiener. For a long time now, the powerful and ambitious had been scrambling for association with San Francisco. That era was over.

In the first months after his inauguration, Daniel Lurie made the rounds in earnest, often making a half dozen appearances or more in a single day, dashing about as if he were still trying to get elected. No new business was too small to earn a visit from the mayor at its opening, no street party or conference too minor to merit his attention. He was amiable and attentive, willing to talk to anyone for at least a minute, and the goodwill that flowed from that was helpful as he assembled his staff and organized his priorities. The cheery but technocratic approach was a welcome relief from the political histrionics of recent years. There was in fact a consensus of sorts on some commonsense steps to address drug dealing and property crime, and Lurie had little trouble winning broad support for his early initiatives.

"There was a cynicism about politics for a long time," said Jesse Blout, the developer who'd once worked for Gavin Newsom and helped get the Mid-Market revitalization underway. "We needed that boyish optimism."

The good news for Lurie was that the artificial intelligence industry was sparking a rebound in the office market, which would accelerate over the course of the year. Dan Kingsley, the developer who'd led the retrofit of SoMa, was working on plans for a huge tower at the dormant Oceanwide site. The Transamerica Pyramid was successfully bringing in high-dollar tenants, though developer Michael Shvo's backers were looking at steep losses and appeared ready to push him aside. Apartment rents were suddenly spiking upwards too as the AI money flooded in.

Picking up where they'd left off in the Internet era, venture capitalists were again smashing records for the sheer volume of dollars they were pouring into the technology industry. Homegrown OpenAI was opening up a lead over Google and other rivals, and by the summer of 2025 was worth an astonishing $500 billion. Young entrepreneurs were once again flocking to San Francisco, some for stints at Y Combinator and others on their own, animating the coffee shops and coworking spaces with eager talk of cap tables and total addressable market. Crypto was back creating fortunes out of thin air.

At Sixth and Market, the sixteen-story building where the Burning Man

Project once had its offices had been bought in 2024 by a trio of thirty-somethings from Berlin, who had ambitions for a tech-centric live-work commune. It wasn't quite squatting a condemned warehouse in SoMa, but in the best tradition of young creatives remaking unwanted real estate, they'd gotten a deal, paying just $11 million for a property that had been worth $62 million less than a decade earlier. "If the market weren't so soft, no one would've taken a meeting with 'these crazy guys from Germany,'" cofounder Christian Nagel told the *SF Standard*. "But given the number of vacancies, people were super open." Now they were recruiting tenants for what they were calling the Frontier Tower, with each floor organized around a theme such as biotech, crypto or AI.

There were other signs of healing. The San Francisco Art Institute campus, with its Diego Rivera mural, was bought by Laurene Powell Jobs, who planned to turn it into something like what it had been in its earliest days—a center for resident artists and experimentation, rather than a degree-granting school. A few stores were returning to Union Square; the Ferry Building was jammed with new food outlets and the tourists who were finally coming back.

"The city's on fire!" Ben Davis, who'd made the Bay Lights happen, enthused in September 2025. The bridge lights were ready for their encore, bigger and better than ever, in 2026.

The national media narrative was shifting too. Not long ago, it was all about *doom loop*. Now it was all about *comeback*.

Yet it was hard to avoid the feeling that the city didn't have the same spirit, or at least not the type of spirit, that it had in the decades before the pandemic. Long blocks that once boasted lively nightlife, along Polk Street, and parts of SoMa, Union Square and the Tenderloin, were now all but shuttered by 10:00 p.m. Slim's was gone, and most of the Eleventh Street nightclubs were struggling at best. There wasn't a late-night dining scene like the one at The Globe—which I'd observed from my old office at *The Industry Standard* in the late 1990s—where you could roll in for an excellent steak or plate of pasta at midnight, if you could get a table. The Great American Music Hall was barely making it: Not only was it tough to get crowds, especially with the poor state of the Tenderloin, but the kids weren't drinking anymore. Whatever else one might say about that, it was terrible for the live entertainment business. "They come in and eat some gummies and they're good, they don't buy any drinks," said Jonathan

Nelson, the head of the venue's ownership group. Business had picked up a little in 2025, Nelson added, though it was still a profit-free operation. Around the corner, the old Phoenix Hotel that Chip Conley had made into a rock 'n' roll destination was closing, and would likely be torn down; the neighborhood was much bleaker than it had been when the Phoenix opened back in the late 1980s, Conley lamented, and big improvements looked a long way off.

Part of the dynamic was that more and more social life was moving into private spaces, whether it be the houses and terraces of wealthy residents or upscale precincts like Shack15. With all the new AI money, the private party business was booming, according to local caterers and event planners; public bars and restaurants and nightclubs, not so much. When Manny Yekutiel did an "ask-me-anything" session at his café in early 2025, a young man who'd grown up in the city wanted to know something. "I'm from San Francisco, and I'm pretty embarrassed at how it feels like in every other city, the gay scene is cooler and more fun and just way more comfortable. They say the gay scene here is dead, and it seems like that," he said.

Yekutiel, citing his own considerable knowledge on the topic, quickly assured him that it wasn't the case, though you might need to know the right parties to go to. Either way, it was a sign of the times.

There was a single-mindedness about the new wave of AI entrepreneurs, many of them very young even by the standards of tech startups, that could be either inspiring or frightening, depending on your perspective. The average age of Y Combinator entrepreneurs in 2025 was just twenty-four, down from thirty in 2022, and teams of nineteen- and twenty-year-olds weren't uncommon. AI was going to change the world posthaste, and the pressure on smart, young would-be entrepreneurs to get in the game right away was intense.

"They're being pulled and pushed at once," said Garry Tan, the Y Combinator president. The push, he said, came from the reality that fat jobs at big tech companies paying a few hundred thousand a year, once the obvious route for smart young techies, were disappearing fast for all but the rare AI specialists. And the pull? "Their friends and their contemporaries and the smartest people they knew from school are starting companies that go from zero to $15, $20 million a year in revenue in less than twelve months." Waiting to graduate from college was to risk being left behind.

(In the fall of 2025, Y Combinator announced a new program where successful applicants could defer their attendance if they wanted to finish school.)

National surveys showed that college students were prioritizing money-making much more than they had in the past, and it was hard to blame them, given how hard the country had shifted to rewarding wealth above everything else. But the money-chase could be a very one-dimensional pursuit: This generation of twentysomethings was very, very different from previous ones that had shaped San Francisco. The music and arts scene had lost a lot of its energy. The social activism and street protests that were long part of the rhythm of the city were only occasionally in evidence, even with the provocations of the second Trump administration. Small bands of activists tried to stand up for people being arrested by ICE agents at the immigration court downtown, but there was nothing like the mass movements against the Gulf War back in the 1990s and the Iraq War in the 2000s, or even the more recent outpouring around Black Lives Matter. Democrats across the country were in retreat from ambitious social policies after the shellacking of 2024. Young people were moving to the right. Both were palpable in San Francisco.

Libertarian-oriented techies who had drifted to Trump were in for a surprise in 2025 as the president began inserting himself into every corner of business and society, often in extralegal fashion. The tech community, always very pious about free markets and the dangers of government interventions, was startlingly mute. It was a truism that American technological leadership was tied to rule-of-law democracy and well-regulated markets, not to mention free trade and free movement of people. Yet the corporate elite, including the card-carrying Democrats atop most of the big tech and media companies, acquiesced to Trump's authoritarian assault on science, free speech, public health, immigrants and the integrity of government with nary a whimper of protest. So did most of San Francisco's startup community and its venture capital backers. Even Marc Benioff, who'd built his personal brand and that of his company on liberal values, declared in the fall of 2025 that he thought Trump was doing a "great job" and urged him to send the National Guard to San Francisco. Local politicos and business leaders were taken aback, but Salesforce faced an existential threat from new AI tools, and Benioff apparently saw benefit in aligning with the president. (Benioff declined to be interviewed for this book, though he sent an email defending Prop C.)

The alarming acquiescence of so many powerful people to Trump's mob-like bullying and dubious presidential decrees was matched only by the breathtaking hypocrisy of the right, which had spent four years raging about government overreach under Joe Biden only to smirk as their man pursued state capitalism and put the federal government at the service of his personal vendettas. For a lot of the billionaires, piling up more money faster seemed to remain the top priority.

Esther Dyson, the PC Forum impresario, investor and early EFF backer who'd gotten to know most of the industry leaders personally, was withering in her assessment of the turn to Trump and the quick abandonment of long-standing industry values. "People focus way too much on money and getting rich and not enough on fundamental moral principles, and they're going to regret it," said Dyson, who had angel investments in many Internet companies including Google, Facebook, Flickr and Square, and had served on the board of John Gilmore's Cygnus Solutions. She attributed the shift in behavior to the thick layers of attendants and yes-men that envelope the super-rich. "They're surrounded by people who don't dare to tell them the truth. First money and power seduce. Then they corrupt."

It was a cruel paradox for Jen Pahlka and others who'd worked in the Obama administration that the US Digital Service—the flagship of their efforts to improve government through technology—was the instrument that Elon Musk seized control of in the early days of the Trump administration to carry out his slash-and-burn campaign against the federal government. It wasn't by chance: USDS had been set up deliberately to have its hands in every part of the executive branch, even if those hands were supposed to handle tools other than chainsaws.

For Pahlka, it was upsetting to see what Musk was doing with her handiwork, but that wasn't her only reaction. "The reporters all called everyone who was involved early on and asked the same question, 'Aren't you horrified by what they're doing?'" Pahlka recounted. "And we were like sure, we're horrified. We're also jealous. If we had been given anywhere near that kind of license, the good we would have done with it..."

The Obama-era USDS was a great idea that never achieved its potential—the digital technology exemplar of a very large problem in Democratic politics. The party and its leaders were very good at coming up with smart and noble initiatives and very bad at turning them into effective programs. In the Obama administration battles over fiefdoms, "everyone tried to kill USDS," Pahlka said, though she took some pride in their having failed.

Pahlka was now advocating ideas associated with the "abundance" theory popularized by journalists Ezra Klein and Derek Thompson, which held that the problem with Democratic liberalism lay in its failure to make things happen on the ground—like building new housing—and show itself to be a party of progress. Tech, in fact, was integral to any program aimed at building things faster, and generally Getting Things Done, and that was never more true than in the AI era. Abundance was seized upon enthusiastically by centrist San Francisco Democrats, especially the pro-housing YIMBYs, as an alternative to the right's version of "techno-optimism." But it was still a long way from being a movement.

Daniel Lurie's honeymoon as San Francisco mayor was coming to an end as summer turned to fall. A citywide rezoning plan that would allow tall buildings on the West Side for the first time was moving forward, but plenty of battles remained. The homelessness and fentanyl crises were as serious as ever, and the old political divisions, unsurprisingly, were emerging once again. Aaron Peskin was still in the arena, staking out a prominent role in the fight against the zoning plan.

Nancy Pelosi, now eighty-five, announced her retirement from Congress in November, after many months of dithering, with Scott Wiener the front-runner to take her seat, though he was no shoo-in. Manny Yekutiel was running for supervisor, but as a moderate in the Castro district, and facing a progressive former aide to Pelosi, he was no shoo-in either.

Gavin Newsom, for his part, executed a remarkable turn, leaning into the media and communications skills that had made him so successful. Early in 2025, he launched a podcast featuring MAGA figures like Steve Bannon, apparently judging that his best path to the White House lay in building bridges to the dominant movement of the moment. But as Trump's extremism, cruelty and corruption enraged Democrats as never before, Newsom pivoted to the resistance, organizing a statewide ballot measure to greenlight the gerrymandering of California's congressional districts after Texas had heeded Trump's call to do the same. To their credit, California Democrats years earlier had led a "good government" initiative to turn redistricting over to an independent commission, but in the Trump era of power over principle, that had come to look like unilateral disarmament. Newsom's measure won in a landslide.

Unlikely as it seemed that Newsom could win a national election, especially after all his hometown, and his state, had gone through since Covid, as 2026 dawned he was being crowned a frontrunner in the coming presidential race. Kamala Harris by all appearances was gearing up for another run too. San Francisco's political superstars, despite all, weren't yet ready for their obituaries.

Among the Web pioneers of South Park and Multimedia Gulch, there were complicated feelings about how their seminal contributions to the rise of the Internet, and the transformation of the city, had turned out. Brian Behlendorf offers a joking apology for the banner ad and the tracking "cookie," which he'd had an important role in inventing as part of the founding team at HotWired. He's appalled by the "surveillance economy" that it helped seed. At the same time, he's proud of the way in which the idea of collaborative, open-source software has become foundational to almost every dimension of computing and the Internet.

John Battelle says a crucial mistake at HotWired was building the advertising system as a separate bit of technology—called an ad server—rather than it being integrated with the editorial content. "My idea was editorial purity—we don't want to mix our chocolate and our peanut butter. So they built an ad server, they built Apache, they built the whole shooting match, and open-sourced it, and editorial was detached from advertising and identity, and context was detached from advertising. Then Facebook put it back together: they knew every single person and knew who they were, they knew the context they were in, and they used the same systems for advertising as they did for content. That was the end of the open Web."

Battelle and many others are drawn to the idea of a far more decentralized Internet, where control lies with the people on the "edge," rather than with centralized platforms like Facebook and Apple and Google. That's closer to what the Internet promised in its early days.

John Gilmore, the EFF cofounder, believes effective "micro-transaction" technologies were a major missed opportunity; if people had a way to pay for things on the Internet a few pennies at a time, then advertising, with all its attendant tracking and manipulations, might not have become the dominant source of revenue for Internet businesses.

Matt Mullenweg, the WordPress cofounder, is still a true believer in the first principles of the independent Web: "To truly be free with technology, you need to have technology that you can own, and open source comes with a Bill of Rights. It comes with your ability to see how it works and what it's doing with your data. You can run it yourself." Though it can mean different things in the context of AI, open-source software is very much a part of the latest wave. It's embedded in all manner of big-company products and guided by sizable nonprofit foundations, including the Linux Foundation, the Apache Foundation, the Mozilla Foundation and the Wikimedia Foundation, the keeper of Wikipedia—all based in San Francisco. (Tensions can still lurk in open-source agreements, though, and Mullenweg in 2025 was fighting a heated legal battle with a major WordPress partner.)

It was poetic justice that Craig Newmark ultimately earned a massive fortune from Craigslist. The site hasn't changed much and still only charges for some real estate and job listings, but against all odds it's retained its enormous scale, and with just a few dozen employees, the profits (though not publicly disclosed) are substantial. After retiring from the company in 2018, Newmark has become a major philanthropist, with a foundation that's distributed $450 million so far to causes including journalism education, support for veterans, and, more recently, Internet security, particularly the problem of online scams. Though mostly living in New York, he bought a different house in the same San Francisco neighborhood he'd lived in for decades, Cole Valley, and was spending a week or two a month there. He's still forced to rebut the idea that he's responsible for the death of the newspaper business.

Brewster Kahle, of the Internet Archive, was continuing his life's work of preserving the independent Web, and much else, lest it be lost to history. The Wayback Machine, as it was called, had become a crucial resource, since it was frequently the case that when a website went offline, all the content from that website was gone forever. The old church in the Richmond district was a shrine of sorts to an earlier vision of a free and open Internet, with humming servers arrayed like altars around the sanctuary. A big warehouse across the Bay in the city of Richmond held container-loads of artifacts that were waiting to be archived. Yet the Internet Archive was also facing a mortal threat from lawsuits by the publishing industry, which didn't want books (like this one) to be copied and offered electronically for free. He believes the promise of the digital economy has been strangled by the oligopolies. "The regulatory structure didn't allow us to have a decent marketplace," he said. "What's missing is the government creating a level playing field."

Kahle was there in June 2025 for a remarkable reunion, organized by Ken Goffman, aka R. U. Sirius, the founder of *Mondo 2000*. Held at an old theater on Mission Street that now housed the Gray Area Foundation, a center for tech in the arts, Goffman called it a "cybernetic ritual" and a veritable who's who of early tech subcultures was there. Mark Pauline of Survival Research Labs had a robot prowling around the crowd. Lee Felsenstein, the PC pioneer and key creator of the Community Memory project, and John Law, the urban explorer and Burning Man cofounder, were among those onstage, though the goings-on were barely audible amid the din of all the friends who hadn't seen each other in a long time. At a table along the side, V. Vale, who'd been publishing offbeat political tracts and techno–science fiction for fifty years, had books and pamphlets for sale.

There were plenty of laments that day about how the big ideas around creating empowering new virtual worlds hadn't panned out, and the worrying prospect of what might come from Trump. Yet there was a certain optimism too—tech tended to draw the optimists, and it was hard to put away.

The cultural changes in the city and its tech industry—both richer than they'd ever been, but riven with uncertainties about their future—were vividly reflected in the woes of Burning Man. The event wasn't sold out in 2024 for the first time in fifteen years, and the organization was forced to undertake an emergency fundraising drive, which brought a lot of carping from veterans about too much money being spent on spreading the gospel at the expense of Black Rock City itself. It didn't sell out in 2025, either, though about seventy thousand people still made their way to the playa for what turned out to be a rough year of windstorms and rain showers.

The AI kids weren't going to Burning Man. They didn't have the time with all the grinding, and anyway that hippie stuff was for the "olds." Over the past few years, just 15 percent of Burning Man attendees were in their twenties, compared with more than 30 percent as recently as 2015. In an ominous first, a man was murdered at the 2025 Burn, found in a pool of blood at a campsite while the climactic festival of fire was underway.

With the defeat of ideas about a more enlightened workplace that once infused the new economy—and with companies taking a more authoritarian approach to their employees as they mimicked what was coming from the Trump White House—it wasn't a surprise that Burning Man had begun to evoke an angry backlash of its own, even beyond the standing criticisms of its self-indulgence and environmental hypocrisy. In an over-the-top Twitter screed in the fall of 2024, set off by a post from another venture capitalist suggesting that psychedelics kill the entrepreneurial drive, Marc Andreessen denounced Burning Man as "hard drugs in the desert" that had led to San Francisco becoming "Burning Man 24x7x365 with predictably devastating results." He went on. "What starts with pot and acid, moves to speed and heroin, and ends in madness and death," he declared, with all the nuance of an angry 1950s middle school teacher. He did get a lot of pushback from people who said they'd benefited from hallucinogens, and numerous commenters, led by Airbnb's Brian Chesky, pointed out that Steve Jobs believed his LSD trips were crucial to his creative genius.

Andreessen threw down the gauntlet: "The transition from the Jobs Way to the Musk Way in U.S. startup culture is fascinating to watch. The founders are switching from artistic UI [user interface] hippies to weight-lifter hard executors."

Andreessen cast this last bit as "overheard in Silicon Valley," but he himself had been working overtime to make such a transition happen. In the summer of 2025, the chatter among startups wasn't about preparations for Burning Man; it was about "9-9-6"—a term from the Chinese tech world that referred to working from 9 in the morning to 9 at night, 6 days a week. A *Wall Street Journal* headline on the new San Francisco zeitgeist captured it well: "AI Startup Founders Tout a Winning Formula: No Booze, No Sleep, No Fun."

I'd gone to Burning Man for the second time in 2024, and the weather was gorgeous. Since I was a hanger-on at my camp and didn't have a project of my own, I volunteered to work at what was called Playa Info, a tent along the broad, curving road called the Esplanade that separated Black Rock City from the open expanses of the playa. The main business there, it turned out, was lost smartphones: At peak times they came in at the rate of about ten an hour. There was a solid procedure developed over the years, where the phones, usually dead, would be powered up to make their lock screen visible and then be photographed, the resulting images loaded into a database. My job was to greet people who arrived looking for a lost phone, show them how to browse the pictures, and then direct them on how to retrieve it if they recognized their screen.

Never in my entire life have I gotten so much appreciation for doing so little.

Burners of every type were united in their ability to lose their phones—dazed young women in scanty clothes who'd been dancing all night; confused first-timers with dowdy Bermuda shorts and bad sunburns; bossy young men who'd fancied themselves shamans for a week and couldn't believe they'd done something so dumb. Some were near tears; some were anxious. Others weren't particularly upset but were going through the motions. Their reactions if they found their phone, though, tended to be very similar: a squeal of joy, a little jumping about and then a big hug for me, accompanied by a tumble of words expressing love and gratitude. Burning Man was a huggy place anyway, and I can state with confidence that when it comes to good hugs and fleeting-but-genuine affection, you can't do any better than returning lost smartphones at Burning Man.

The only trouble is, smartphones didn't belong at Burning Man in the first place. Their presence was corrosive.

There's hardly any cell service in the Black Rock Desert, and at first, people carried their phones primarily for the camera. As a result there was a lot less nudity over the years, and a gradual loss of the feeling that you were in a special, separate, even secret place. That was followed by the growing number of attendees who, determined to stay connected, began to bring satellite links, which were getting cheaper all the time.

The idea of Burning Man becoming just another Coachella, with crowds focused on their Instagram feeds rather than letting their inner artist roar, was anathema to almost everyone involved. What would be the point of that? Yet just as the city of San Francisco often had little ability to control the changes that big money and technology brought, so Burning Man found itself helpless in the face of trends and behaviors that undermined the spirit of the event, but couldn't be stopped. The very act of taking a picture cuts against the idea of being "present," fully experiencing a moment. If you're trying to imagine and inhabit a different kind of world for a week, being cut off from the Internet is all but required. There's been a lot of talk about how money has ruined Burning Man, gentrifying it with RVs and expensive amenities, but the creep of technology and the Internet onto the playa—a fun challenge when Scott Beale and Brian Behlendorf were doing the first webcasts and websites back in the mid-1990s—had dulled the novelty and the early ethos of the event too, with an impact at least as big as the dollars.

In San Francisco, there are similar dynamics at play. Those who lament the changes that tech and the Internet have brought to the city point to the big money squeezing out the artists and the renegades. Yet the technology itself is bringing even more monumental upheavals to urban life, in San Francisco and around the world. Cities draw their magic from human contact and physical connection, and lose some of their reason for being when so many interactions are mediated by a smartphone. The Internet has loosened connections to physical places and can make cities less essential, less dynamic and less fun.

The digital technology revolution has indeed been a story for the ages, as Louis Rossetto foresaw many decades ago, and every bit as fascinating as he promised, if not as liberating. As a journalist, I've always considered it my job to be skeptical of grandiose claims about the inherent virtues of technological progress and those who are celebrating it, or the ability of venture-backed tech startups to "change the world." in an intentional and positive way. The emptiest cliché of the era is that new technologies are morally neutral, neither good nor bad—it all depends on how we use them, and to what ends. San Francisco's Internet economy, though it didn't start out that way, was ultimately structured around capitalist success to the exclusion of almost everything else, and took that to a level that few could have imagined even a generation ago. As the next boom takes hold, we can only hope that the city preserves a whiff of the creative idealism that's always been at the heart of what makes San Francisco great.

Acknowledgments

When I started this book, my first, I was looking forward to the solo nature of the endeavor after many years of running newsrooms, and I've enjoyed that aspect of the work very much. Yet this project is ultimately a collaboration too, and it's been an honor to have the support and engagement of so many talented friends and mentors, so many book professionals, and so many people who took the time to share their stories and ideas.

It was a major stroke of good fortune for me when Karen Wickre, a veteran journalist and PR pro who knows everyone in town, offered to be my researcher—a title that doesn't begin to describe her contributions. She's made numerous important introductions and offered countless good suggestions, in addition to the research and fact-checking, and has been a reader and sounding board throughout. It's been a pleasure to have her as a partner.

This book might not have happened without the late Esmond Harmsworth, who on the recommendation of my friend Adam Lashinsky took my call when I had only a basic treatment. He immediately got what I was trying to do, asked all the right questions, and not long after agreed to be my agent. He and his colleague, Jen Marshall, then spent many months coaching me through the proposal, editing it and then selling it for a worthy price.

Like everyone who knew him, I was shocked and extremely saddened by Esmond's untimely death last year. My deepest thanks to Jen and to Aevitas Creative, where Esmond was a partner, for picking up the ball amid the tragedy of Esmond's passing and ably representing me since.

This story is largely based on interviews, more than 200 people in all, with a full list on the following pages. I sincerely appreciate everyone's contributions very much, and I wanted to call out a few folks who went above and beyond in offering their time and helping me understand events. They include John Battelle, Brian Behlendorf, Cindy Cohn, Chip Conley, Doug Dalton, Esther Dyson, Ted Egan, Sean Elsbernd, Christin Evans, John Gilmore, Phil Ginsburg, Marian Goodell, Chris Gruwell, Del Harvey, Mary Jung, Sharky Laguana, Chris Larsen, John Law, Summer Lee, Michael Mikel, Jonathan Moscone, Jonathan Nelson, Craig Newmark, Tom O'Connor, Tim O'Reilly, Mark Pincus, Jennifer Pahlka, Paul Paradis, Jonathan Steuer and Greg Suhr.

I'm delighted to have the chance to thank the many longtime friends and collaborators who I've learned so much from over the years, and who were incredibly helpful as I tackled this somewhat daunting project.

Marty Baron gave me my first "big media" break when he hired me as an intern at the *Los Angeles Times* and then tapped me to cover the tech industry. He taught me a lot about how to write and report with clarity and integrity, and it's been a privilege to have his counsel and friendship over the years.

I met Michael Wolff when I hired him as a columnist at *The Industry Standard*, and I've appreciated his comradeship and peerless media instincts ever since. From the start of this effort, he's offered generous feedback and invaluable guidance on the ins and outs of the book business.

Brad Feld was a crucial investor in my startup, New West, and has remained an unwavering backer of my pursuits, even though I lost his money. He's also an author and a big reader himself, and I hugely appreciate his comments and fact-checks of the manuscript.

My longtime friend Thomas Goetz has been with me throughout the book process, offering sharp journalistic insight and encouragement, as well as helpful suggestions on the writing. Lessley Anderson, Cory Johnson, and John Temple talked me through many things over many months. Susie Davis offered important introductions to the cultural underground and kindly reviewed some passages. She and her husband, Marc Mowrey, have supported me personally in ways too numerous to count, and will always have my gratitude.

Maryann Thompson, the only person who worked for me at *The Industry Standard* and then again at *The San Francisco Standard* twenty-five years later, volunteered to help run down some of the data in this book. Even more importantly, her integrity in standing up for me in a difficult professional situation is something I will never forget.

My staunch friend and pro bono attorney, Edward Hernstadt, has devoted an embarrassing number of hours to my employment and media issues over the years and never steered me wrong. Without his generosity, many of my career adventures might not have played out so well.

I'm very thankful to Laura Fraser for her review of the manuscript and friendship over many years. Lawrence Comras has been ahead of the curve and a lifelong interlocutor on technology and society. Steedman Hinckley and Eric Sack have been there for me for many decades. I only wish the treasured members of our posse who left us too soon, John Moynihan and Janet Allon, were here to share these moments.

Jordan Shlain, whom I am very lucky to count as my friend, doctor and occasional collaborator, has done much to keep me healthy over the years while also offering shrewd insights on technology and introductions to his remarkable network.

I'm indebted to a number of other friends and associates who read portions of the manuscript, made introductions or otherwise helped in some fashion, including Diane Anderson, Vince Bielski, Michael Barba, Jeanne Carstensen, Bob Cohn, Frances Dinkelspiel, Dave Eggers, Jim Evans, Steve Fainaru, Annie Gauss, Stacey Foreman, Heather Grossmann, Katie Hafner, Reyhan Harmanci, Scott James, Gary Kamiya, Josh Koehn, Bonnie Lin, John Markoff, Sarah McBride, Joe Menn, Meaghan Mitchell, Dan Morain, Dan Newman, Alexei Oreskovic, Riley Rant, Gary Rivlin, Jim Schachter, Gerry Shih, Joshua Slayton, David Sjostedt, Matt Smith, Anna Tong and Lance Williams.

My gratitude to Joe Eskenazi for his close read on some of the local politics sections; there are few who know the history better. Many thanks also to Mark Gimein and Michael Parsons for their thorough reading and input.

A hat tip to Chrissy Farr, a former Reuters colleague, who helped crystallize my vague ideas over coffee one day when she declared, "You should write a book about the city, San Francisco in the tech era."

Much appreciation to Eric Newcomer and the crew at his Substack, and to Damon Darlin and Judy Lin of KFF Health News, who collectively offered me fun and rewarding part-time editing work that helped everything add up. My thanks as well to Jørn Lyseggen and the very kind staff at Shack15, where I spent many, many hours interviewing and writing.

This book was inspired and informed by so many great writers and reporters who came before me, and I can only begin to acknowledge all the influences. I've been asked more than once if I was aiming for a sequel to David Talbot's *Season of the Witch*, and if it isn't exactly that, I do hope it's close to as entertaining and educational as that seminal work. Gray Brechin's *Imperial San Francisco*, an underappreciated masterpiece that tells the city's history through its architecture and monuments, sparked my passion too.

In addition to the journalists cited above, I also wanted to call out the work of Felicity Barringer, Nick Bilton, Denise Caruso, Chris Carlsson, J. K. Dineen, Brian Doherty, Conor Dougherty, John Heilemann, Heather Knight, Steven Levy, Ryan Mac, Katharine Mieszkowski, James Reginato, Scott Rosenberg, Elizabeth Lesly Stevens, Brad Stone, Brad Wieners and Gary Wolf.

My editor at Atria Books, Yaniv Soha, has been a writer's dream—encouraging, knowledgeable and helpful, with a good eye for separating the essential from the dispensable. I didn't quite know what to expect in the editing process and couldn't have hoped for better. My appreciation also to everyone on the team at Atria and Simon & Schuster, including Abby Mohr, Debbie Norflus, Erin Kibby, Sonja Singleton, Esther Paradelo, Emma Van Deun and Kate Napolitano.

Many thanks to Betsy Streeter for her delightful map, and to Erfert Fenton for her diligent proofreading.

An extra-special shout-out to my sister, Leslie Weber, who has been not only a loving and supportive sibling but a shrewd reader of the manuscript too. My brother David Weber has been a stalwart ally. And much love and appreciation to my parents, Sophie Weber and Joe Weber, who in addition to everything else helped instill in me a love of reading and writing.

My crucial lifeline and partner throughout this project and beyond has been my wife, Karen Taylor. Her love and support, and her tolerance of the single-minded focus a book like this requires, have meant everything to me. So has the love of our children, Shannon, Joshua and Zachary. I am eternally grateful.

INTERVIEWS 2023–2025

Sachin Agarwal, Angela Alioto-Piers, Diane Anderson, Lessley Anderson, Sunny Angulo, Indira Allegra, Tiffany Apczynski, John Avalos, Ed

Axelsen, Sara Fenske Bahat, Caroline Barlerin, Nathan Ballard, Carl Bass, John Battelle, Scott Beale, Brian Behlendorf, Mariane Bekker, Michael Birch, Steven Blumenfeld, Jay Bradshaw, London Breed, David Bressie, Willie Brown, Mark Buell, Steve Buss, David Campos, Chris Carlsson, Jeanne Carstensen, Denise Caruso, Tim Chang, Jay Cheng, David Chiu, Amy Cohen, Greg Cohn, Cindy Cohn, Lawrence Comras, Chip Conley, Ron Conway, Linda Corso, Dick Costolo, Suzannah Cowell, Colin Crowell, Deborah Cullinan, Doug Dalton, Chris Daly, Ben Davis, Susie Davis, Frances Dinkelspiel, Matt Dorsey, Bevan Dufty, Esther Dyson, Ted Egan, Theo Ellington, Jason Elliott, Sean Elsbernd, Joe Eskenazi, Christin Evans, Mike Farrah, Brad Feld, Ana Teresa Fernández, Laura Foote, Craig Forman, Renee Franzwa, Mark Frauenfelder, Jennifer Friedenbach, Julia Friedlander, Brad Garlinghouse, Sean Garrett, Gillian Gillett, John Gilmore, Phil Ginsburg, Ken Goffman, Jason Goldman, Marian Goodell, Matt Graves, Terry Gross, Eric Grossberg, Chris Gruwell, Karen Gullo, Ed Harrington, Del Harvey, Will Hearst, Susan Hirsch, Peter Hirschberg, Minnie Ingersoll, Mike Isaac, Mark Jacobson, Scott James, Eric Jaye, Tom Jennings, Alicia John-Baptiste, Cory Johnson, EJ Jones, Nick Josefowitz, Mary Jung, Brewster Kahle, Matt Kallman, Gary Kamiya, Mitch Kapor, Jane Kim, Dan Kingsley, Amy Kirsch, Will Kreth, Christine Kristen, Margot Kushell, Sharky Laguana, Chris Larsen, Alex Lash, John Law, Summer Lee, Chris Lehane, Mark Leno, Steven Levy, Jeremy Liew, Janice Li, Bonnie Lin, Candace Locklear, Autumn Looijen, Jørn Lyseggen, Charles Malet, Om Malik, Rafael Mandleman, Stuart Mangrum, John Markoff, Joe McCarthy, Cary McClelland, Hamish McKenzie, Gabriel Metcalf, Jane Metcalfe, Michael Mikel, Meaghan Mitchell, Jonathan Moscone, Matt Mullenweg, Jonathan Nelson, Dan Newman, Craig Newmark, Rudy Nothenberg, David Noyola, Tom O'Connor, Tim O'Reilly, Chikai Ohazama, David Owen, Jen Pahlka, Carli Paine, Paul Paradis, David Pescovitz, Aaron Peskin, Mark Pincus, Marc Porat, Dean Preston, Siva Raj, Tim Redmond, Clint Reilly, Susan Dyer Reynolds, David Rodriguez, Chris Roeder, Zach Rosen, Scott Rosenberg, Ben Rosenfield, Louis Rossetto, Todd Rufo, Elliot Schrage, Jim Schachter, Rob Schmitt, Del Seymour, Randy Shaw, Nathan Shedroff, Jordan Shlain, Tiffany Shlain, Sara Shortt, Anne Shulock, Sam Singer, David Sjostedt, Derek Slater, Josh Slayton, Larry Smarr, Jack Song, Cari Spivack, Jonathan Steuer, Greg Suhr, Kara Swisher, Garry Tan, Alex Tourk, Trevor Traina,

Merredith Treaster, Nancy Tung, Bob Wachter, Marc Weber, Scott Wiener, Margit Wennmachers, Patrick Wolff, Karen Wickre, Ev Williams, Jim Wunderman, Manny Yekutiel

A handful of additional interview subjects preferred to remain anonymous.

Notes

Chapter 1: South Park and the Birth of the Web

11 *Growing up in the Los Angeles suburb*: Author interview, Brian Behlendorf, June 12, 2024.

11 *They were powered*: "History of Unix," Wikipedia, https://en.wikipedia.org/wiki/History_of_Unix.

12 *Launched in the late 1970s*: Philip Elmer-DeWitt, "First Nation in Cyberspace," *Time*, December 6, 1993, archived at https://web.archive.org/web/20210408023213/https://kirste.userpage.fu-berlin.de/outerspace/Internet-article.html.

12 *That was the same year there was a coup in Russia*: Behlendorf, interview.

13 *This new genre of electronic dance party*: Samantha Durbin, "Acid, Dance, Unity: What Happened to the '90s Bay Area Rave Scene?," *Bold Italic*, June 12, 2019, https://thebolditalic.com/acid-dance-unity-what-happened-to-the-90s-bay-area-rave-scene-50a5320b7001; see also Shari Buck, "During the first San Francisco dot-com boom, techies and ravers got together to save the world," Quartz, July 20, 2022, https://qz.com/1045840/during-the-first-san-francisco-dot-com-boom-techies-and-ravers-got-together-to-save-the-world.

13 *Behlendorf started to think*: Behlendorf, interview; see also Stephanie Buck, "Raves + Tech = San Francisco," Medium, https://medium.com/timeline/raves-tech-san-francisco-727d1e0d224d.

13 *A couple of years earlier*: "History of the World Wide Web," Wikipedia, https://en.wikipedia.org/wiki/History_of_the_World_Wide_Web.

14 *A grassy oval*: Brock Keeling, "A Brief History of South Park," Curbed SF, May 20, 2020, https://sf.curbed.com/2020/5/20/21264976/south-park-history-timeline-sf-san-francisco.

14 *Yet something new was starting to simmer*: Author interview, Will Kreth, December 11, 2024.

15 *"I walk in and there were three guys . . ."*: Author interview, Michael Mikel, May 4, 2024; see also *Jasmine Technologies*: Steve Sande, "Throwback Thursday: That 'Affordable' $2,795 300MB Hard Drive," AppleWorld Today, May 14, 2015, https://appleworld.today/throwback-thursday-that-affordable-2795-300mb-hard-drive/.

16 *Tom Jennings, a hacker, skateboarder*: Author interview, Tom Jennings, February 13, 2024.

17 *John Battelle, later a prolific digital media entrepreneur*: Author interview, John Battelle, December 21, 2023.
18 *Brand, who grew up in Illinois*: John Markoff, *Whole Earth: The Many Lives of Stewart Brand* (Penguin Press, 2022). https://www.penguinrandomhouse.com/books/554161/whole-earth-by-john-markoff/.
18 *Barlow recounted his experience*: John Perry Barlow, "A Not Terribly Brief History of the Electronic Frontier Foundation," EFF.org, November 8, 1990, https://www.eff.org/pages/not-terribly-brief-history-electronic-frontier-foundation#main-content.
19 *Survival Research Laboratories*: "Illusions of Shameless Abundance," YouTube, 1989, https://www.youtube.com/watch?v=1Cp7aD0q63g/.
19 *Joegh Bullock created the Anon Salon*: Sam Whiting, "Joegh Bullock, San Francisco party pioneer who helped build Craigslist and Burning Man, dies at 71," *San Francisco Chronicle*, June 24, 2022, https://www.sfchronicle.com/bayarea/article/Joegh-Bullock-San-Francisco-party-pioneer-who-17261331.php.
20 *ToonTown and 1991 New Year's Eve rave*: Cameron Holbrook, "26 photos capturing the blissful essence of San Francisco's 90s rave scene," Mixmag, June 19, 2018, https://mixmag.net/feature/26-photos-capturing-the-blissful-essence-of-san-franciscos-90s-rave-scene/.
21 *A lanky, blue-eyed man*: Author interview, John Gilmore, December 30, 2024.
23 *A pioneering MIT spin-out*: Author interview, Brewster Kahle, July 3, 2024.
23 *For Louis Rossetto, an itinerant editor*: Author interview, Louis Rossetto, June 16, 2025.
24 *Rossetto and Metcalfe made the move*: Rossetto, interview; see also Gary Wolf, *Wired: A Romance* (Penguin Random House, 2003), https://www.penguinrandomhouse.com/books/193249/wired-a-romance-by-gary-wolf/.
24 *The neighborhood could be*: Author interview, Jane Metcalfe, July 1, 2025.
25 *Some of their ideas, and their future writers*: Author interview, Ken Goffman, March 26, 2025.
25 *Battelle had grown up in Pasadena*: Author interview, John Battelle, December 21, 2023.
27 *A company later known as Macromedia had set up shop*: Trip Gabriel, "Gurus of Multimedia Gulch," *New York Times*, September 4, 1994, https://www.nytimes.com/1994/09/04/style/gurus-of-multimedia-gulch.html.
27 *One of those companies, Vivid Studios*: Author interview, Nathan Shedroff, July 1, 2024.
28 *"We were trying to define digital culture"*: Author interview, Daniel Carter, January 24, 2025.
28 *He developed an especially close bond*: Willie Brown, *Basic Brown: My Life and Our Times* (Simon & Schuster, 2006).
29 *Will Hearst III, the grandson*: "Riptide: An oral history of the epic collision between journalism and digital technology, 1980 to the present," videotaped interview, Shorenstein Center on Media, Politics and Public Policy, Harvard University, April 2013. https://www.digitalriptide.org/person/will-hearst/.

29 *"My primary source of information was the* Bay Guardian*"*: Author interview, Sharky Laguana, November 29, 2023.

32 *Now he'd been tapped to create an electronic expression*: Author interview, Jonathan Steuer, June 21, 2024; see also *Jonathan Steuer and Cyberorganic*: Links.net, "Jonathan Steuer biography," https://links.net/vita/cyb/jss/.

32 *Rossetto had been jolted into action*: Rossetto, interview.

34 *Clark persuaded Andreessen to leave*: Brian McCullough, "20 Years On: Why Netscape's IPO Was the Big Bang of the Internet Era," *Internet History* podcast, August 27, 2015, https://www.Internethistorypodcast.com/2015/08/20-years-on-why-netscapes-ipo-was-the-big-bang-of-the-Internet-era/.

35 *Tim O'Reilly, a computer book publisher*: Author interview, Tim O'Reilly, February 12, 2025.

36 *Steuer's childhood friend and now-housemate Jonathan Nelson*: Author interview, Jonathan Nelson, October 3, 2024.

36 *When HotWired launched in the fall of 1994*: Author interview, Jonathan Nelson, July 9, 2025.

37 *Netscape launched its Web browser*: Adam Lashinsky, "Netscape IPO 20-year anniversary," *Fortune*, August 9, 2015, https://fortune.com/2015/08/09/remembering-netscape/.

37 *In August 1995, Netscape sold stock*: W. Joseph Campbell, "Memorable moment in an exceptional year: The Netscape IPO of 1995," *1995 Blog*, August 7, 2020, https:/1995blog.com/2020/08/07/memorable-moment-in-an-exceptional-year-the-netscape-ipo-of-1995/.

Chapter 2: The "Motor City" of the Internet

39 *Willie Brown, dapper as always*: "San Francisco Mayoral Inauguration," C-SPAN, January 8, 1996, https://www.c-span.org/video/?69289-1/san-francisco-mayoral-inauguration.

39 *Nancy Pelosi was stuck in a Washington snowstorm*: "Brown's Inauguration Draws Star-Studded Crowd to San Francisco," *Los Angeles Times*, January 9, 1996.

39 *Art Agnos, adored by the left*: California State Archives, Oral History of Art Agnos, vol. 2, https://archives.cdn.sos.ca.gov/oral-history/pdf/oh-agnos-art-v2.pdf.

40 *Brown's career was nothing less than a tour de force*: Willie Brown, *Basic Brown: My Life and Our Times* (Simon & Schuster, 2006); see also Elizabeth Lesly Stevens, "The Power Broker," *Washington Monthly*, July 7, 2012, https://washingtonmonthly.com/2012/07/07/the-power-broker/.

41 *The target of more than one federal corruption investigation*: Mary Curtius, "Corruption Scandal in SF Grows," *Los Angeles Times*, September 23, 1999, https://www.latimes.com/archives/la-xpm-1999-sep-23-mn-13313-story.html.

41 *More seriously, Brown saw the investigations as unethical*: Brown, *Basic Brown*.

42 *The mayor's office seemed like an obvious step*: Author interview, Willie Brown, July 18, 2024.

42 *Brown retained the speakership even after Republicans won the majority*: B. Drummond Ayres Jr., "California Speaker Frustrates G.O.P. One Last Time," *New*

York Times, June 6, 1995, https://www.nytimes.com/1995/06/06/us/california-speaker-frustrates-gop-one-last-time.html.

42 *Brown didn't love the more mundane aspects of retail politics*: Maura Dolan and Norma Kaufman, "SF Parties as Brown Is Sworn In," *Los Angeles Times*, January 9, 1996, https://www.latimes.com/archives/la-xpm-1996-01-09-mn-22626-story.html.

42 *Brown, a legendary clothes horse*: Brown, *Basic Brown.*

42 *On inauguration day*: Brown, interview.

42 *Brown seemed to be trying to convince himself*: Mayor Willie Brown 1996 swearing-in ceremony video, KPIX Archive, January 9, 1996, https://www.youtube.com/watch?v=1xPfIg-aPTo.

45 *Ted Egan had written an academic paper*: Trip Gabriel, "Gurus of Multimedia Gulch," *New York Times*, September 4, 1994, https://www.nytimes.com/1994/09/04/style/gurus-of-multimedia-gulch.html.

46 *The fledgling industry didn't have a local political agenda*: Brown, interview.

48 *Megan Smith, who'd met Rielly*: Author interview, Megan Smith, March 18, 2024.

48 *General Magic was ahead of its time*: Smith, interview; see also Matt Maude and Sarah Kerruish, directors, *General Magic* documentary, 2018, https://www.generalmagicthemovie.com/.

49 *Rossetto had tried to take the parent company public*: Dawn Kawamoto, "Wired pulls plug on IPO," CNET, October 25, 1996, https://www.cnet.com/tech/tech-industry/wired-pulls-plug-on-ipo/; also Rossetto, interview.

50 *Just around the corner from* Wired: Author interview, Fran Maier, May 2, 2024.

51 *The dot-com boom hadn't quite started yet*: Author interview, Mark Pincus, October 22, 2024.

52 *Newmark was already steeped in the Internet*: Author interview, Craig Newmark, October 23, 2024.

53 *Cindy Cohn, a young lawyer from Michigan*: Author interview, Cindy Cohn, May 7, 2025.

54 *The young Internet industry, led by the EFF, rose up in protest*: Cohn, interview; see also Parker Higgins, "The Web's First Blackout Protest: The CDA, 20 Years Later," EFF, February 23, 2016, https://www.eff.org/deeplinks/2016/02/webs-first-blackout-protest-cda-20-years-later.

55 *While Gilmore and Barlow and some other EFF backers*: Cohn, interview; author interview, Mitch Kapor, October 9, 2024.

56 *John Law stood forty-five feet above the salt flats*: Burning Man 1996, Trippingly.net, 1996. See also Brad Wieners, "An Oral History of Burning Man, the Biggest, Weirdest, Most Clothing-Optional Desert Carnival on the Planet," *Outside*, August 24, 2012, https://www.outsideonline.com/adventure-travel/destinations/hot-mess/; Scott Beale, "The John Law Burning Man 'Exit Interview,'" *Laughing Squid* blog, 1996; author interview, John Law, February 16, 2024.

56 *Eight thousand people had gathered*: Caveat Magister, "A Brief History of Who Ruined Burning Man," *Burning Man Journal*, October 6, 2016, https://journal.burningman.org/2016/10/philosophical-center/tenprinciples/a-brief-history-of-who-ruined-burning-man/.

56 *Small wooden man burned on Baker Beach*: Brewster Kahle's Blog, "Zone Trip #5: Inventing a Desirable Physical World in Pandemic August 2020," April 22, 2020, https://brewster.kahle.org/2020/04/22/zone-trip-5-inventing-a-desirable-physical-world-in-pandemic-august-2020/.

56 *Law and his friend Michael Mikel*: Law, interview. See also Gilmore, interview.

57 *A woman named P. Segal*: Trippingly, "P Segal, Cacophonist and Influencer," https://www.trippingly.net/burning-man-musings/p-segal.

57 *On the Friday ahead of Labor Day weekend*: Author interview, Michael Mikel, May 4, 2024.

57 *What they called a Zone Trip*: Kahle, "Zone Trip #5."

58 *Kahle tied the knot in the Black Rock Desert*: Kahle, "Zone Trip #5."

59 *A SoMa DJ named Craig Ellenwood*: Kirsten Weisenburger, "Burning Man's First Sound Camp," *Burning Man Journal*, November 17, 2021, https://journal.burningman.org/2021/11/black-rock-city/tales-from-the-playa/burning-mans-first-sound-camp/.

59 *"There were all kinds of gutter punks"*: Author interview, Candace Locklear, May 7, 2025.

59 *. . . by 1996 it was a $35 ticket*: Justin Berton, *SFGate*, "Burning Man Tries to Cope with Cash," August 19, 2007, https://www.sfgate.com/news/article/BURNING-MAN-TRIES-TO-COPE-WITH-CASH-2509458.php.

60 *That was the final straw for Law*: Wieners, "An Oral History."

60 *A T-shirt began to circulate*: Wieners, "An Oral History."

61 *For Law, the choice was clear*: Law, interview, July 10, 2025.

62 *She'd grown up in Ohio*: Author interview, Marian Goodell, January 17, 2025.

63 *Harvey died*: "Larry Harvey, founder and leader of Burning Man, dies in SF," *SF Gate*, April 30, 2018, https://www.sfgate.com/bayarea/article/Larry-Harvey-founder-Burning-Man-dead-after-stroke-12871983.php.

63 *"He was just an amazing guy"*: Author interview, Terry Gross, May 14, 2025.

66 *Scott Beale . . . helped create the first live webcast*: Author interview, Scott Beale, September 17, 2024; see also Trippingly, "Burning Man 1993: Article from Silicon Valley Metro," February 27, 2019, https://www.trippingly.net/burning-man-musings/2019/2/27/burning-man-1993-article-from-silicon-valley-metro.

66 *In August 1998, the multicolored logo on the home page of Google*: Google Doodles, https://doodles.google/doodle/burning-man-festival/.

Chapter 3: The Bible of the Dot-Com Boom

71 *Netscape had gone public just sixteen months*: Chart, "1996's Internet IPOs," *Wall Street Journal*, December 9, 1996, https://www.wsj.com/articles/SB850102729387399500.

72 *A San Francisco outfit called The Globe*: "TheGlobe.com," Wikipedia, https://en.wikipedia.org/wiki/TheGlobe.com.

72 *In 1999 alone, nearly 300 Internet firms went public*: David Westenberg, "Internet IPOs Conclude Sensational 1999," Wilmer Hale Publications, January 1, 2000, https://www.wilmerhale.com/en/insights/publications/Internet-ipos-conclude-sensational-year-in-1999-january-2000.

72 *Frank Quattrone, the leading investment banker in the sector*: "Frank Quattrone," Wikipedia, https://en.wikipedia.org/wiki/Frank_Quattrone.

72 *Even Mayor Willie Brown*: Lance Williams and Chuck Finnie, "How the Mayor Really Built His Stock Portfolio," *San Francisco Examiner*, April 6, 2000, https://www.sfchronicle.com/news/article/how-mayor-really-built-his-stock-portfolio-3066136.php.

73 *"I missed the whole dot-com bubble"*: Pincus, interview.

73 *Unemployment in the city had fallen*: Chart: Unemployment Rate in San Francisco, Federal Reserve Economic Data, https://fred.stlouisfed.org/series/CASANF0URN.

74 *People were coming from all over the country*: Author interview, Ev Williams, June 3, 2024.

75 *Craig Newmark had been laid off*: Newmark, interview.

75 *Trevor Traina's startup*: Author interview, Trevor Traina, March 13, 2025.

77 *CEO Jonathan Nelson was buying out the leases*: Nelson, interview.

77 *Installing the signs would fall to John Law and Michael Mikel*: Law, interview.

78 *Nelson already had Goldman Sachs lined up*: Nelson, interview.

78 *The great-grandson of Isaiah Hellman*: "Warren Hellman, 77, investor who loved bluegrass, dies," Peter Lattman, *New York Times*, December 19, 2011, https://www.nytimes.com/2011/12/20/business/warren-hellman-dies-at-77-ex-lehman-president-and-music-festival-founder.html.

79 *The annoying wrinkle that it wasn't in fact our company*: John Battelle, Revisionist History at IDG, Searchblog, March 29, 2007.

80 *For city residents who weren't interested*: Laguana, interview.

81 *A wave of prosperous young dot-commers was soon pushing rents*: "30+ Years of Housing Market Cycles in the San Francisco Bay Area: Recessions, Recoveries, Booms, Bubbles & Adjustments (Sometimes Crashes)," Patrick Carlisle, Compass Real Estate, updated November 2021, https://www.bayareamarketreports.com/trend/3-recessions-2-bubbles-and-a-baby.

82 *We had a five-hour hearing*: Author interview, Dan Kingsley, February 11, 2025.

83 *Chris Daly had come to [the] Mission*: Author interview, Chris Daly, April 25, 2024.

84 *Resistance to development in the Mission*: Eva Knowles, "Mission Yuppie Eradication Project," FoundSF, 2024, https://www.foundsf.org/Mission_Yuppie_Eradication_Project.

84 *Crime had fallen significantly since the early 1990s*: Author interview, Greg Suhr, July 15, 2024.

85 *Chris Carlsson, a flinty writer and activist*: Author interview, Chris Carlsson, October 11, 2023.

86 *"We dressed up like big computer heads"*: Author interview, Laura Fraser, July 20, 2025.

86 *On this day in 1992*: Carlsson, interview.

88 *Welch and his allies would sour on Brown*: Matthew Robinson and Brian C. Anderson, "Willie Brown Shows How Not to Run a City," *City Journal*, Autumn 1998, https://www.city-journal.org/article/willie-brown-shows-how-not-to-run-a-city.

89 *"My job was to go and talk to these old Italian ladies"*: Author interview, Aaron Peskin, June 8, 2024.

89 *Brown has his own story*: Brown, interview.

90 *Willie Brown compliments Aaron Peskin*: Mission Local, "So Many Speedo Jokes at Aaron Peskin's Birthday Roast," June 2024, https://missionlocal.org/2024/06/so-many-speedo-jokes-at-aaron-peskins-birthday-roast/.

90 *Industry Standard shutdown*: Editorial, *New York Times*, "Economic Tombstones: The Industry Standard Unplugged," August 18, 2001, https://www.nytimes.com/2001/08/18/opinion/economic-tombstones-the-industry-standard-unplugged.html.

96 *The* Wall Street Journal *ran a long, well-reported story*: Matthew Rose and Suzanne McGee, "Publisher to Close Industry Standard as Ad Slump Claims Another Victim," *Wall Street Journal*, August 17, 2001, https://www.wsj.com/articles/SB997994475396278436; and also Sharon Waxman, "Final Public Offering," *Washington Post*, August 26, 2001, https://www.washingtonpost.com/archive/lifestyle/2001/08/27/final-public-offering/8d31fca6-f067-477d-88c6-76f1a956ab41/.

96 *Some thirty thousand people moved out*: Anastasia Hendrix, "Census: Bay Area counties shrink / Santa Clara, S.F. lead U.S. in drop in population," *SFGate*, April 17, 2003, https://www.sfgate.com/bayarea/article/Census-Bay-Area-counties-shrink-Santa-Clara-2654647.php.

97 *Lycos, a Boston startup that was trying to rival Yahoo*: "Lycos acquires Wired Digital," *WIRED*, October 6, 1998, https://www.wired.com/1998/10/lycos-acquires-wired-digital/.

97 *Rossetto and Metcalfe seized the moment*: David Carr, "The Coolest Magazine on the Planet," *New York Times*, July 27, 2003, https://www.nytimes.com/2003/07/27/books/the-coolest-magazine-on-the-planet.html; see also Wolf, "Wired."

97 *Over at Organic, Jonathan Nelson acted sooner than most*: Nelson, interview.

Chapter 4: Gavin Newsom Meets Web 2.0

97 *Newsom was a child of the city's political and financial royalty*: Seema Mehta, Ryan Menezes, and Maloy Moore, "Gavin Newsom and Old Money," *Los Angeles Times*, September 7, 2018, https://www.latimes.com/projects/la-pol-ca-gavin-newsom-san-francisco-money/.

104 *Ann and Gordon bought*: James Reginato, *Growing Up Getty: The Story of America's Most Unconventional Dynasty* (Gallery Books, 2002), 203.

104 *PlumpJack was just getting ramped up*: Brown, interview.

107 *The next year*: Author interview, Tom O'Connor, November 19, 2024.

107 *Kimberly Guilfoyle would make a name for herself*: Jaxon Van Derbeken, "Newsmaker Profile: Kimberly Guilfoyle and James Hammer," *SFGate*, May 6, 2001, https://www.sfgate.com/politics/article/NEWSMAKER-PROFILE-Kimberly-Guilfoyle-and-James-2924603.php.

107 *The burgeoning Willie Brown machine*: Mehta, Menezes, and Moore, "Gavin Newsom and Old Money."

107 *A seat on the Board of Supervisors came open*: Elizabeth Lesly Stevens, "The Power Broker," *Washington Monthly*, July 7, 2012, https://washingtonmonthly.com/2012/07/07/the-power-broker/.

108 *High-level alliances*: Author interview, Ed Harrington, May 7, 2024.

108 *Brown also pushed through a charter reform measure*: Lance Williams, "THE MAYOR'S LEGACY: WILLIE BROWN / FBI's 5-year City Hall probe yields few convictions amid cries of political meddling," *San Francisco Chronicle*, January 4, 2004, https://www.sfgate.com/politics/article/the-mayor-s-legacy-willie-brown-fbi-s-5-year-2832559.php.

108 *Brown also packed his office*: Robinson and Anderson, "Willie Brown Shows How Not to Run a City"; Mary Curtius, "Corruption Scandal in S.F. Grows," *Los Angeles Times* archives, September 23, 1999, https://www.latimes.com/archives/la-xpm-1999-sep-23-mn-13313-story.html.

108 *Brown appointed Harris to two state commissions*: Michael Finnegan, "Kamala Harris was shaped by the crucible of San Francisco Politics," *Los Angeles Times*, January 21, 2019, https://www.latimes.com/politics/la-na-pol-kamala-harris-san-francisco-20190121-story.html.

109 *She reached out to Mark Buell*: Author interview, Mark Buell, January 30, 2025.

109 *Numerous local tech companies going out of business*: Federal Reserve Economic Data, "Unemployment Rate in San Francisco City/County, CA," https://fred.stlouisfed.org/series/CASANF0URN.

109 *Williams ended up running Blogger by himself*: Nick Bilton, *Hatching Twitter: A True Story of Money, Power, Friendship, and Betrayal* (Portfolio, 2013).

111 *Justin Hall, a teenager from Pennsylvania*: Justin Hall, "Jonathan Steuer," Justin Hall's Personal Site, Links.net, https://links.net/vita/cyb/jss/.

111 *Matt Mullenweg . . . had created an open-source blogging platform called WordPress*: Author interview, Matt Mullenweg, April 10, 2025.

113 *It was just a very generative time*: Mullenweg, interview.

113 *"I learned so much from Craig," says Pincus*: Author interview, Mark Pincus, October 22, 2024.

114 *SF LAN, an early effort to provide free internet connectivity*: Brewster Kahle's blog, "Zone Trip #5: Inventing a Desirable Physical World in Pandemic," April 22, 2020, https://brewster.kahle.org/2020/04/22/zone-trip-5-inventing-a-desirable-physical-world-in-pandemic-august-2020/.

115 *Pincus . . . was an early investor in Facebook*: Pincus, interview.

116 *The Newsom-Gonzalez matchup*: SF Gate, "Mayoral hopefuls come out swinging in debate," https://www.sfgate.com/politics/article/Mayoral-hopefuls-come-out-swinging-in-debate-2512314.php.

115 *Newsom won*: John Wildermuth, "Mayoral hopefuls come out swinging in debate," *SFGate*, November 12, 2003, https://www.sfgate.com/politics/article/Mayoral-hopefuls-come-out-swinging-in-debate-2512314.php.

119 *In what was quickly dubbed "Camp Agnos"*: California State Archives, State Government Oral History Program, "Art Agnos Oral History," 2002, 2003, https://archives.cdn.sos.ca.gov/oral-history/pdf/oh-agnos-art-v2.pdf.

122 *An online system called SFGov*: CivLab/SF Gov, https://sfgov.civlab.org/.

123 *An even bigger problem was Aaron Peskin*: Author interview, Minnie Ingersoll, July 25, 2024; see also "SF Supervisor Aaron Peskin's Message to Newsom," *SF Weekly*, March 5, 2008, https://www.sfweekly.com/archives/sf-supervisor-aaron-peskins-message-to-newsom-quit-attacking-me/article_d044f2c2-3b81-591e-ae50-89cc0fefe326.html.

123 *Dubbed Web 2.0*: Author interview, Tim O'Reilly, February 12, 2025; John Battelle, January 17, 2025.

126 *He'd debuted Foo Camp*: *Justinsomnia* blog, "Foo Camp," August 20, 2005, https://justinsomnia.org/2005/08/foo-camp/; see also John Battelle, "When geeks go camping, ideas hatch," CNN, January 10, 2004, https://www.cnn.com/2004/TECH/ptech/01/09/bus2.feat.geek.camp/; see also author interview, Scott Beale, September 17, 2024.

126 *They sold the company to AOL*: Author interview, Michael Birch, October 29, 2024; see also Michael Arrington, "AOL Buys Bebo for $850 Million," TechCrunch, March 13, 2008, https://techcrunch.com/2008/03/13/aol-buys-bebo-for-750-million/.

127 *It would earn the nickname the "Y-Scraper"*: Quora, "What is the Y scraper," https://www.quora.com/What-is-the-Y-scraper.

127 *Larsen was on to his next venture, Prosper Marketplace*: Author interview, Chris Larsen, February 23, 2024.

128 *Key participants had met at Marian Goodell's house*: Author interview, Marian Goodell, January 17, 2025; see also *StudioAlpha* podcast, https://www.youtube.com/watch?v=j6TrEHe1a8c.

130 *Pepe and his opera*: Kirsten Weisenburger, "Meet the DJs Who Started Burning Man's First Sound Camp," *Burning Man Journal*, November 17, 2021, https://journal.burningman.org/2021/11/black-rock-city/tales-from-the-playa/burning-mans-first-sound-camp/; Trippingly, "Burning Man 1993 Article from Silicon Valley Metro," February 27, 2019, https://www.trippingly.net/burning-man-musings/2019/2/27/burning-man-1993-article-from-silicon-valley-metro.

130 *"The grant program was pretty much based on trust"*: Author interview, Christine "Ladybee" Kristen, September 20, 2024; see also https://journal.burningman.org/author/ladybee/.

131 *The rave camps*: Cameron Holbrook, "26 Photos Capturing the Blissful Essence of San Francisco's '90s Rave Scene," Mixmag, June 19, 2018, https://mixmag.net/feature/26-photos-capturing-the-blissful-essence-of-san-franciscos-90s-rave-scene/; Samantha Durbin, "Acid, Dance, Unity: What Happened to the '90s Bay Area Rave Scene," *Bold Italic*, June 12, 2019, https://thebolditalic.com/acid-dance-unity-what-happened-to-the-90s-bay-area-rave-scene-50a5320b7001.

132 *The company did not*: "Burning Man Tries to Cope with Cash," *SFGate*, August 19, 2007, https://www.sfgate.com/news/article/BURNING-MAN-TRIES-TO-COPE-WITH-CASH-2509458.php; Brad Wieners, "An Oral History of Burning Man," *Outside*, August 24, 2012, https://www.outsideonline.com/adventure-travel/destinations/hot-mess/.

132 *Quite a surprise for Law*: Author interview, John Law, July 20, 2025.

Chapter 5: The Twitter Tax Break

135 *A fledgling podcasting company called Odeo*: Nick Bilton, *Hatching Twitter: A True Story of Money, Power, Friendship, and Betrayal* (Portfolio, 2013), 15–22.

135 *Jason Goldman joined them*: Bilton, *Hatching Twitter*, 22–25.

135 *It stole the show at the SXSW tech festival*: Scott Beale, "Twitter is Super-Active at SXSW Interactive," Laughing Squid, March 12, 2007, accessed via Internet Archive; Bilton, *Hatching Twitter*, 76–84.

136 *Facebook . . . made a bid to buy Twitter*: Kara Swisher, "When Twitter Met Facebook: The Acquisition Deal That Fail-Whaled," *Wall Street Journal*/AllThingsD, November 24, 2008, https://allthingsd.com/20081124/when-twitter-met-facebook-the-acquisition-deal-that-fail-whaled/.

136 *The city's economy contracted*: Federal Reserve Economic Data, "Unemployment Rate in San Francisco County/City, CA," https://fred.stlouisfed.org/series/CASANF0URN.

137 *"Texas Hold 'Em Poker" on Facebook*: Sarah Lacy, "The Real History of Twitter," TechCrunch, October 4, 2006; David Kirkpatrick, *The Facebook Effect* (Simon & Schuster, 2010), 201–203.

137 *Airbnb was getting its feet under it*: "Just Announced: Airbnb Raises $7.2MM Series A Round and Releases iPhone App," *Business Insider*, November 10, 2010, https://www.businessinsider.com/airbnb-raises-72mm-series-a-round-and-releases-iphone-app-2010-11.

137 *Salesforce, founded in 1999 by Marc Benioff*: Marc Benioff and Carlye Adler, *Behind the Cloud* (Jossey-Bass, 2009), 23–45.

138 *Mark Zuckerberg mocked it as a "clown car"*: Ben Mezrich, *The Accidental Billionaires* (Doubleday, 2009).

138 *Twitter closed its third round of funding*: "Twitter Closes Its $100 Million Round," TechCrunch, September 25, 2009, https://techcrunch.com/2009/09/25/twitter-closes-its-100-million-round/.

139 *Williams wanted to find a permanent home for Twitter*: Author interview, Ev Williams, June 3, 2024; also referenced in multiple *San Francisco Chronicle* articles about the Twitter headquarters decision.

139 *1355 Market was an odd but alluring structure*: Jay C. Barmann, "Twitter Building History: From Furniture Exchange to Tech HQ," Curbed SF, January 22, 2019, https://sf.curbed.com/2019/1/22/18184279/twitter-building-bio-history-origins-architect-furniture-exchange.

140 *Beginning in the mid-1990s*: Author interview, Linda Corso, February 10, 2025.

141 *A friend had tipped him off*: J. K. Dineen, "Mr. Buy-and-hold," *San Francisco Business Times*, July 29, 2007, https://www.bizjournals.com/sanfrancisco/stories/2007/07/30/focus1.html.

141 *Willie Brown, who met*: Brown, interview.

142 *Peskin recalls*: Peskin, interview.

142 *Dworman was a regular*: Author interview, Eric Grossberg, September 3, 2025.

142 *There were special events*: Michael Dougan and Julie Chao, "Raunchy 'Ritual,'" *San Francisco Examiner*, May 8, 1997, https://www.sfgate.com/news/article/RAUNCHY-RITUAL-3120012.php; see also "Mary Curtius and Maria L. La

Ganga, "Party Proves Even San Francisco Can Blush," *Los Angeles Times*, May 8, 1997, https://www.latimes.com/archives/la-xpm-1997-05-08-mn-56756-story.html.

143 *And he had some ideas*: Grossberg, interview.

143 *Twitter in the early days wanted to "save the world"*: Author interview, Chris Roeder, March 18, 2024.

144 *The city at the time had a payroll tax*: Alejandro Lazo, "Tax Breaks for Twitter Bring Benefits and Criticism," *Wall Street Journal*, April 29, 2016, https://www.wsj.com/articles/tax-breaks-for-twitter-bring-benefits-and-criticism-1461947597.

145 *It was a really scary time*: Author interview, Jason Elliott, August 14, 2024.

145 *The city had offered a time-limited payroll tax break*: San Francisco Controller's Office reports on tax increment financing for biotech companies, 2000–2005.

145 *They'd met decades ago*: Brown, interview.

145 *At the Furniture Mart, Brown led a meeting*: Author interviews with multiple participants who requested anonymity, 2019–2020.

146 *When Matz proposed the idea*: Daly, interview.

146 *He was replaced by Dick Costolo*: Amir Efrati, "Twitter Switches Chief Executives," *Wall Street Journal*, October 4, 2010, https://www.wsj.com/articles/SB10001424052748704631504575532260428606800?gaa_at=eafs&gaa_n=ASWzDAinNOm1w6j4YnyCc5Bk8n0NiAYomnsjb8RIG-8CpyYDiSCOH9oIz4iCZHGKZA4%3D&gaa_ts=68ddd94a&gaa_sig=bsjnytGT5T_IpeDfunU_79r5CX5tJQMYZoJQepKL8iTlN8nz2yZMfptDoHHcx7AV_gZw6Anlcv429kG0wg5FDg%3D%3D.

146 *Dick Costolo too*: Author interview, Dick Costolo, July 22, 2024.

147 *Board of Supervisors considered candidates for interim mayor*: "The Inside Story of How Ed Lee Became Mayor," *San Francisco Magazine*, March 2011.

149 *Chiu liked Lee*: Author interview, David Chiu, March 31, 2025.

149 *He retreated to Newcom's office*: Author interview, Bevan Dufty, June 20, 2024.

150 "*To present an opportunity to a 'person of color'*": Gerry Shih, *New York Times*, January 6, 2011, https://www.nytimes.com/2011/01/07/us/07bcmayor.html.

150 *Chris Daly raged*: "Twitter Loves Chris Daly's 'It's on like Donkey Kong' Threat," *SF Examiner*, accessed via Wayback Machine, January 5, 2011, https://www.sfexaminer.com/news/twitter-loves-chris-dalys-its-on-like-donkey-kong-threat/article_1e024986-9df1-5fe5-8057-152a352296aa.html.

150 *He called Doug Shorenstein*: Author interview, Charles Malet, September 10, 2025.

150 *There was horsetrading*: Author interview, Jane Kim, April 22, 2024.

152 *The Board approved the tax break*: John Coté, "S.F. Controller Hails 'Twitter Tax Break' as rousing success," *San Francisco Gate*, October 15, 2014, https://www.sfgate.com/bayarea/article/S-F-controller-hails-Twitter-tax-break-as-5851498.php.

152 *Mark Pincus's game company*: Pincus, interview.

152 *Pincus' lobbyist*: Author interview, Chris Gruwell, April 15, 2025.

153 *It wasn't even clear*: Gary Rivlin, *The Godfather of Silicon Valley* (Random House, 2001).

154 *Conway lived in the city*: Author interview, Ron Conway, February 22, 1994.

155 *Conway said it was he and Warren Hellman*: Conway, interview.

155 *A priceless video*: Elizabeth Flock, "Ed Lee dubbed '2 Legit 2 Quit' in new political ad by M.C. Hammer and Biz Stone (video)," *Washington Post*, October 25, 2011, https://www.washingtonpost.com/blogs/blogpost/post/ed-lee-dubbed-2-legit-2-quit-in-new-political-ad-by-mc-hammer-and-biz-stone-video/2011/10/25/gIQAfy9EGM_blog.html.

Chapter 6: The Media Shift

157 *Hellman had become the leader*: Elizabeth Lesly Stevens, "The Power Broker," *Washington Monthly*, July 7, 2012, https://washingtonmonthly.com/2012/07/07/the-power-broker/.

157 *Underwriting Hardly Strictly Bluegrass*: https://www.hellmanfoundation.org/hardly-strictly-bluegrass.html.

157 *Back in the day*: Author interview, Rudy Nothenberg, January 5, 2024.

159 *That wouldn't be enough*: Redmond, interview.

160 *Criminally prosecuting Backpage.com*: Michael Finnegan, "Kamala Harris was shaped by the crucible of San Francisco politics," *Los Angeles Times*, January 21, 2019, https://www.latimes.com/politics/la-na-pol-kamala-harris-san-francisco-20190121-story.html.

160 *John Battelle had seen the promise*: Battelle, interview.

160 *Kara Swisher, a former*: Author interview, Kara Swisher, June 24, 2025.

160 *He took to calling the business a "platform for change"*: Benioff, Marc, "*Trailblazers: The Power of Business as the Greatest Platform for Change*," Crown Currency, October 15, 2019.

165 *Benioff's grandfather, Marvin Lewis*: Peter Hartlaub, "Marc Benioff, Salesforce, and the monorail-loving SF supervisor who inspired them," *San Francisco Chronicle*, October 15, 2019, https://www.sfchronicle.com/oursf/article/Before-Marc-Benioff-blazed-trails-grandfather-14516074.php.

165 *The leukemia that would be diagnosed a few months later*: David R. Baker, "Warren Hellman, financier and philanthropist, dies at 77," *San Francisco Chronicle*, December 19, 2011, https://www.sfgate.com/news/article/Warren-Hellman-financier-and-philanthropist-2411646.php.

167 *Garrett Camp, the Canadian*: Brad Stone, *The Upstarts: Uber, Airbnb and the Battle for the New Silicon Valley* (Little, Brown, 2017).

168 *Kalanick's relentless drive*: Mike Isaac, *Super Pumped: The Battle for Uber* (W. W. Norton, 2019).

168 *Hayashi in fact had been exploring*: Elliott, interview.

168 *Uber agreed to stop marketing*: Stone, *Upstarts*.

170 *Airbnb was far from the only place*: Stone, *Upstarts*.

170 *Hotels in San Francisco tended to be extremely expensive*: Sam Altman, "Airbnb and San Francisco," Sam Altman's blog, September 28, 2015, https://blog.samaltman.com/airbnb-and-san-francisco.

170 *San Francisco . . . clearly had the authority to set lodging rules*: Carolyn Said, "Supes back 'Airbnb Law' to allow short-term rentals, with limits," *SFGate*,

October 8, 2014, https://www.sfgate.com/news/article/Supervisors-approve-Airbnb-law-5807858.php.

171 *Conway's investing style involved finding ways*: Rivlin, *Godfather*, p.

172 *Lee branded San Francisco the "tech capital of the world"*: Joe Eskenazi, "How Mayor Ed Lee remade San Francisco in Big Tech's image," *Verge*, December 13, 2017, https://www.theverge.com/2017/12/13/16771912/mayor-ed-lee-san-francisco-silicon-valley-tech-jobs.

172 *Twitter moved eight hundred employees into . . . Market Square*: Jay C. Barmann, "How the Internet reshaped Market Street's Art Deco monolith," *Curbed San Francisco*, April 6, 2020, https://sf.curbed.com/2019/1/22/18184279/twitter-building-bio-history-origins-architect-furniture-exchange.

172 *Lee toured the site*: Robin Wilkey, "Twitter Headquarters Opens: San Francisco Mid-Market Office Revealed," *HuffPost*, June 13, 2012, https://www.huffpost.com/entry/twitter-headquarters-opens-san-francisco_n_1594237.

172 *"Job creation is everything," declares Conway*: Conway, interview; see also Joe Eskenazi, "How Mayor Ed Lee remade San Francisco," *The Verge*, December 13, 2017, https://www.theverge.com/2017/12/13/16771912/mayor-ed-lee-san-francisco-silicon-valley-tech-jobs.

173 *Zendesk, a Danish customer service startup*: Sarah Klearman, "Zendesk's longtime Mid-Market headquarters listed for sale," *San Francisco Business Times*, April 10, 2024, https://www.bizjournals.com/sanfrancisco/news/2024/04/04/989-market-for-sale-asb.html.

173 *The exemption excluded all of the large office properties*: Author interview, Ted Egan, April 23, 2024.

173 *New restaurants began appearing on Market Street*: Carolyn Alburger, "As Twitter Tax Break Nears Its End, Mid-Market Restaurants Feel Glimmer of Hope," Eater San Francisco, September 19, 2018, https://sf.eater.com/2018/9/19/17862118/central-market-tax-exclusion-restaurants-post-mortem-future.

174 *Zendesk had tapped*: Author interview, Tiffany Apczynski, February 14, 2024.

174 *South Park was being colonized by venture capital firms*: Carmel DeAmicis, "Has South Park Finally Become the New Sand Hill Road?" Vox, April 13, 2015, https://www.vox.com/2015/4/13/11561376/has-south-park-finally-become-the-new-sand-hill-road.

175 *The Netscape Moment*: "Netscape IPO 20-year anniversary," *Fortune*, August 9, 2015, https://fortune.com/2015/08/09/remembering-netscape/; see also Brian McCullough, "20 Years On: Why Netscape's IPO Was the 'Big Bang' of the Internet Era," *Internet History Podcast*, August 7, 2015, https://www.Internethistorypodcast.com/2015/08/20-years-on-why-netscapes-ipo-was-the-big-bang-of-the-Internet-era/.

Chapter 7: The Big Boom

179 *Unemployment falling by more than half*: Federal Reserve Economic Data, Unemployment Rate in San Francisco City/County, CA, https://fred.stlouisfed.org/series/CASANF0URN.

179 *Twitter would go public in 2013*: Sarah McBride and Alexei Oreskovic, "Twitter IPO highlights big money friction in San Francisco," Reuters, November 7, 2013, https://www.reuters.com/article/twitter-ipo-sanfrancisco/twitter-ipo-highlights-big-money-friction-in-san-francisco-idINDEE9A60H220131107; John Shinal, "With Twitter IPO—changes are coming to Market Street," *USA Today*, October 24, 2013, https://www.usatoday.com/story/tech/columnist/shinal/2013/10/24/twitter-ipo-san-francisco/3183865/.

180 *The risk-loving Japanese entrepreneur*: Katie Benner, "Masayoshi Son's Grand Plan for Softbank's $100 Billion Vision Fund," *New York Times*, October 10, 2017.

180 *The firm's first*: "Meet Our Newest Portfolio Company," Andreessen Horowitz, July 24, 2011, https://a16z.com/announcement/meet-our-newest-portfolio-company-airbnb/.

181 *This whole blitz-scaling idea*: O'Reilly, interview.

182 *Paradis, who grew up in a French-speaking family*: Author interview, Paul Paradis, December 16, 2024.

182 *"We wanted to have a building"*: Author interview, Merredith Treaster, December 16, 2024.

182 *In 2010 Salesforce had spent*: "Salesforce.com Purchases Site for New San Francisco Global Headquarters," Salesforce.com, November 1, 2010. https://www.salesforce.com/news/press-releases/2010/11/01/salesforce-com-purchases-site-for-new-san-francisco-global-headquarters/.

185 *Shorenstein, after acquiring the Twitter building*: Emily Landes, "Shorenstein pays off $400M loan on Twitter Building," The Real Deal, March 6, 2023, https://therealdeal.com/san-francisco/2023/03/06/shorenstein-pays-off-400m-loan-on-twitter-building/.

186 *By the middle of the decade*: Evgenia Peretz, "Blue Bloods and Billionaires," *Vanity Fair*, September 24, 2013, https://www.vanityfair.com/news/2013/10/pacific-heights-real-estate?srsltid=AfmBOoolfHzz3dyndmLxA1Lwr7WPim4WUZOr3nsed-HCbdmNL5Jvxzcm; see also author interview, Traina.

187 *Jack Dorsey, fiddling with a few associates*: Levy, Steven.

187 *Birch recounted later*: Author interview, Michael Birch, October 29, 2024.

188 *Long-established clubs for the upper crust*: Katie Sweeney, "Coterie by the Bay: Step Inside Six of San Francisco's Private Clubs," *Nob Hill Gazette*, December 1, 2023, https://www.nobhillgazette.com/places/step-inside-six-of-san-franciscos-private-clubs/article_ae9b56a2-87ea-11ee-b8e8-6b20a7e834fb.html.

189 *Jamie Zawinski, a well-known programmer*: Evany Thomas, "From Netscape to Nightclub," *Wired*, July 16, 2001.

189 *Nelson had also run into issues*: Nelson, interview.

190 *Doug Dalton, another early-Netscape alum*: Author interview, Doug Dalton, March 7, 2025.

190 Buell, interview.

193 *The gaudy excess*: "Sean Parker's Lavish Big Sur Wedding," *Vanity Fair*, August 5, 2014.

194 *A signature case of the hazards*: Ellen Huet and Patricia Hurtado, "Founders of Sexual Wellness Startup One Taste Goes on Trial," May 5, 2025.

195 *Apache founder Brian Behlendorf*: Behlendorf, interview.

196 *The event sold out for the first time*: Justin Berton, "Burning Man tries to cope with cash," *SFGate*, August 19, 2007, https://www.sfgate.com/news/article/BURNING-MAN-TRIES-TO-COPE-WITH-CASH-2509458.php.

196 *The rich could fly in*: Nick Bilton, "A Line Is Drawn in the Desert," *New York Times*, August 20, 2014, https://www.nytimes.com/2014/08/21/fashion/at-burning-man-the-tech-elite-one-up-one-another.html; Gregory Ferenstein, "Why Silicon Valley billionaires are obsessed with Burning Man," Vox, August 22, 2014, https://www.vox.com/2014/8/22/6050625/why-silicon-valley-billionaires-are-obsessed-with-burning-man.

196 *Chang had grown up*: Author interview, Tim Chang, May 13, 2025.

197 *Took amenities up a notch*: Author interview, Chip Conley, January 14, 2025.

197 *Conley would play a key role*: Conley, interview.

198 *A camp called Caravancycle*: Author interview, David Rodriguez and Suzanne Cowell, May 7, 2025.

199 *Even as it matured*: burning man history and timeline, https://burningman.org/about/history/brc-history/afterburn/2011-2/financial_chart/.

199 *"I used to scoff"*: Locklear, interview.

199 *The spirit of Burning Man*: Author interview, Mike Farrah, November 15, 2024.

199 *Davis had first visited San Francisco as a teenager*: Author interview, Ben Davis, December 19, 2024.

201 *The unassuming library patron, it turned out*: Andy Greenberg, "Undercover Agent Reveals How He Helped the FBI Trap Silk Road's Ross Ulbricht," *Wired*, January 14, 2015.

202 *At First, Larsen said later, the main concern was a simple one*: Larsen, interview.

203 *That's what spurred Jen Pahlka*: Author interview, Jen Pahlka, June 27, 2025.

204 *O'Reilly had begun eyeing*: O'Reilly, interview.

204 *Behlendorf was soon drawn*: Author interview, Brian Behlendorf, March 21, 2025.

205 *She did just that*: Pahlka, interview.

Chapter 8: The Buses and the Backlash

207 *He recalled years later*: Author interview, David Campos, March 15, 2025.

209 *In the fall of 2012 he plunked down*: Emily Landes, "Mark Zuckerberg sells SF home for $31M," The Real Deal, July 21, 2022, https://therealdeal.com/san-francisco/2022/07/21/mark-zuckerberg-sells-sf-home-for-31m/.

209 *A city whose key industry used to stand a bit apart*: Reyhan Harmanci, "The Heart of Silicon Valley Is Now in the Mission," BuzzFeed News, October 16, 2012, https://www.buzzfeednews.com/article/reyhan/the-heart-of-silicon-valley-is-now-in-the-mission.

210 *Campos called for a moratorium*: Campos, interview.

210 *He wasn't a fan of the Twitter tax break*: Campos, interview; see also Alejandro Lazo, "Tax Breaks for Twitter Bring Benefits and Criticism," *Wall Street Journal*, April 29, 2016, https://www.wsj.com/articles/tax-breaks-for-twitter-bring-benefits-and-criticism-1461947597.

210 *A proposed ten-story apartment building*: "Eight-Year Community Fight Leads to 16th Street Plaza Set to Feature a 100% Affordable Housing Development," MEDA SF, March 28, 2022, https://medasf.org/eight-year-community-fight-leads-to-16th-street-plaza-set-to-feature-a-100-affordable-housing-development/.

210 *A designer by trade*: Author interview, Cari Spivack, October 1, 2024.

212 *With unemployment topping 10%*: "Unemployment Rate in San Francisco County/City, CA," Federal Reserve Economic Data, https://fred.stlouisfed.org/series/CASANF0URN.

212 *Ed Lee and tech industry*: Joe Eskenazi, "How Mayor Ed Lee remade San Francisco in Big Tech's image," The Verge, December 13, 2017, https://www.theverge.com/2017/12/13/16771912/mayor-ed-lee-san-francisco-silicon-valley-tech-jobs.

213 *Gillian Gillett, who was in charge*: Author interview, Gillian Gillett, September 20, 2024.

215 *Shortt had been active in housing rights causes*: Author interview, Sara Shortt, May 5, 2025.

216 *The buses represented a huge class divide*: Leslie Dreyer, as quoted in Cary McClelland, *Silicon City* (W. W. Norton, 2018), 92.

216 *Google bus protests begin*: "Twitter IPO Highlights Big Money Friction in San Francisco," Reuters, November 7, 2013, https://www.reuters.com/article/twitter-ipo-sanfrancisco/twitter-ipo-highlights-big-money-friction-in-san-francisco-idINDEE9A60H220131107.

216 *First protest, on Valencia St*: Sarah McBride, "Google bus blocked in San Francisco gentrification protest," Reuters, December 9, 2013, https://www.reuters.com/article/technology/google-bus-blocked-in-san-francisco-gentrification-protest-idUSBRE9B818J/.

217 *Gillett had succeeded*: Gillett, interview.

218 *Summer Lee was among the many local artists*: Author interview, Summer Lee, June 9, 2024.

218 *The Art Institute had nourished generations*: Margaret Crane, "The Rise, Fall and Reinvention of the San Francisco Art Institute," *East of Borneo* blog, April 3, 2024, https://eastofborneo.org/articles/the-rise-fall-and-reinvention-of-the-san-francisco-art-institute/.

219 *It was an unexplored area*: Author interview, Ana Teresa Fernández, January 29, 2025.

219 *Los Angeles beckoned*: Evan Minsker, "Ty Segall and John Dwyer on Why So Many Musicians Are Leaving San Francisco for LA," Pitchfork, January 15, 2014, https://pitchfork.com/thepitch/210-ty-segall-and-john-dwyer-on-why-so-many-musicians-are-leaving-san-francisco-for-la/.

220 *"They're totally happy with the idea"*: Author interview, Gordon Knox, March 24, 2025.

220 *Marc Benioff had given heavily to schools and hospitals*: Joe Garofoli, "Marc Benioff challenges Bay Area's tech leaders to give more," *SFGate*, March 7, 2014, https://www.sfgate.com/politics/joegarofoli/article/marc-benioff-challenges-bay-area-s-tech-leaders-5296188.php.

220 *Matt Mullenweg, the WordPress founder*: Mullenweg, interview.
221 *His vision went well beyond*: Knox, interview.
221 *Lee said later*: Lee, interview.
222 *Chesky said his company wanted to be a good actor*: Andy Kessler, "Brian Chesky: The 'Sharing Economy' and Its Enemies," *Wall Street Journal*, January 17, 2014, https://www.wsj.com/articles/SB10001424052702304049704579321001856708992.
222 *Chris Lehane, a veteran of*: Author interview, Chris Lehane, August 7, 2025.
223 *Airbnb spent more than $8 million*: "Airbnb Spends $8 Million Lobbying Against San Francisco Ballot Initiative," Reuters, November 1, 2015, https://www.huffpost.com/entry/airbnb-san-francisco-proposition-f_n_56366676e4b0c66bae5cc3b6#.
223 *Legislation that . . . legalized Airbnb but imposed some rules*: Carolyn Said, "Supes back 'Airbnb law' to allow short-term rentals, with limits," *SFGate*, October 8, 2014, https://www.sfgate.com/news/article/Supervisors-approve-Airbnb-law-5807858.php.
223 *The Supervisors passed another ordinance*: "Supervisors Approve Strict New Limits on Airbnb-ing in San Francisco," SocketSite, November 16, 2016, https://socketsite.com/archives/2016/11/supervisors-approve-strict-new-airbnb-limits-in-san-francisco.html.
223 *Preliminary ruling in our favor*: Peskin, interview.
224 *Lee pointed to studies*: Lee, interview.
224 *The official count crossing 7,500*: Beth Spotswood, "2017 San Francisco 'Homeless Census' Reveals That Despite Numbers, Things Are Worse, Not Better," *SFist*, June 26, 2017, https://sfist.com/2017/06/26/2017_san_francisco_homeless_census/.
225 *"We had a meeting in our offices"*: Apczynski, interview.
226 *"Community development agreements"*: "Twitter IPO," Reuters, November 7, 2013, https://www.reuters.com/article/twitter-ipo-sanfrancisco/twitter-ipo-highlights-big-money-friction-in-san-francisco-idINDEE9A60H220131107.
226 *"When I got there"*: Author interview, Caroline Barlerin, May 23, 2024.
226 *Dorsey would later give*: *Tenderloin Community Benefit District* blog, "Twitter CEO awards Code Tenderloin $1.6M to continue post-COVID impact efforts," June 30, 2020, https://tlcbd.org/blog/2020/06/30/twitter-code-tenderloin-donation/.
227 *"People really had their hearts in it"*: Author interview, Amy Cohen, February 28, 2024.
227 *Seymour said later*: Author interview, Del Seymour, June 17, 2024.
227 *Jane Kim says her one regret*: Kim, interview.
228 *Harvey had been hired back in 2007*: Author interview, Del Harvey, February 21, 2024.
229 *Jack Dorsey himself*: James Cook, "Twitter Co-Founder Explains Why He Protested Against Police Brutality," *Business Insider*, November 20, 2014, https://www.businessinsider.com/jack-dorsey-explains-why-he-took-part-in-the-ferguson-protests-2014-11.

231 *Even Marc Andreessen*: Rich Jaroslavsky, "CodeCon '17 Was Tech Versus Trump," *Observer*, June 2, 2017, https://observer.com/2017/06/code-conference-review-donald-trump/.

232 *Another contrarian was Louis Rossetto*: Author interview, Louis Rossetto; see also Nick Gillespie, "Louis Rossetto Is Wired for Optimism," Reason, May 2018, https://reason.com/2018/04/26/louis-rossetto-is-wired-for-op/.

Chapter 9: The Late Empire

233 *He championed the heartwarming adventures of Batkid*: Angela Watercutter, "Make-A-Wish Foundation Turns San Francisco Into Gotham for 5-Year-Old Batkid," *Wired*, November 15, 2013, https://www.wired.com/2013/11/san-francisco-batkid-wish/.

233 *He had a real interest in local history*: Author interview, John Law, February 18, 2025.

233 *An A-list celebrity*: Author interview, Francis Tsang, December 20, 2024.

233 *Traveling in style*: Matier and Ross, "Globe-trotting mayor makes himself scarce in SF," *San Francisco Chronicle*, September 10, 2017, https://www.sfchronicle.com/bayarea/matier-ross/article/Globe-trotting-mayor-makes-himself-scarce-in-SF-12185508.php.

234 *Unemployment . . . was almost nonexistent*: Federal Reserve Economic Data, Unemployment Rate in San Francisco County/City, CA, https://fred.stlouisfed.org/series/CASANF0URN.

234 *Lee was proud of his record on housing*: "Mayor Lee Creates Housing for All San Francisco Residents," Office of the Mayor, 2014, https://www.sfgov.org/mayoredlee/housing-for-residents#.

235 *His friend Randy Shaw*: Carlsson, interview.

235 *Now down to just 5 percent*: Wikipedia, https://en.wikipedia.org/wiki/African_Americans_in_San_Francisco.

235 *Suhr handed in his resignation*: Alex Emslie and Lisa Pickhoff White, KQED News, "SFPD Chief Suhr Resigns After Series of Fatal Officer-Involved Shootings," May 19, 2016, https://www.kqed.org/news/10961193/suhr-resigns-after-series-of-fatal-officer-involved-shootings.

236 *The first city to sue the president*: Thomas Fuller, "San Francisco Sues Trump Over 'Sanctuary Cities' Order," *New York Times*, January 31, 2017, https://www.nytimes.com/2017/01/31/us/san-francisco-lawsuit-trump-sanctuary-cities.html.

236 *Statements of solidarity from tech CEOs*: Samantha Masunaga, "CEOs speak out about Trump's travel ban," *Los Angeles Times*, January 30, 2017, https://www.latimes.com/business/la-fi-ceo-tweets-trump-20170130-story.html#.

237 *Clinton's campaign*: "Why did Hillary Clinton turn down embedded help from Facebook?" ePolitics.com, October 10, 2017, https://www.epolitics.com/2017/10/10/hillary-clinton-turn-embedded-help-facebook/.

237 *They reported that a political consulting firm*: Nick Confessore, "Cambridge Analytica and Facebook: The Scandal and the Fallout So Far," *New York Times*, April 4, 2018, https://www.nytimes.com/2018/04/04/us/politics/cambridge-analytica-scandal-fallout.html.

237 *Aaron Peskin's return to Board of Supervisors*: "Aaron Peskin Returns to San Francisco Board of Supervisors," *Bay City News*, November 4, 2015, https://www.nbcbayarea.com/news/local/incumbent-san-francisco-board-of-supervisor-aaron-peskin-retains-seat/2075632/.

238 *$100 billion in market value*: Rupert Neate, "Over $119bn wiped off Facebook's market cap after growth shock," *Guardian*, July 26, 2018, https://www.theguardian.com/technology/2018/jul/26/facebook-market-cap-falls-109bn-dollars-after-growth-shock.

238 *Om Malik, a highly respected tech blogger*: Author interview, Om Malik, June 18, 2025.

238 *The blogging revolution*: Battelle, interview.

239 *Google and Facebook were eating*: Author interview, Kara Swisher, June 24, 2025.

239 *One of the curious things*: Rebecca Solnit, "Get off the Bus," *London Review of Books*, February 20, 2014; see also Rebecca Solnit, "In the Shadow of Silicon Valley," *London Review of Books*, February 28, 2024.

240 *That was not what Battelle had in mind*: Author interview, John Battelle, June 11, 2025.

241 *Investigating Willie Brown*: Mary Curtius, "Corruption Scandal in S.F. Grows," *Los Angeles Times*, September 23, 1999, https://www.latimes.com/archives/la-xpm-1999-sep-23-mn-13313-story.html; Matthew Robinson and Brian C. Anderson, "Willie Brown Shows How Not to Run a City," *City Journal*, Autumn 1998, https://www.city-journal.org/article/willie-brown-shows-how-not-to-run-a-city.

241 *Other skeletons loomed*: Michael Barba, "Years after FBI probe, SF pay-to-play case ends in plea deals," *San Francisco Examiner*, April 2, 2019, https://www.sfexaminer.com/archives/years-after-fbi-probe-sf-pay-to-play-case-ends-in-plea-deals/article_e237b9e3-ee87-5c51-adff-6892e6844ae2.html.

241 *His heart stopped beating*: Joe Eskenazi, "How Mayor Ed Lee remade San Francisco in Big Tech's image," The Verge, December 13, 2017, https://www.theverge.com/2017/12/13/16771912/mayor-ed-lee-san-francisco-silicon-valley-tech-jobs.

242 *Breed had gotten her start as an intern in Brown's office*: C. W. Nevius, "London Breed is S.F.'s election shocker," SF Gate, November 10, 2012, https://www.sfgate.com/bayarea/nevius/article/London-Breed-is-S-F-s-election-shocker-4025223.php.

242 *"I'm from the 'hood,"*: Madeleine Johnston, "London Breed: How personal hardships inspired a political career," The Lowell, January 31, 2020, https://thelowell.org/8248/features/london-breed-how-personal-hardships-inspired-a-political-career/.

243 *He frequented in search of her brother*: Author interview, Greg Suhr, July 15, 2024.

243 *O'Connor says it was clear*: Author interview, Tom O'Connor, November 19, 2024.

243 *I did feel protected*: Author interview, London Breed, January 6, 2025.

243 *An expletive-filled rant*: Heather Knight, "London Breed cites her experience as path to transforming San Francisco," *San Francisco Chronicle*, April 7, 2018, https://www.sfchronicle.com/news/article/london-breed-cites-experience-transforming-sf-12806970.php.

244 *Breed would narrowly defeat Mark Leno in the special election*: Heather Knight, "It's a really big deal that SF elected London Breed," *San Francisco Chronicle*, June 13, 2018, https://www.sfchronicle.com/news/article/It-s-a-really-big-deal-that-SF-elected-London-12991871.php.

245 *She developed what would be known as Prop C*: Author interview, Jennifer Friedenbach, December 20, 2025.

245 *Evans had grown up in the East Bay*: Author interview, Christin Evans, July 9, 2024.

248 *She exhorted him to support Prop C*: https://x.com/christinevans.

248 *Marc Benioff's on Prop C*: Vox, "Marc Benioff and other tech CEOs are fighting over a San Francisco homeless tax," November 14, 2018, https://www.vox.com/2018/11/14/18093170/marc-benioff-homeless-tax-prop-c-tech-ceos.

249 *Crushing and credibility-destroying decision*: Joe Eskenazi, "Mayor London Breed's huge political fumble on Prop. C," Mission Local, October 9, 2018, https://missionlocal.org/2018/10/mayor-london-breeds-huge-political-fumble-on-prop-c/.

249 *Stripe, with one thousand employees*: Roland Li, "2nd most valuable U.S. startup to leave SF as city loses another headquarters," *San Francisco Chronicle*, October 23, 2019, https://www.sfchronicle.com/business/article/2nd-most-valuable-U-S-start-up-to-leave-SF-as-14558067.php.

250 *Overdose deaths had doubled*: Christian Leonard and Yoohyun Jung, "48 people died in August in San Francisco from accidental overdoses," *San Francisco Chronicle*, updated September 16, 2025, https://www.sfchronicle.com/projects/san-francisco-drug-overdose-deaths/#.

250 *A program called LEAD*: David Sjostedt, "A mother, an officer, a heartbreaking death," *San Francisco Standard*, October 21, 2021.

251 *The move backfired badly*: Joe Eskenazi, "Suzy Loftus' appointment as DA 24 hours before voting commenced was a crass political flex," Mission Local, October 17, 2019, https://missionlocal.org/2019/10/suzy-loftus-appointment-as-da-24-hours-before-voting-commenced-was-a-crass-political-flex/.

253 *Corruption charges*: Mark Fiore and Joe Fitzgerald Rodriguez, "San Francisco's Unfolding Web of Corruption: A Cartoon Interactive," KQED, March 17, 2021, https://www.kqed.org/news/11859677/san-franciscos-unfolding-web-of-corruption-a-cartoon-interactive.

253 *Brown's favor-trading style*: Lance Williams and Chuck Finnie, "Willie Brown Inc.—How SF's mayor built a city based on 'juice' politics," *SFGate*, April 29, 2001, https://www.sfgate.com/politics/article/Willie-Brown-Inc-How-S-F-s-mayor-built-a-city-2926278.php.

254 *Nuru and Recology pushed it too far*: Joe Eskenazi, "Recology to reimburse San Franciscans some $100M following City Attorney investigation," Mission Local, March 4, 2021.

Chapter 10: Covid Comes to Town

259 *Breed quickly declared a state of emergency*: Maura Dolan, "London Breed, a mayor who scolds and empathizes, is San Francisco's face of the coronavirus crisis," *Los Angeles Times*, April 20, 2020, https://www.latimes.com/california/story/2020-04-20/coronavirus-san-francisco-mayor-london-breed.

259 *Breed issued the "shelter-in-place" order*: Angela Hart and Anna Maria Barry-Jester, "The Inside Story of How the Bay Area Got Ahead of the Covid-19 Crisis," California Healthline, April 21, 2020, https://californiahealthline.org/news/the-inside-story-of-how-the-bay-area-got-ahead-of-the-covid-19-crisis/.

260 *"There was a lot of pushback"*: Breed, interview; See also Dolan, "London Breed, a mayor who scolds and empathizes," *Los Angeles Times,* April 20, 2020, https://www.latimes.com/california/story/2020-04-20/coronavirus-san-francisco-mayor-london-breed.

260 *"San Francisco a national model in fighting the pandemic"*: Russell Berman, "The City That Has Flattened the Coronavirus Curve," *Atlantic*, April 12, 2020, https://www.theatlantic.com/politics/archive/2020/04/coronavirus-san-francisco-london-breed/609808/.

260 *San Francisco executed by far the most successful initial response*: Daniel Duane, "San Francisco Was Uniquely Prepared for Covid-19," *Wired*, August 11, 2020, https://www.wired.com/story/san-francisco-uniquely-prepared-covid-19/.

261 *Dean Preston, the Democratic Socialist*: Author interview, Dean Preston, February 26, 2025.

262 *Mullenweg noted*: Mullenweg, interview.

262 *Salesforce . . . began shedding some of its San Francisco real estate*: Sarah Klearman, "Salesforce shrank its San Francisco office footprint by more than 40% last year," *San Francisco Business Times*, March 12, 2024, https://www.bizjournals.com/sanfrancisco/news/2024/03/12/salesforce-occupied-office-san-francisco.html.

262 *Anything but safe havens*: "Project Home: Neighbors of San Francisco Shelter-In-Place Hotels Complain of Increased Crime," CBS News, November 12, 2020, https://www.cbsnews.com/sanfrancisco/news/sf-shelter-in-place-hotels-increase-911-calls-crime/.

262 *Aby Rosen, a big-spending social climber*: Alex Williams, "Making His Life the Party," *New York Times*, May 29, 2013, https://www.nytimes.com/2013/05/30/fashion/aby-rosen-is-the-life-of-the-party.html.

263 *Jack Dorsey told Twitter employees . . . that they could work from home*: Brock Keeling, "Now that Twitter employees can work at home forever, what's to become of its headquarters?" *SF Curbed*, May 13, 2020, https://sf.curbed.com/2020/5/13/21256351/twitter-building-headquarters-work-home-employees-sf.

263 *Harvey recalled*: Harvey, interview.

264 *Return-to-office metrics*: Hannah Kanik, "San Francisco office attendance stands at less than half of pre-pandemic levels," *San Francisco Business Times*, February 7, 2024, https://www.bizjournals.com/sanfrancisco/news/2024/02/07/san-francisco-office-return-to-work-hybrid-wfh.html.

264 *Who accounted for 35 percent*: Author interview, Ted Egan, April 23, 2024.

264 *Told CBS News*: "Project Home: Neighbors of San Francisco Shelter-in-Place Hotels Complain of Increased Crime," CBS News, November 12, 2020, https://www.cbsnews.com/sanfrancisco/news/sf-shelter-in-place-hotels-increase-911-calls-crime/.

264 *Now the Market Square complex was all but empty*: Kanik, "San Francisco office attendance."

264 *Now it would serve some of the neighborhood's down and out*: Kristina Shevory, "Twitter Helps Revive a Seedy San Francisco Neighborhood," *New York Times*, November 1, 2013, https://www.nytimes.com/2013/11/02/business/twitter-helps-revive-a-seedy-san-francisco-neighborhood.html.

267 *"They don't trust children"*: Dalton, interview.

268 *Sunset resident*: Author interview, Patrick Wolff, December 19, 2024.

270 *Wolff had gotten ahold of the spreadsheet*: Wolff, interview.

270 *The renaming committee had indeed gone off the deep end*: Joe Eskenazi, "The San Francisco School District's renaming debacle has been a historic travesty," Mission Local, January 28, 2021, https://missionlocal.org/2021/01/the-san-francisco-school-districts-renaming-debacle-has-been-a-historic-travesty/.

270 *Breed called it "offensive"*: "San Francisco Board of Education Set to Reverse Controversial School Renaming Plan," CBS News, April 6, 2021, https://www.cbsnews.com/sanfrancisco/news/san-francisco-school-board-set-to-reverse-controversial-school-renaming-plan/.

270 *Both were single parents*: Author interview, Autumn Looijen and Siva Raj, February 4, 2025.

274 *Netscape's Open Directory Project*: Adam Lashinsky, "Netscape IPO 20-year anniversary: Read Fortune's 2005 oral history of the birth of the web," *Fortune*, August 9, 2015, https://fortune.com/2015/08/09/remembering-netscape/.

274 *A half dozen unprovoked attacks*: Joe Fitzgerald Rodriguez and Han Li, "Why High-Profile Attacks on SF's Asian Communities Rarely Lead to Hate Crime Charges," KQED News, June 2, 2022, https://www.kqed.org/news/11915634/why-high-profile-attacks-on-sfs-asian-communities-rarely-lead-to-hate-crime-charges.

274 *His only big donation*: Megan Cassidy, "These are the ultra-wealthy donors pouring money into the Chesa Boudin recall battle," *San Francisco Chronicle*, May 4, 2022, https://www.sfchronicle.com/sf/article/chesa-boudin-recall-17052312.php.

275 *On St. Patrick's Day*: Author interview, Mary Jung, March 6, 2025.

275 *Xiao Zhen Xie had been punched*: Rodriguez and Li, "Why High-Profile Attacks."

276 *Neighbors for a Better San Francisco*: Joe Rivano Barros, "BigMoneySF: How one group quickly became the 800-pound gorilla of San Francisco politics," Mission Local, April 24, 2024, https://missionlocal.org/2024/04/bigmoneysf-how-one-group-quickly-became-the-800-pound-gorilla-of-san-francisco-politics/.

Chapter 11: Recalls and Recriminations

279 *He'd bankrolled*: Joe Eskenazi, "Who is Michael Moritz, and what does he want for San Francisco?" Mission Local, October 30, 2024, https://missionlocal.org/2024/10/michael-moritz-san-francisco-togethersf-sf-elections-crankstart/.

279 *The son of German Jews*: https://en.wikipedia.org/wiki/Michael_Moritz.

280 *His political group*: Kate Selig, "New Venture Here/Say Media won't disclose who its donors are," Mission Local, January 29, 2021, https://missionlocal.org/2021/01/who-is-funding-here-say-media-the-founders-refuse-to-say/.

282 *The San Francisco startup Substack*: Author interview, Hamish McKenzie, August 21, 2025.

283 *Among the Substackers*: Shellenberger, Michael, "San Fransicko."

283 *Sacks, a South African native*: Zoe Bernard, "First, All-In Red-Pilled Billionaires. Now It's Coming for Everyone Else," *Vanity Fair*, September 25, 2025.

283 *The* All-In *podcast*: Alex Carter, "*All-In* Podcast Hosts' Net Worth: Chamath, Jason, David & Sacks," Capitaly VC, April 10, 2025, https://www.capitaly.vc/blog/all-in-podcast-hosts-net-worth-chamath-jason-david-sacks.

283 *Calacanis had successfully*: David Li, "What one investor saw in Uber when it was valued at only $5 million," Medium, December 19, 2018.

284 *Ricci Wynne*: Garrett Leahy, "Ricci Wynne produced child sexual abuse images, feds say," *San Francisco Standard*, March 18, 2025, https://sfstandard.com/2025/03/18/ricci-wynne-indicted-producing-child-sexual-abuse-images/.

285 *For all the heated political drama*: Christian Leonard and Yoohyun Jung, "San Francisco Drug Overdose Deaths," *San Francisco Chronicle*, Jan. 16, 2026. https://www.sfchronicle.com/projects/san-francisco-drug-overdose-deaths/.

285 *Mayor Breed declared a state of emergency for the Tenderloin*: John Coté, "S.F. hails 'Twitter tax break' as rousing success,": *SFGate*, November 1, 2013 (updated November 3, 2014), https://www.sfgate.com/bayarea/article/S-F-controller-hails-Twitter-tax-break-as-5851498.php.

286 *The reason for that*: Anna Tong and Josh Koehn, "DA Boudin and fentanyl: Court data shows just 3 drug dealing convictions in 2021 as immigration concerns shape policy," *San Francisco Standard*, May 17, 2022.

286 *Breed said later that she had explored*: Author interview, London Breed, June 18, 2025.

292 *Tensions would come to a boil*: Gilmore, interview; Cohn, interview.

293 *A generational change in the local tech industry*: Joel Umanzor, "Republicans unlikely to feel safe in San Francisco, Gallup poll shows," *San Francisco Standard*, August 24, 2023, https://sfstandard.com/2023/08/24/just-over-half-of-americans-said-they-feel-safe-in-san-francisco/.

293 *Tent encampments covered long stretches of sidewalks*: Sarah Klearman, "Showplace Square building that once housed Airbnb's HQ headed for foreclosure," *San Francisco Business Times*, March 26, 2024, https://www.bizjournals.com/sanfrancisco/news/2024/03/26/airbnb-rhode-island-showplace-square-foreclosure.html.

297 *The school board recall and the Boudin recall*: Barros, "BigMoneySF," Mission Local, April 24, 2024.

297 *Jack Dorsey had returned as CEO*: Kaplan, Michael, "Inside the 'odd, eccentric' life of 'checked-out Twitter CEO Jack Dorsey," *The New York Post*, Nov. 29, 2021.

298 *Musk made a $44 billion offer*: Saul Sugarman, "The many hot takes on Twitter moving out of San Francisco," *Bold Italic*, August 7, 2024, https://thebolditalic.com/the-many-hot-takes-on-twitter-moving-out-of-san-francisco-er-x-3cb8ec948490.

Chapter 12: Doom Loop

301 *But it was too late*: "Bob Lee murder case: Nima Momeni trial, timeline, what to know," NBC Bay Area, October 7, 2024, https://www.nbcbayarea.com/news/local/bob-lee-murder-nima-momeni-trial/3673231/.

301 *Starting to tip over into panic*: Joel Umanzor, "Republicans unlikely to feel safe in San Francisco, Gallup poll shows," *San Francisco Standard*, August 24, 2023, https://sfstandard.com/2023/08/24/just-over-half-of-americans-said-they-feel-safe-in-san-francisco/.

303 *Musk himself*: https://x.com/elonmusk/status/1687663413877714944.

304 *Shoots him dead on the sidewalk*: Nuaia Bishari, "Killed over '$14 of candy': Banko Brown's death is a tragedy of San Francisco's making," *San Francisco Chronicle*, May 4, 2023, https://www.sfchronicle.com/opinion/article/shoplifting-crime-san-francisco-banko-brown-candy-18073871.php.

305 *Most properties had lost 60 to 75 percent of their worth*: "InvestSF San Francisco Commercial Property Statistics," InvestSF, https://www.investsf.com/investsf-san-francisco-commercial-property-statistics/; Alan Rappeport, "Cities Face Cutbacks as Commercial Real Estate Prices Tumble," *New York Times*, March 14, 2024, https://www.nytimes.com/2024/03/14/business/commercial-real-estate-tax-revenue.html.

305 *"We can no longer operate"*: Burke Williams Spa, https://www.burkewilliams.com/san-francisco-spa-closure#.

305 *At designer boutiques*: "Thieves smash car into Union Square Dior Store," ABC News 7, October 11, 2024, https://abc7news.com/post/thieves-smash-car-dior-store-san-francisco-union-square-steal-thousands-dollars-worth-merchandise/15418259/.

305 *Business tax collections were plunging*: Danielle Echeverria, "Businesses in most S.F. neighborhoods are still struggling to bring back customers," *San Francisco Chronicle*, September 25, 2024, https://www.sfchronicle.com/bayarea/article/sales-tax-neighborhood-recovery-19779849.php.

306 *A new term encapsulated the worst fears: "Doom Loop"*: Eliza Relman, "Midwest America Cities Are Facing a Downtown Crisis, Urban Doom Loop," *Business Insider*, June 22, 2023, https://www.businessinsider.com/midwest-america-cities-downtown-crisis-office-apocalypse-urban-doom-loop-2023-6; Roland Lee and Laura Waxmann, "S.F.'s fight against the doom loop weighed down by stark economic realities," *San Francisco Chronicle*, October 8, 2024, https://www.sfchronicle.com/sf/article/doom-loop-economy-19807505.php.

306 *An outdoor meeting*: "San Francisco Mayor booed in UN Plaza, brick thrown in crowd," May 23, 2023, https://www.kron4.com/news/bay-area/san-francisco-mayor-booed-in-un-plaza-brick-thrown-in-crowd/.

307 *She was done*: April 2023 WIRED podcast: https://www.wired.com/story/have-a-nice-future-podcast-1/.

308 *The San Francisco Art Institute would shut down*: Margaret Crane, "The Rise, Fall and Reinvention of the San Francisco Art Institute," East of Borneo, April 3, 2024, https://eastofborneo.org/articles/the-rise-fall-and-reinvention-of-the-san-francisco-art-institute/.

310 *Salesforce was among those pulling back*: Sarah Klearman, "Salesforce shrank its San Francisco office footprint by more than 40% last year," *San Francisco Business Times*, March 12, 2024, https://www.bizjournals.com/sanfrancisco/news/2024/03/12/salesforce-occupied-office-san-francisco.html.

311 *A large, illuminated X sign*: Janie Har and Haven Daley, "'X' logo installed atop Twitter building, spurring San Francisco to investigate permit violation," Associated Press, July 28, 2023, https://apnews.com/article/twitter-san-francisco-building-x-elon-musk-4e0ae2a3b1b838b744bb2dc494f5b23c.

311 *Sold in 2023 for five cents on the dollar*: Jillian D'Onfro, "SF office sells for a stunning 90% discount from 2016 price," *SF Gate*, April 24, 2024, https://www.sfgate.com/local/article/995-market-st-san-francisco-office-market-19420409.php.

312 *Waiting for the world's richest man*: Author interview, Sean Elsbernd, April 9, 2025.

313 *Musk moved the company to Texas*: Laura Waxmann, "Elon Musk: X headquarters will move from S.F. to Austin, Texas," *San Francisco Chronicle*, July 16, 2024, https://www.sfchronicle.com/sf/article/elon-musk-x-headquarters-moving-to-texas-19577613.php; Holly Secon, "X's SF Landlord Is Suing over Unpaid Rent, Seeking $13.6 Million," SFist, February 14, 2024, https://sfist.com/2024/02/14/x-sf-hq-landlord-is-suing-over-unpaid-rent-seeking-13-6-million/; Jaimie Ding and Brian Contreras, "Laid-off Twitter workers feared meager severance deals. Elon Musk just set the bar even lower," *Los Angeles Times*, January 10, 2023, https://www.latimes.com/business/story/2023-01-10/musk-twitter-layoffs-severance-lawsuits.

314 *That would have required Apple*: Elsbernd, interview.

315 *A tawdry family drama also spilled into public view*: Michael Kaplan, "Dianne Feinstein leaves stunning properties to her billionaire husband's feuding kids," *New York Post*, September 29, 2023, https://nypost.com/2023/09/29/dianne-feinsteins-stunning-properties-and-feuding-family/.

316 *A lot of older Burners would be calling it quits*: Justin Berton, "Burning Man Tries to Cope with Cash," SF Gate, August 19, 2007, https://www.sfgate.com/news/article/BURNING-MAN-TRIES-TO-COPE-WITH-CASH-2509458.php; Brad Wieners, "An Oral History of Burning Man, the Biggest, Weirdest, Most Clothing-Optional Desert Carnival on the Planet," *Outside Online*, August 24, 2012, https://www.outsideonline.com/adventure-travel/destinations/hot-mess/.

319 *My journalistic sensibilities were shocked*: Aja Romano, "What happened at Burning Man: The flameout, explained," Vox, September 6, 2023, https://www.vox.com/culture/2023/9/6/23861675/burning-man-2023-mud-stranded-climate-change-playa-foot.

Chapter 13: Regime Change

321 *He leased a restaurant space*: Author interview, Manny Yekutiel, March 3, 2024; see also Caleb Persham, "New Cafe Opens for Food, Drinks, and Political Activism in the Mission," SF Eater, November 2, 2018, https://sf.eater.com/2018

/11/2/18053288/mannys-opens-mission-cafe-political-food-drink-engagement.

324 *Yerba Buena Center for the Arts*: Lily Janiak, "Conflict over S.F. and YBCA's guaranteed income for artists shows tension in movement for racial equity," *San Francisco Chronicle*, June 1, 2021, https://datebook.sfchronicle.com/entertainment/conflict-over-s-f-and-ybcas-guaranteed-income-for-artists-shows-tension-in-movement-for-racial-equity.

324 *She'd face an uphill climb*: Rachel Swan, "London Breed, mayor: San Francisco, drugs, homelessness, crime, economy, election," *SF Standard*, June 20, 2023, https://sfstandard.com/2023/06/20/london-breed-mayor-san-francisco-drugs-homelessness-crime-economy-election/.

324 *London Breed was the moderate centrist*: Heather Knight, "It's a really big deal that SF elected London Breed as mayor," *San Francisco Chronicle*, June 13, 2018, https://www.sfchronicle.com/news/article/It-s-a-really-big-deal-that-SF-elected-London-12991871.php.

325 *A key part of their election strategy*: "Big Money SF: How One Group Quickly Became the 800-Pound Gorilla of San Francisco Politics," Mission Local, April 2024, https://missionlocal.org/2024/04/bigmoneysf-how-one-group-quickly-became-the-800-pound-gorilla-of-san-francisco-politics/.

325 *Conway boasted*: Conway, interview.

325 *Ramping up his profile*: https://www.sfchronicle.com/politics/joegarofoli/article/dnc-san-francisco-bash-19666567.php.

326 *Willie Brown had spent a career*: Matthew Robinson and Brian C. Anderson, "Willie Brown Shows How Not to Run a City," *City Journal*, Autumn 1998, https://www.city-journal.org/article/willie-brown-shows-how-not-to-run-a-city.

327 *His campaign and the associated PAC*: SF Ethics Commission Campaign Finance Dashboards, https://sfethics.org/ethics/2023/12/campaign-finance-dashboards-november-5-2024.html.

328 *There was no small amount of head-scratching*: Noah Kirsch, "Transamerica Pyramid Sale Closes for $650 Million, One of the Largest Commercial Deals Since Covid," *Forbes*, October 28, 2020, https://www.forbes.com/sites/noahkirsch/2020/10/28/transamerica-pyramid-sale-closes-for-650-million-the-largest-commercial-deal-since-covid/.

329 *Shvo and his backers remained undeterred*: "InvestSF San Francisco Commercial Property Statistics," 2011–2021, https://www.investsf.com/investsf-san-francisco-commercial-property-statistics/.

330 *Since leaving Apple in 2019*: Jony Ive Properties.

330 *Ive was the kind of Mogul*: Aaron Peskin comments.

331 *A product dubbed ChatGPT*: Coverage of OpenAI's November 30, 2022, ChatGPT release.

331 *Altman was now firmly in control*: Kari Paul, "OpenAI reinstates CEO Sam Altman to board after firing and rehiring," *Guardian*, March 8, 2024, https://www.theguardian.com/technology/2024/mar/08/openai-sam-altman-reinstated; see

also Keach Hagey, "The Secrets and Misdirection Behind Sam Altman's Firing from OpenAI," *Wall Street Journal*, March 28, 2025, https://www.wsj.com/tech/ai/the-real-story-behind-sam-altman-firing-from-openai-efd51a5d.

332 *The city was capturing*: Eli Tan, "The Insider's Guide to San Francisco's AI Boom," *New York Times*, August 4, 2025.

333 *It was here that Garry Tan*: Author interview, Garry Tan, August 25, 2025.

333 *The rhetoric around how AI*: Newcomer Cerebral Valley Ai Summit, NovemberXX, 2023.

334 *The second tech boom*: Heather Knight, "Projects: Mid-Market and the 'Twitter tax break': Vision and reality in San Francisco's tech corridor," *San Francisco Chronicle*, 2019, https://projects.sfchronicle.com/2019/mid-market/; Alejandro Lazo, "Tax Breaks for Twitter Bring Benefits and Criticism," *Wall Street Journal*, April 29, 2016, https://www.wsj.com/articles/tax-breaks-for-twitter-bring-benefits-and-criticism-1461947597.

334 *So had an essay*: Paul Graham, Founder Mode, September 2024, https://paulgraham.com/foundermode.html.

335 *A forlorn reminder of the former glory days of 'multimedia gulch'*: Douglas Fruehling, "LinkedIn announces new round of layoffs across Bay Area offices," *San Francisco Business Times*, May 29, 2025, https://www.bizjournals.com/sanfrancisco/news/2025/05/29/linkedin-announces-layoffs-across-bay-area.html.

336 *San Francisco's population had mostly recovered*: Douglas Sams, "SF sees big slowdown in population loss," *San Francisco Business Times*, April 25, 2025.

337 *Harris was a San Francisco politician through and through*: Michael Finnegan, "Kamala Harris was shaped by the crucible of San Francisco politics," *Los Angeles Times*, January 21, 2019, https://www.latimes.com/politics/la-na-pol-kamala-harris-san-francisco-20190121-story.html.

338 *His circle would pour tens of millions*: Gabe Greschler, "Kamala World: The Bay Area elites powering Harris' campaign for president," *San Francisco Standard*, July 23, 2024, https://sfstandard.com/2024/07/23/kamala-harris-president-san-francisco-campaign-donors/.

341 *An especially surprising turn came from Mark Pincus*: Author interview, Mark Pincus, October 22, 2024.

343 *By elbowing Breed out*: Thomas Fuller and Conor Dougherty, "San Francisco Ousts a Mayor in a Clash of Tech, Politics and Race," *New York Times*, January 24, 2018, https://www.nytimes.com/2018/01/24/us/san-francisco-mayor-breed-farrell.html.

343 *Breed was facing a scandal of her own*: Aldo Toledo, "S.F. Nonprofit scandals: New legislation promises tougher oversight amid $1.7 billion in contracts," *San Francisco Chronicle*, March 12, 2024, https://www.sfchronicle.com/sf/article/sf-nonprofit-scandals-city-hall-oversight-18926219.php; see also Mark Fiore and Joe Fitzgerald Rodriguez, "San Francisco's Unfolding Web of Corruption: A Cartoon Interactive," KQED, March 17, 2021, https://www.kqed.org/news/11859677/san-franciscos-unfolding-web-of-corruption-a-cartoon-interactive.

343 *Closet Trump supporters began to emerge*: David Jeans, "Here Come the Anti-Woke Venture Capitalists," *Forbes*, December 23, 2024, https://www.forbes.com/sites/davidjeans/2024/12/23/anti-woke-venture-capital-new-founding/; Ali Breland, "Even the Koch Brothers Weren't This Brazen," *Atlantic*, December 9, 2024, https://www.theatlantic.com/technology/archive/2024/12/tech-billionaires-trump-administration/680930/.

Epilogue

355 *A new dawn was at hand*: "The inauguration of Daniel Lurie," *San Francisco Chronicle* YouTube video, January 8, 2025, https://www.axios.com/2025/05/28/ai-jobs-white-collar-unemployment-anthropic.

356 *"There was a cynicism about politics"*: Author interview, Jesse Blout, June 26, 2025.

357 *Nagel told the* SF Standard: Zara Stone, "'Market Street is coming back to life': The tech utopia trying to revitalize Mid-Market," *San Francisco Standard*, May 18, 2025, https://sfstandard.com/2025/05/18/first-look-frontier-tower-san-francisco-vertical-village/.

357 *Bought by Laurene Powell Jobs*: Laura Waxmann, "S.F. Art Institute bought by nonprofit backed by Laurene Powell Jobs. Here's their plan," *San Francisco Chronicle*, February 29, 2024, https://www.sfchronicle.com/realestate/article/s-f-art-institute-bought-laurene-powell-jobs-18693505.php.

357 *"They don't buy any drinks"*: Nelson, interview.

358 *"Ask-me-anything" session at his café*: Manny's Café, June 24, 2025.

358 *Average age of Y Combinator entrepreneurs*: Natallie Rocha, "The 20-Somethings Are Swarming San Francisco's A.I. Boom," *New York Times*, August 4, 2025, https://www.nytimes.com/2025/08/04/technology/ai-young-ceos-san-francisco.html.

358 *"They're being pulled and pushed"*: Author interview, Garry Tan, August 26, 2025.

360 *Withering in her assessment*: Author interview, Esther Dyson, May 14, 2025.

360 *For Pahlka, it was upsetting*: Pahlka, interview.

360 *Progressives like Dean Preston*: Preston, interview.

362 *A crucial mistake at HotWired*: Battelle, interview.

362 *It's actually more important*: Mullenweg, interview.

363 *A remarkable reunion*: Gulf of Deliria at the Gray Area Foundation, https://grayarea.org/event/come-to-the-gulf-of-deliria/.

364 *Wasn't sold out in 2024*: Sam Mondros, "'Incredibly unusual': Burning Man ticket sales dry up after sloppy year," *San Francisco Standard*, August 8, 2024, https://sfstandard.com/2024/08/08/burning-man-tickets-rain-heat-weather/.

364 *Seventy thousand people made their way*: Aidan Vaziri, "Burning Man 2025 census: Middle-aged, highly educated, mostly white," *San Francisco Chronicle*, August 31, 2025, https://www.sfchronicle.com/entertainment/article/burning-man-2025-census-demographics-21024056.php.

364 *Just 15 percent of Burning Man attendees*: Vaziri, "Burning Man 2025 Census," *San Francisco Chronicle*.

365 *Andreessen denounced Burning Man*: Marc Andreessen (@pmarca), Twitter (now X), September 24, 2024, https://x.com/pmarca/status/1838544274167861513; https://x.com/pmarca/status/1839694268799324649.

364 *They shut down the school*: Gennady Scheyner, "Chan Zuckerberg Initiative announces it will close East Palo Alto school," *Palo Alto Weekly*, April 21, 2025, https://www.paloaltoonline.com/education/2025/04/21/the-primary-school-to-shut-down-at-end-of-school-year/.

Bibliography

Benioff, Marc, with Carlye Adler. *Behind the Cloud: The Untold Story of How Salesforce .com Went from Idea to Billion-Dollar Company-and Revolutionized an Industry.* Wiley-Blackwell, 2009.

Benioff, Marc. *Trailblazer: The Power of Business as the Greatest Platform for Change.* Crown, 2019.

Berlin, Leslie. *Troublemakers: Silicon Valley's Coming of Age.* Simon & Schuster, 2017.

Bilton, Nick. *Hatching Twitter: A True Story of Money, Power, Friendship and Betrayal.* Portfolio, 2013.

Brechin, Grey. *Imperial San Francisco: Urban Power, Earthly Ruin.* University of California Press, 1999.

Brown, Willie, with P.J. Corkery. *Basic Brown: My Life and Our Times.* Simon & Schuster, 2008.

Caro, Robert. *The Power Broker: Robert Moses and the Fall of New York.* Vintage, 1975.

Conger, Kate, and Ryan Mac. *Character Limit: How Elon Musk Destroyed Twitter.* Penguin Press, 2024.

DeLeon, Richard. *Left Coast City: Progressive Politics in San Francisco, 1975-1991.* University of Kansas Press, 1992.

Doherty, Brian. *This Is Burning Man.* Little, Brown and Company, 2004.

Dougherty, Conor. *Golden Gates: The Housing Crisis and the Reckoning of an American Dream.* Penguin Books, 2021.

Evans, Kevin, John Law, and Carrie Galbraith. *Tales of the San Francisco Cacophony Society.* Last Gasp Publishing, 2013.

Hafner, Katie. *The Well: A Story of Love, Death & Real Life in the Seminal Online Community.* Carroll & Graf, 2001.

Isaac, Mike. *Super Pumped: The Battle for Uber.* W. W. Norton & Company, 2019.

Livingston, Jessica. *Founders at Work: Stories of Startups' Early Days.* Apress, 2008.

Kirkpatrick, David. *The Facebook Effect: The Inside Story of the Company That Is Connecting the World.* Simon & Schuster, 2010.

Markoff, John. *What the Dormouse Said: How the Sixties Counterculture Shaped the Personal Computer Industry.* Penguin Publishing Group, 2005.

Markoff, John. *Whole Earth: The Many Lives of Stewart Brand.* Penguin Press, 2022.

McClelland, Cary. *Silicon City: San Francisco in the Long Shadow of the Valley.* W. W. Norton & Company, 2018.

Mitchell, Lincoln. *San Francisco Year Zero: Political Upheaval, Punk Rock and a Third-Place Baseball Team.* Rutgers University Press, 2019.

Morain, Dan. *Kamala's Way: An American Life.* Simon & Schuster, 2021.

Newsom, Gavin, with Lisa Dickey. *Citizenville: How to Take the Town Square Digital and Reinvent Government.* Penguin Publishing Group, 2014.

Reginato, James. *Growing Up Getty: The Story of America's Most Unconventional Dynasty.* Gallery Books, 2022.

Richardson, James. *Willie Brown: A Biography*. University of California Press, 1996.

Rivlin, Gary. *The Godfather of Silicon Valley: Ron Conway and the Fall of the Dot-coms.* Random House, 2001.

Rosenberg, Scott. *Say Everything: How Blogging Began, What It's Becoming, and Why It Matters.* Crown, 2009.

Rushkoff, Douglas. *Throwing Rocks at the Google Bus: How Growth Became the Enemy of Prosperity.* Portfolio, 2017.

Schuerman, Matthew. *Newcomers: Gentrification and Its Discontents.* University of Chicago Press, 2019.

Shellenberger, Michael. *San Fransicko: Why Progressives Ruin Cities.* Harper, 2021.

Smith, Ben. *Traffic: Genius, Rivalry, and Delusion in the Billion-Dollar Race to Go Viral.* Penguin Books, 2024.

Solnit, Rebecca. *Infinite City: A San Francisco Atlas.* University of California Press, 2010.

Stone, Brad. *The Upstarts: How Uber, Airbnb, and the Killer Companies of the New Silicon Valley Are Changing the World.* Little, Brown and Company, 2017.

Svane, Mikkel, with Carlye Adler. *Startupland: How Three Guys Risked Everything to Turn an Idea into a Global Business.* Jossey-Bass, 2014.

Swisher, Kara. *Burn Book: A Tech Love Story.* Simon & Schuster, 2024.

Talbot, David. *Season of the Witch: Enchantment, Terror, and Deliverance in the City of Love.* Free Press, 2013.

Turner, Fred. *From Counterculture to Cyberculture: Stewart Brand, the Whole Earth Network, and the Rise of Digital Utopianism.* University of Chicago Press, 2008.

Wolf, Gary. *Wired—A Romance.* Random House, 2003.

Wolff, Michael. *Burn Rate: How I Survived the Gold Rush Years on the Internet.* Simon & Schuster, 1999.

Index

M

N

T

About the Author

Jonathan Weber was the *Los Angeles Times*'s first-ever Silicon Valley reporter, in 1990. He helped found *The Industry Standard*, a chronicle of and bellwether for the dot-com boom. He has since held several prestigious posts in media covering the tech industry and Silicon Valley, including overseeing global tech coverage for Reuters. He was most recently the editor in chief of *The San Francisco Standard*.